BOOK OF
EARTH

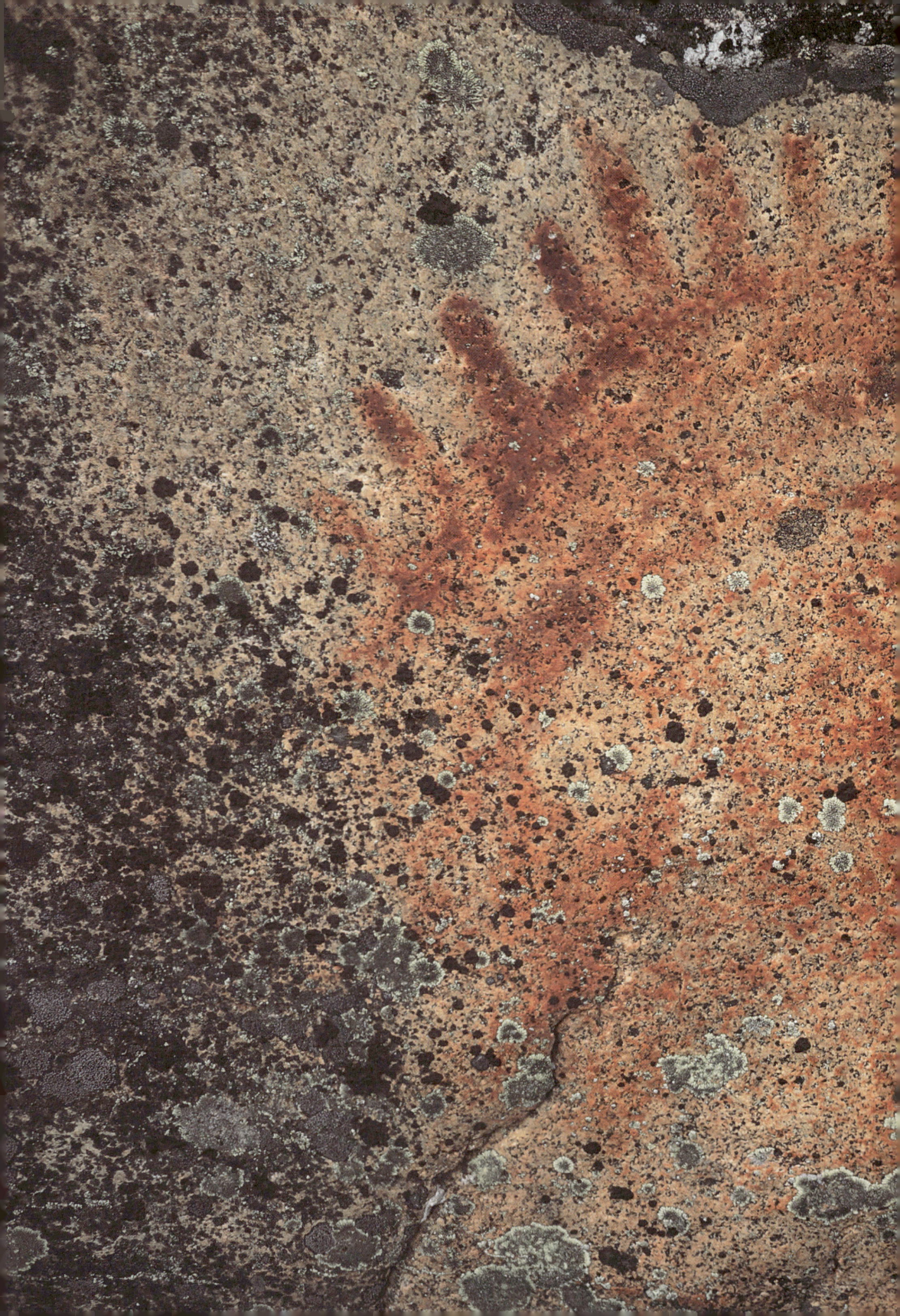

BOOK OF EARTH

A Guide to OCHRE, PIGMENT, and RAW COLOR

HEIDI GUSTAFSON

ABRAMS, NEW YORK

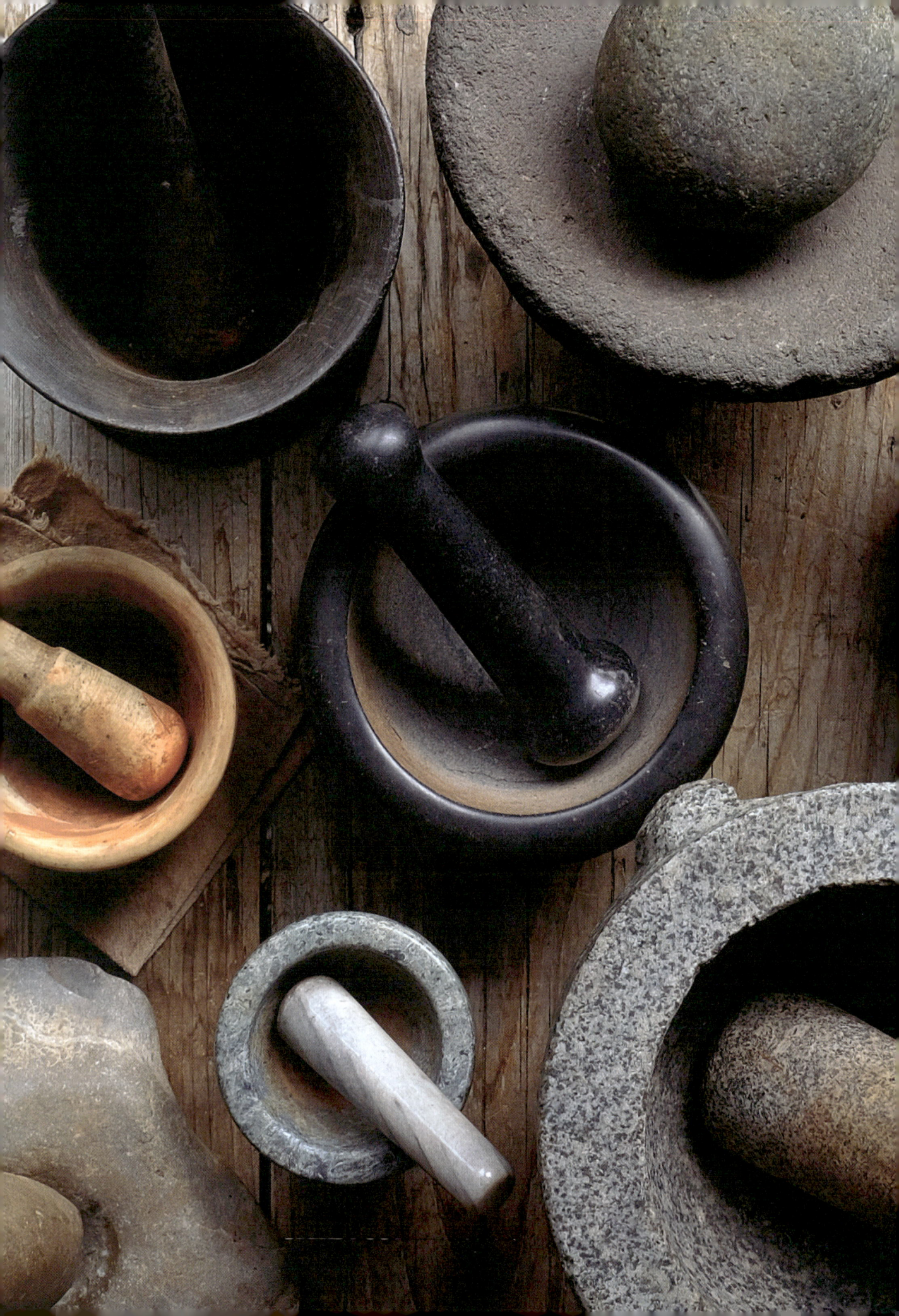

CONTENTS

PREFACE 13

ON OCHRE 17

BIOGENIC OCHRE

VULTURE MUD 39

TERRA SIGILLATA 49

RED OCHRE

SUPERGENE ORE 69

RED? 73

YELLOW OCHRE

SHADOW ORE 91

GREEN EARTH

EARTHGRIP 109

BLUE OCHRE

BLUE CLAY 129

NIGHT SOIL 135

BLACK OCHRE

LODESTONE 147

WHITE EARTHS

BONE DUST 165

FROM ROCK TO PAINT

Key Processes 175

Seed Recipes 193

A Quick Reference Chart 199

NOTES 203

RESOURCES 211

BIBLIOGRAPHY 213

IMAGE CREDITS 219

ACKNOWLEDGMENTS 222

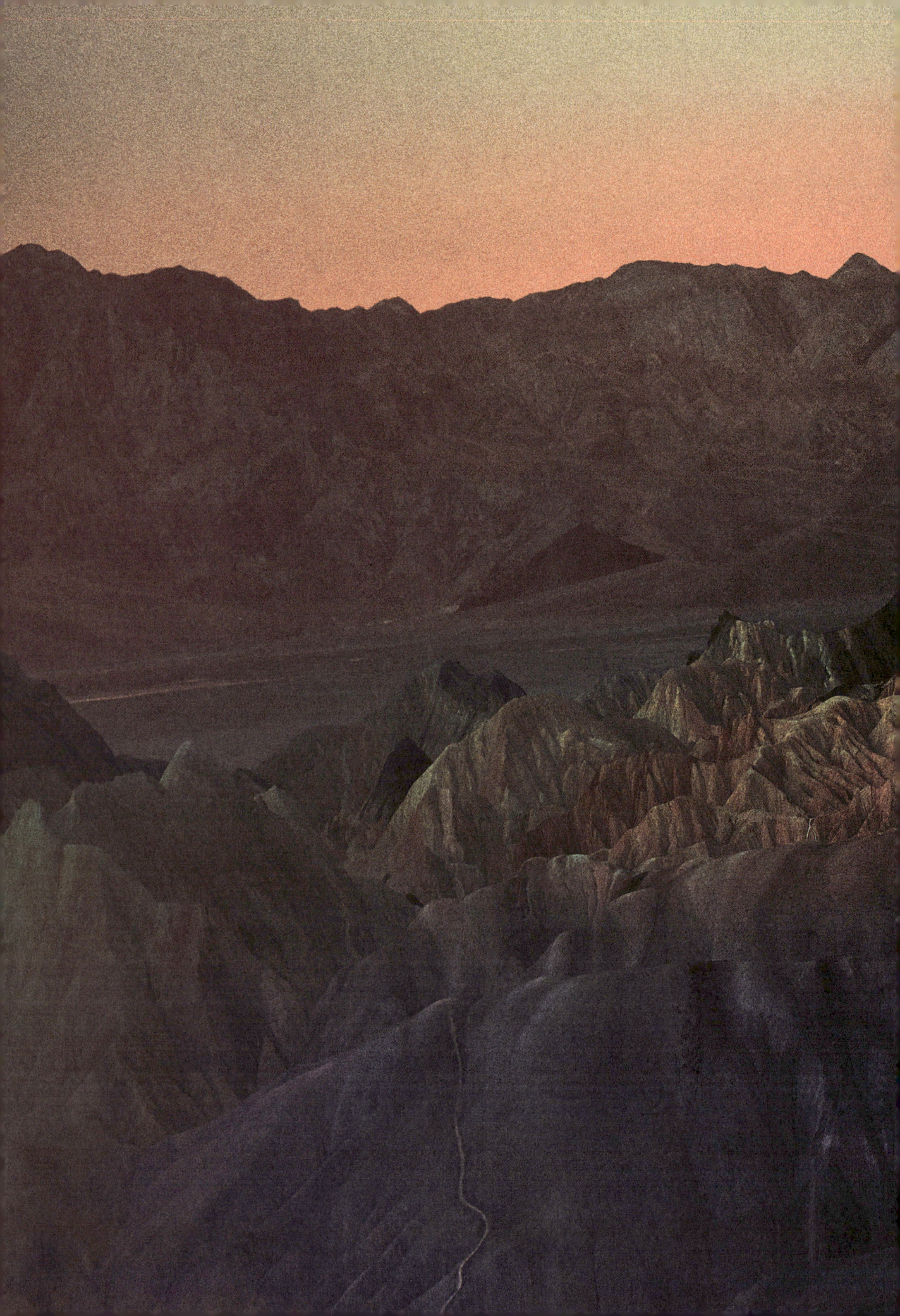

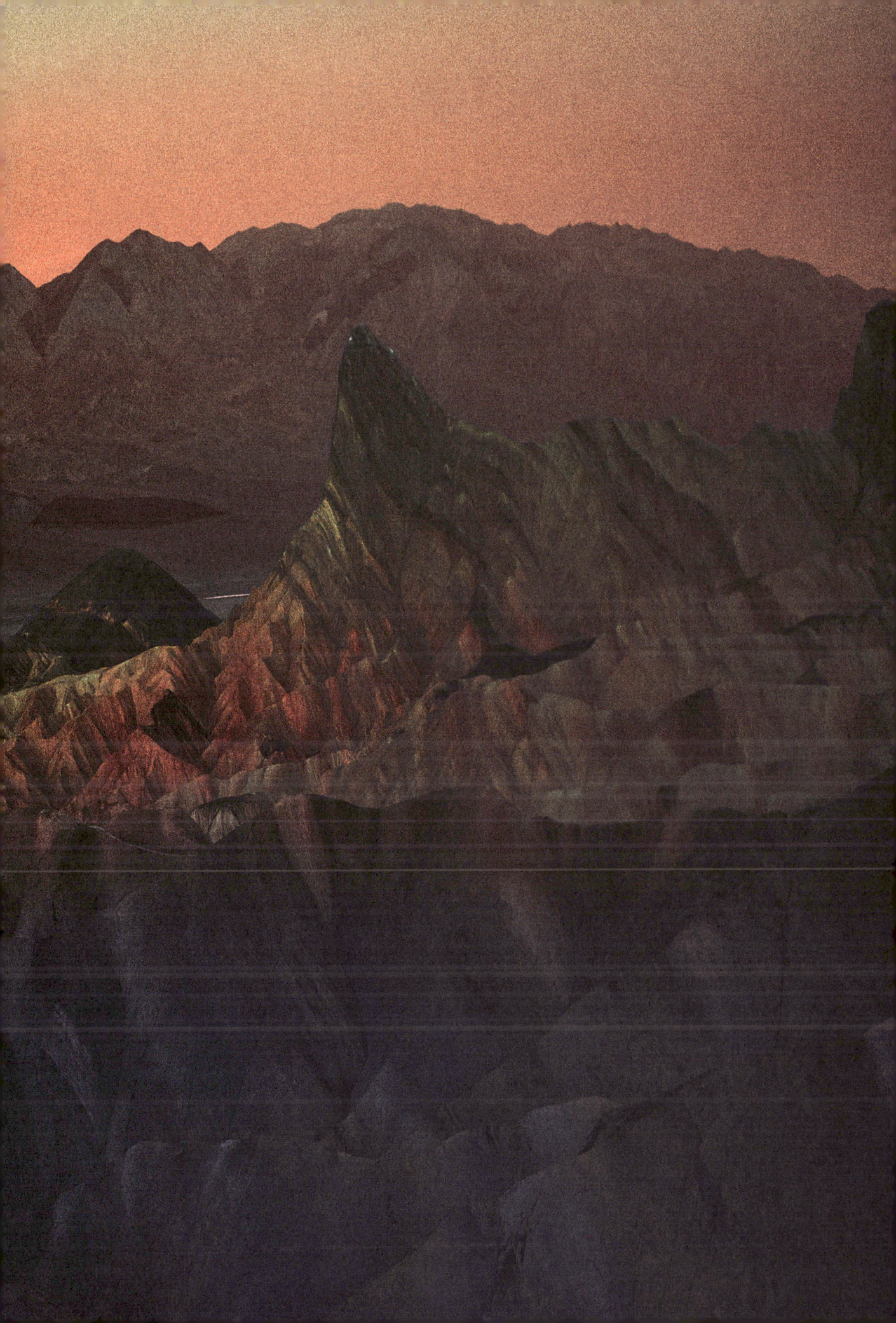

I felt like I was almost looking at a secret
that humans weren't supposed to see this
It's too beautiful . . . like looking into absolute paradise
My first reaction was to look away from it
I actually turned my head.
I thought . . .

I'm not supposed to be looking at this.

—MICHAEL MASSIMINO,
ASTRONAUT ON WITNESSING EARTH FROM SPACE FOR THE FIRST TIME

So the heart breaks
Into small shadows
Almost so random
They are meaningless
Like a diamond
Has at the center of it a diamond
Or a rock
Rock.

—JACK SPICER, *BILLY THE KID*

The nonhuman animal, the rock, the river,
the beach, the wind, and soil: let them be heard,
be represented and representable
in the governance of the earth.

—ELIZABETH POVINELLI, *GEONTOLOGIES*

LAND ACKNOWLEDGMENT

Indigenous and Native peoples, specifically Coast Salish, Nooksack, Lummi, Methow, Duwamish, Ohlone, and Lenape peoples are the traditional caretakers and storytellers of the lands on which the majority of this book was made. I respectfully honor Indigenous and Native stories, traditions, technology, knowledge, and cultures, and celebrate their continuous connection to land, water, sky, and creatures and their stewardship of land across time and space. I seek to be a good guest on traditional Indigenous lands and in the presence of the ancestors, heeding the words of Chief Si'ahl "Seattle" (translated from Lushootseed): "When your children's children shall think themselves alone in the fields, the store, the shop, upon the highway, or in the silence of the pathless woods, they will not be alone. . . . At night when the streets of your cities and villages will be silent and you think them deserted, they will throng with returning hosts that once filled and still love this beautiful land."[1]

DEDICATION

The title *Book of Earth* came to me after my best spiritual friend and Buddhist scholar/practitioner Steven Goodman died from a rare blood iron disorder during my early writing. One morning a few weeks after he died, I thought I heard his voice saying *book of earth . . . book of earth . . .* out of the blue, whispered as if by way of secret afterlife telegraph. I took it to heart and immediately researched the phrase. I found an almost identical historical title, *The Book of the Earth*, naming an elaborate series of ancient Egyptian funerary texts (ca. 1100–1200 BCE).

Unlike books as we know them today—something like this to hold in your hand—funerary texts were stories carved and painted on stone tomb walls. They are seen only when entering dark underground catacombs. Ominous images—a stunning red glowing orb; humongous boats lifting a midnight sky; a double-headed Sphinx spreading into a horizon—would be painted from floor to ceiling surrounding a royal mummy in the center.

These strange and colorful "earth spells" were said to envision, or re-create, the nocturnal path of the Sun below the depths of Earth (a deity named ༄ Aker, earth, ochre). Were they an afterlife survival guide for souls? Dark magic? I imagine a confused dead king or lost soul somehow able to experience the tomb paintings as lifelike magic whose art transforms their death into a luminous cosmos. Death turns from scary to tender, protecting their spirit through a journey in which, if willing, they learn how to conquer heavens and split open a "sky made of iron."

My intentions are to carry this vision in honor of Steven, and to enact a series of earth spells—images that may magnetize and guide through an unknown passage where, like the daily cycling of our Sun, something slowly dawns and calls us back into this world.[2]

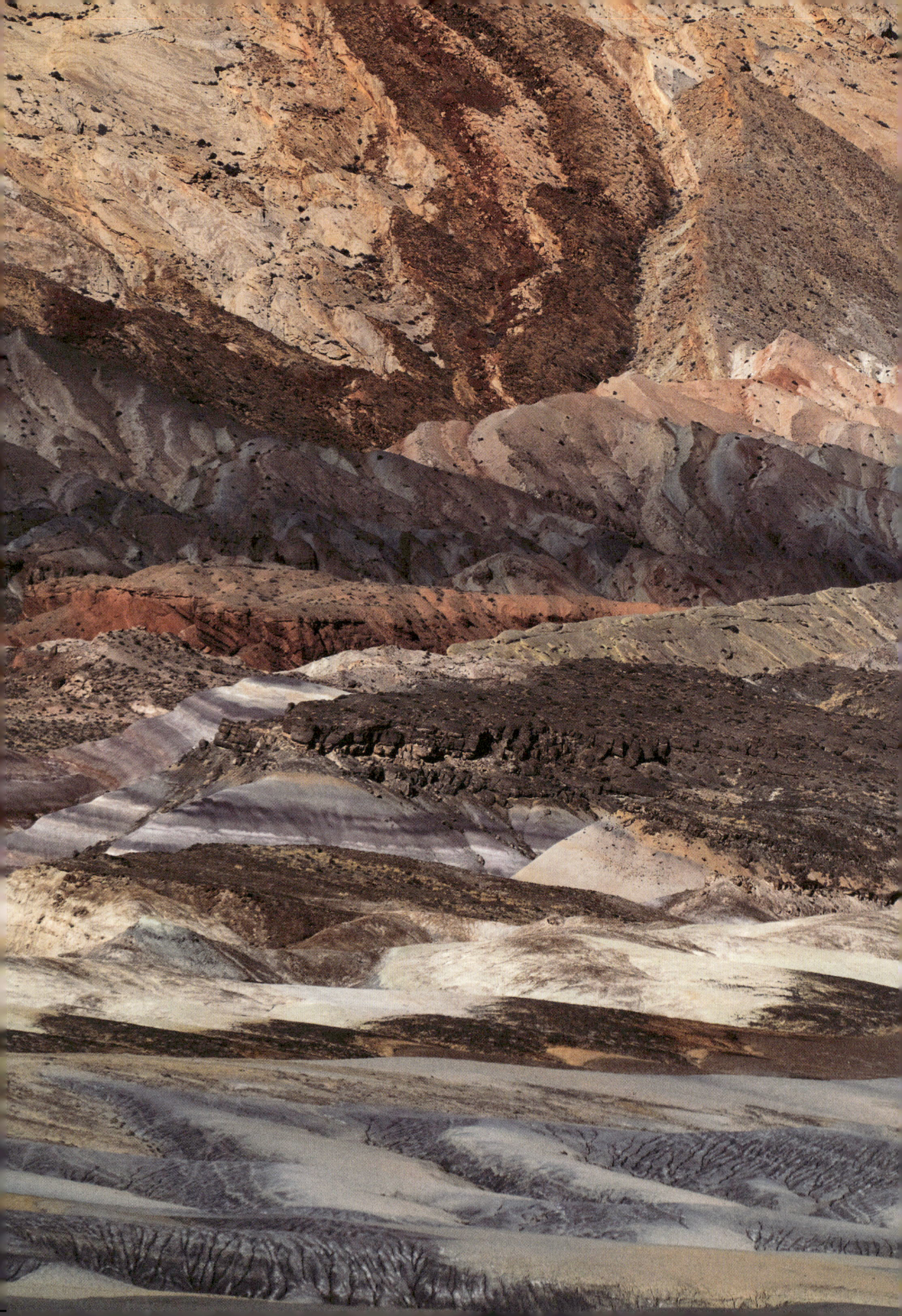

I go inside a place I belong to for the first time.

I see old, worn earthen floors, dovecote niches sculpted by hand.

A wall made entirely of pale marble. An oracular mirror?

Another wall holds shelves full of crystalline minerals,

twinkling in thousands of colors.

They make all art materials, medicines, anything   a voice says

I turn to look to who is speaking.

What a breathtaking being.

I don't remember who they are.

I ask *what's your name?*

Ochre Window.

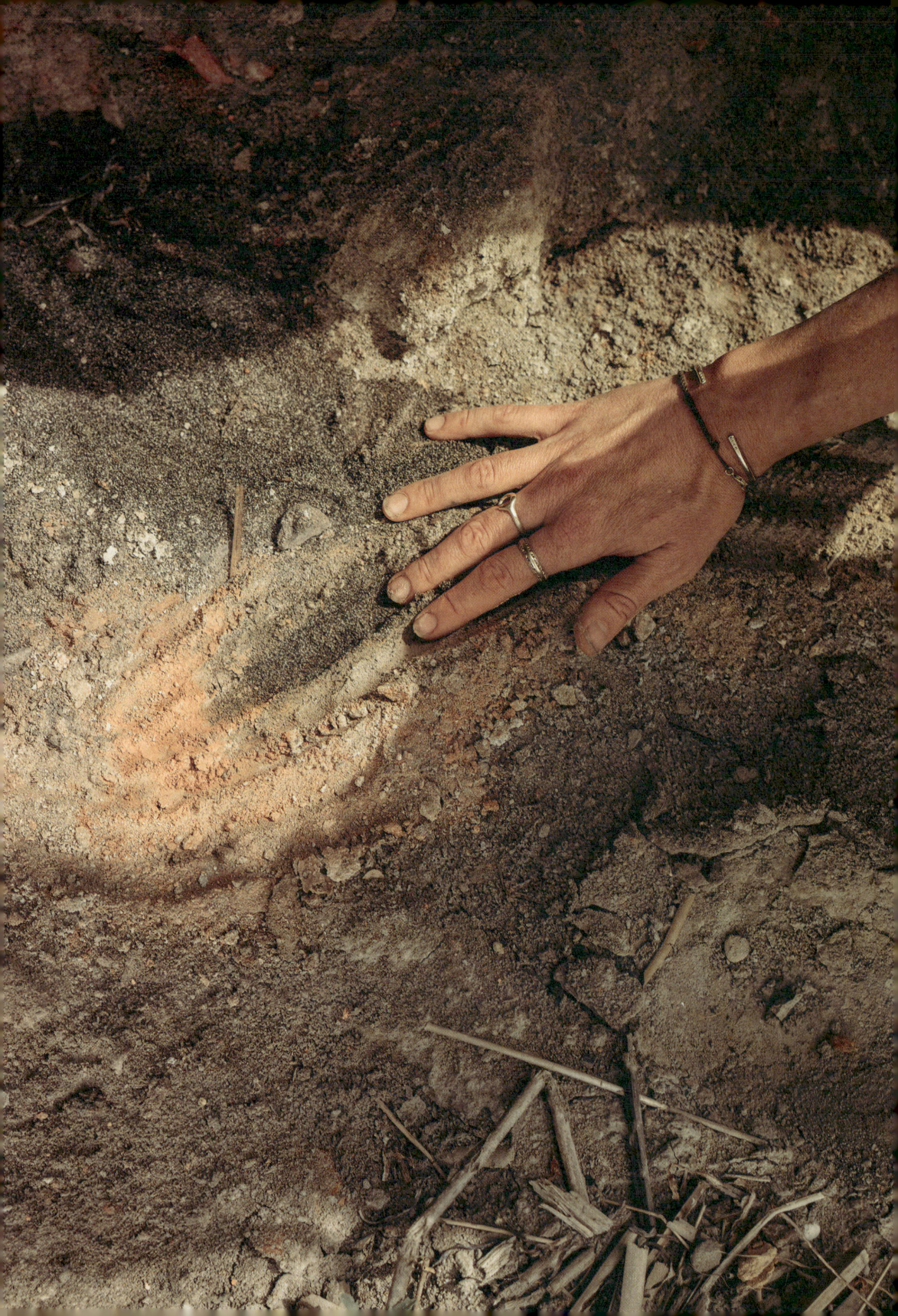

PREFACE

I envisioned *Book of Earth* some years ago in my imaginary inner studio, albeit without a title. The book I originally saw had no words and rocks spoke for themselves. I did not really know what ochres or earth pigments were or why they mattered. A decade later, I now offer this book as one of many attempts to translate what ochres have come to mean to me and how they seek our attention.

Structurally, *Book of Earth* follows humankind's oldest art material, a full spectrum of ochre and earth pigments made with soil, stone, and land, from rock to paint to creative potency. Each chapter reveals a well-known member of the ochre and earth pigment mineral family (the "stuff" that makes color), in their rainbow order: biogenic iron; hematite and specularite; goethite and limonite; celadonite, glauconite, and other ferromagnesian silicates; vivianite; magnetite; and calcium carbonates, along with a brief mention of ferruginous soils.

While walking along with me through swamps, headlands, and soil exposures rich in shimmering earthen pigments, you may find yourself asking: How did she end up talking to and crushing rocks for a living? Or become devoted to humankind's oldest art material in the first place? And why don't these essays follow some obvious linear logic?!?

Ochre and earth pigment are situated at the nexus of huge elemental cycles: a gazillion years of

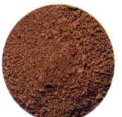

biogenic iron

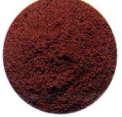

hematite
specularite

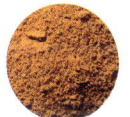

goethite
limonite

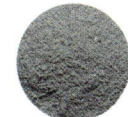

celadonite
glauconite
other ferromagnesian silicates

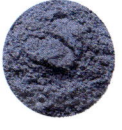

vivianite

magnetite

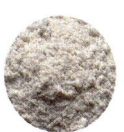

calcium carbonates

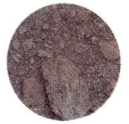

ferruginous soils

13

outer space galaxy creation, a few billion years of geologic and biologic growth on (and of) Earth, and several hundred thousand recent years of human evolution. To explore ochres as a unique and wild *prima materia*, or raw cosmic material, and talk about their influence today, I feel responsible to move fluidly between lenses, scales, places, and kinds of knowledge, often without much forewarning. That said, *Book of Earth* focuses mostly on *matter* and *materiality*, the mythic geologies of color.

My method follows from artistic research, which is a professional discipline that enjoys diverse forms of knowledge and seeks resources from many walks of life. Casual conversation, rigorous collaboration, scribbles, and de-compositions made while on psychedelics, personal experimentation, and devout contemplation are all important forms of research and experience. I deeply value and prioritize insights from dreamwork and inner realms, as well as nonhuman life-forms (rivers, microbes, storms). This method aims to synthesize scientific insight—the current mythic paradigm of a large part of my culture—as accurately and respectfully as possible, with the helpful review of expert colleagues. To me, science, art, spiritual life, and sensuality are each to be honored as masterful ways to intimately engage Earth's creative process, closely observe matter (and spirit), and make meaningful connections between them.

My formal training is fairly interdisciplinary: I was born on a biogenic ochre swamp creek and trained in childhood as a painter. I got an art degree in sculpture, an almost-completed degree in forensic science, a several year co-teaching mentorship in early childhood education (with Baltimore city's legendary educator Ms. Winnie Thomas), graduate research degree in philosophy and religion (with a specialty in philosophy, cosmology, and consciousness), a handful of incredible mentor-friends (whom I'll get to later and in the acknowledgments), and over a decade of non-institutional spiritual training with an inner counsel of beings, including a central guide who is dedicated to teaching and protecting ochre.

I mention teachings from my personal ochre guide, whose name is Ochre Window, where publicly appropriate, just as I try to name other beings (human and otherwise) whom I learn specific knowledge from and who teach me about ochre in extrasensory or non-ordinary ways. For me, insight from spiritual beings and nonhuman life-forms are a foundation from which I can slowly learn and validate teachings through observation, curiosity, shared experience, engaged sense perception, questioning, deep research, and formal experimentation.

Because of my multivalent and open-minded training, in writing and translating the subtle worlds of ochres, I will often change perspectives and scales—as if going from a mythic microscope to a scientific megaphone to a galactic stethoscope—which you may find abrupt, or aimless at first, as I do not follow linear storytelling and instead honor what is real to my lived experience. I tell stories how they appear to me in my heart.

That said, my perspective is limited, and I communicate and make art from within those bounds. Although I wrote *Book of Earth* for any person to explore, this book isn't for everyone. Nor is it an easy, breezy, encyclopedic global trot around the planet to check out colorful places to travel or escape to. While I celebrate and heed diverse brilliance from several places on Earth, my birthplace (the Pacific Northwest, northern Lake Washington, traditional lands of Coast Salish and Duwamish peoples) is an important, central influence. As a descendant of ya'aats' xaatgaay, the term used by the Haida people of the Northwest Coast for "Iron People" (or Europeans), my thoughts and work are informed by my heritage and cultures (Western Euro-American, materialist, artist-alchemist) and shared in my primary languages (Anglo-Saxon/English and images). I respect and celebrate that each different culture, clan, nation, down to each individual person, stone, and their ancestors, may have their own rules, spirituality, timing, and relationship with ochre.

As I've become initiated into the wisdom of ochre, earth/soil pigments, and their beauty over the years, I've come to understand that there are also limits to discussing them. You may (or may not) notice that I make every effort not to disclose images, place, or knowledge shared with me in confidence, nor do I mention specifics involving ochre use from known sacred places not meant to be shared outside of a specific culture as a form of respect and effort at protection, unless given permission or publicly available from proper sources.

Color gathering and making are specialized activities in many cultures, my own included, with certain restrictions for physical, cultural, or spiritual safety and protection. Yet, their discreet influence is a rather big part of our lives. Thus, while I do offer introductory recipes and ways to engage—as a shared creative inheritance to renew, celebrate, steward, and protect—I ask you to consider how to integrate earth pigment into your own specific cultural time and place and ecological bioregion, and with appropriate teachers, with respect, reciprocity, and pleasure.

Personally, my integrity thermometer is ochres themselves: If in doubt about the right relation, I look to and ask ochres and their guardians, in myriad ways. I won't always recognize or hear an answer. So then I follow Robin Wall Kimmerer's teaching from *Braiding Sweetgrass*: "How can we distinguish between that which is given by the earth and that which is not? When does taking become outright theft? I think my elders would counsel that there is no one path, that each of us must find our own way. In my wandering with this question, I've found dead ends and clear openings. Discerning all that it might mean is like bushwhacking through dense undergrowth. Sometimes I get faint glimpses of a deer trail."[1]

—*HG, Sumas, Washington*

PS. For those with an interest in natural color and dyes more broadly, I must prepare you that I only discuss a small sliver of the natural color and pigment worlds. And within the ochres that I do discuss, I am only able to share a handful of tales and uses. If you are not familiar with the difference between earth pigments and natural pigments, here's a short summation: Earth pigments are crushed up from mineral rocks, soil, clay, mud, or other earthen lumps of geological material—generally they contain iron and soil. They're also powdered land, or in Afrikaans, *grondstof*, meaning "ground stuff" or "ground dust." You can find them on the ground (without much digging or mining). Natural pigments, on the other hand, include a far wider umbrella of material, including several other non-iron based stones, crystals, and minerals (orpiment, malachite, jade, etc.); botanical pigments (charcoal, plant extractions, nuts, galls, lakes); biological pigments from animals and insects (cochineal, murex, cow piss); and hydrocarbon-based synthetic pigments (which are, of course, also made from "earth materials").

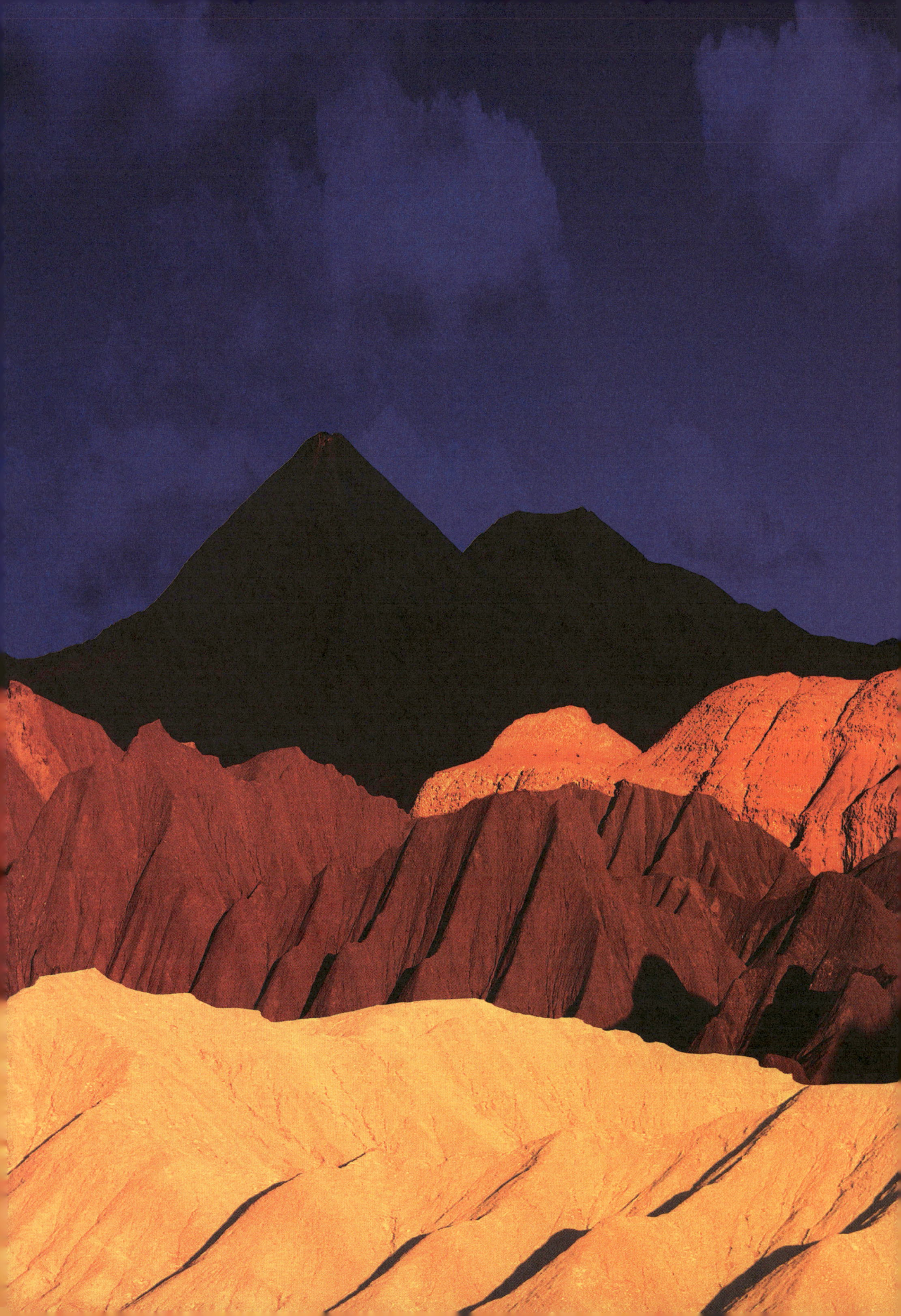

this
is the place
where time
slows down, where
light is collected and flashes
in all the colors of love it is

—CHERYL SAVAGEAU, *ABENAKI POEM*

✳

Iron seeds the imagination.

—NOR HALL, *IRONS IN THE FIRE*

On Ochre

Let's start with the endgame. Why do ochre and earth pigment matter? Because ochre shows up significantly—across humanity's cultural records—in times of crucial rites of passage. Ochres emerge over and over to ferry us through major changes including everyday basics such as birth, puberty, marriage, and death, and also the big unfathomable stuff: large-scale extinctions, weather chaos, climate changes, mass death, radical adaptations, profound loss, creative joy, new divergent forms of expression, evolution: moments that color humanity.

By most accounts, we're officially living through one of these intense threshold passages right now. Renewing bonds with the earth is a crucial and continuous process, especially in tumultuous times. I think this may be why there is a rising tide of people publicly discussing, researching, revitalizing, and protecting ochre pigment knowledge and places. Together many of us wonder: Could ochre and earth pigments help us magnetize creative response in this uncertain phase for us and our kin? Or are we going to turn this beautiful, radiant place full of song and joy and grit and bite into just another red dead orb?[1] As climate writer Elizabeth Kolbert puts it, "the Sixth Extinction will continue to determine the course of life long after everything people have written and painted and built has been ground into dust and giant rats have—or have not—inherited the earth."[2]

Perhaps my read on the importance of ochre sounds implausible and grandiose? I might have agreed with you not that long ago. Ochre? Colorful dirt? Do they really matter? *Really?* How could old paint stones be *that* helpful right now? Shouldn't I be doing something more immediate, more pressing? To me, color-making is ecological and cultural activism. Creative expression and process are a matter of life and death. "The only war that matters is the war against the imagination," as revolutionary poet Diane di Prima bravely put it. "All other wars are subsumed in it . . . and no one can fight it but you & no one can fight it for you."

In order to explore the revolutionary and deep ecological approach to earth pigments, as part of my own fight to nourish exhausted imaginations, I want to begin by introducing what ochre is and does (both today and in the past), and give a bit of historical background.

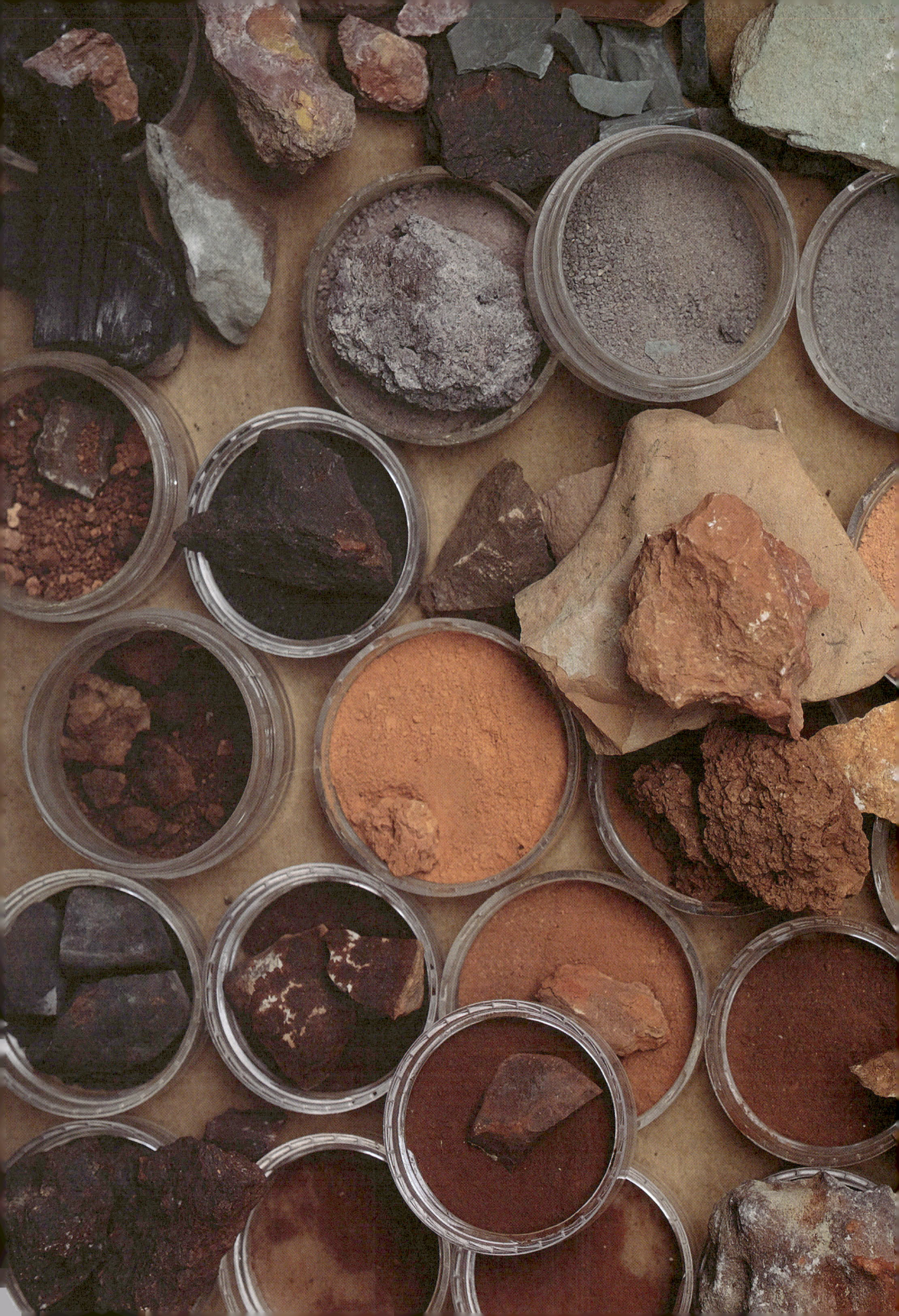

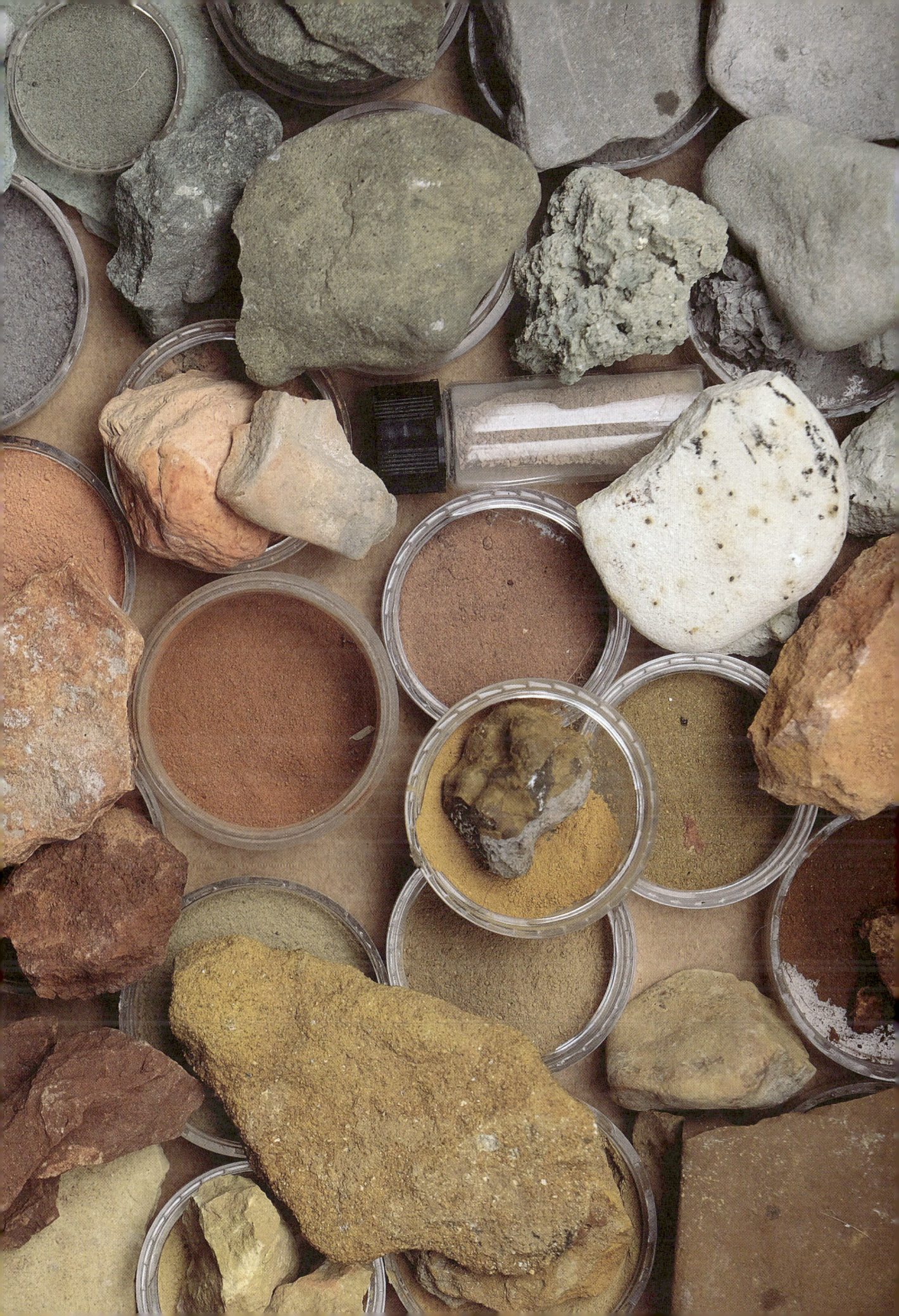

What is ochre?

Ochre, ocher, ocre, ocker, okra, ohkra, ochra, okür, oker, okker, aker—the spoken word itself—is shared among diverse tongues and cultures worldwide, but no one really agrees on what it means. I've heard ochre spoken of in uncountable ways. For some there's a deep awe, fear even; for others it means very little, like a heavy bag of cheap red dirt. To me, ochre is family and kin, a profound guide, companion, and ancestor—a supernatural material. I love American superstar Cardi B's way of saying "*okurrrrrr*," which she defines as, "I ain't know she had all of that in her."[3]

What exactly *is* ochre? What's it made of? What *do* we know about "all of that in her"? And who cares?

Ochre usually conjures a predictable, benign earthy color. A color devoid of mystery or value. A dirty yellow? Musty brown? Or maybe a prehistoric red?

Archeologists use the term *ochre* or "iron earths" to identify humanity's primal colors, used worldwide to draw cave lions and bears, steppe bison, aurochs, giant sloth lemur, and other extinct mammals, as well as affix early symbols, handprints, sprinkles on shells, antlers, bone, graves, and deepest cave ceilings. Painters throughout millennia have used ochres (iron oxides), and it is quite challenging to find paintings lacking in their presence. Red ochre was laid beneath what was gold and gilded, indicating ordinary soil was a background for spiritual life. Modern artists (of all kinds) and builders continue to follow suit today, applying durable, pervasive ochres and earth pigments as central agents for both its function and color in cement, plaster, paints, pastels, pencil, ceramics, textiles, cosmetics, digital printing inks, and on and on.

Research scientists identify ochre as part of the "iron oxide and hydroxide" group of iron-ore minerals, which include sixteen specific mineral types. Geochemists tend to equate ochre more broadly with *all* metal oxides (lead, copper, mercury, etc.). Recent papers in prehistoric archaeometry attempt fresh definitions such as *ochre sensu lato*, meaning "natural materials with an iron mineral chromophore that are culturally significant in symbolic and applied roles."[4]

Iron-ore mining, the largest metal extraction industry on Earth, on the other hand, superstitiously never mentions ochre. I've heard that some geologists go so far as to say, "Ochre isn't real."[5] Yet industrial forecasters claim the "iron oxide pigment market is expected to surpass US\$3.82 billion by 2032,"[6] because of iron oxide pigment's dominant role in the construction, cosmetic, and automotive paint industries. Modern plastic and cosmetic tycoons would probably go bankrupt without the use of ochres to coax "natural" tonal ranges for everything from baby pacifiers to sneakers, lipstick to foundation. "Ochre forms such a part of our identity that it is difficult to disentangle the use of ochre from the human experience," as early sapiens behavior researcher Dr. Elizabeth Velliky sums up.[7]

Complex definitions of ochre are old news to Indigenous and Native knowledge keepers, whose stories define ochre in radically nuanced ways and have for millenia. Places rich in colorful pigments and natural paint materials are often honored and named after the ochre materials present there, signifying their importance. Pigment naming and descriptions also indicate significant ecological knowledge and may be described in great observational detail including textures, smell, function, behavior, binders, "feminine" or "masculine," geologic origin, and, of course, color. Ochre is respected as protected cultural knowledge, related to crucial ancestral land, memory, and ceremony. In some cases, like with the Mirarr clans in Northern Australia, ochre-gathering practices have survived, many in closely held secret, over one thousand generations (more than forty thousand years—talk about sustainable mining). Aboriginal Australians consider ochre sacred (so much so that some books with descriptions of ochre places or practices are restricted from public access), whose material spectrum covers a full rainbow of colors.[8] My friend Mel Ladkin, a female Awabakal painter in Bundjalung Country, calls ochres simply "persons," and some local elders say they "have spirit."[9] Ochres are portals, potencies, life-forms unto themselves.

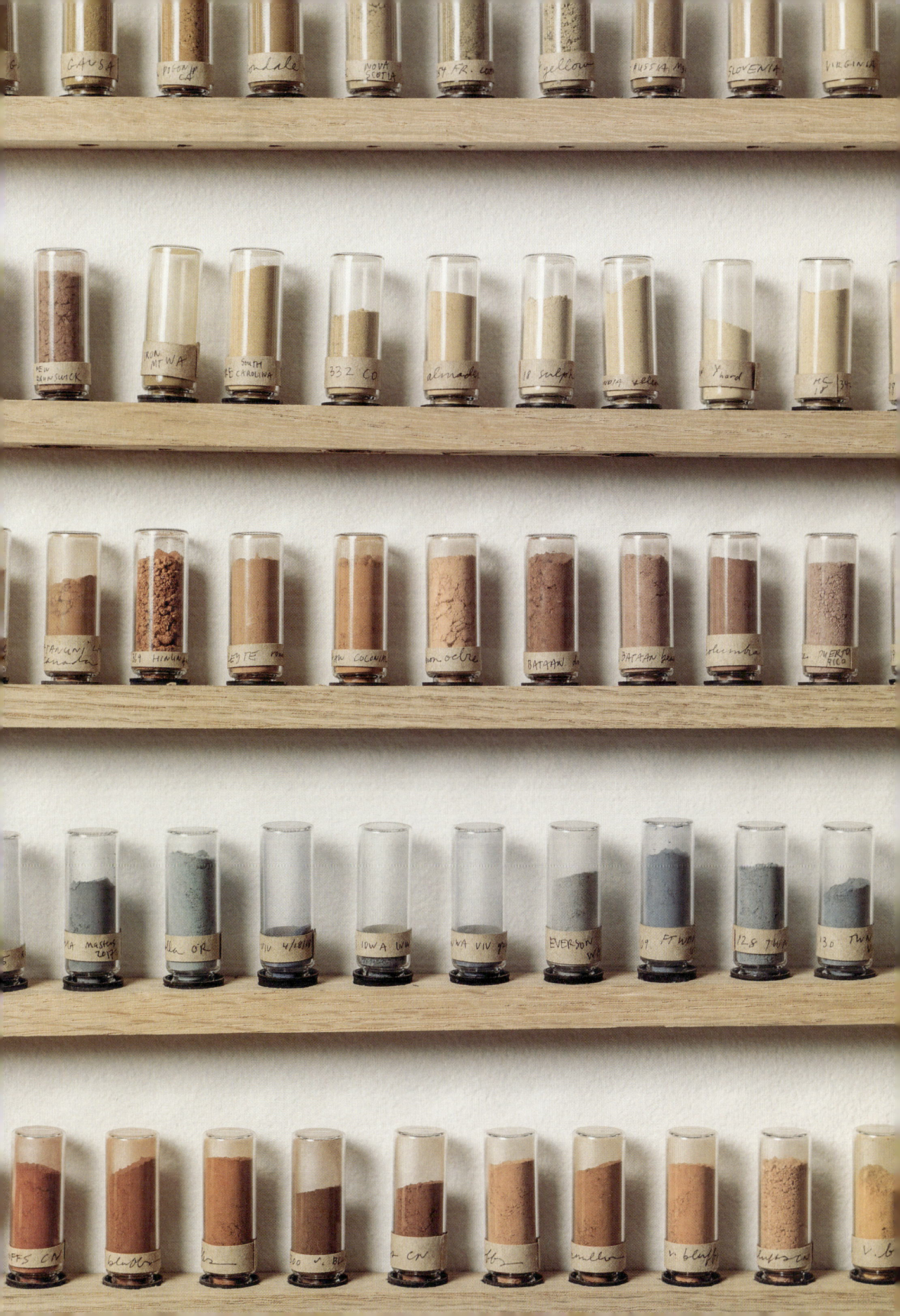

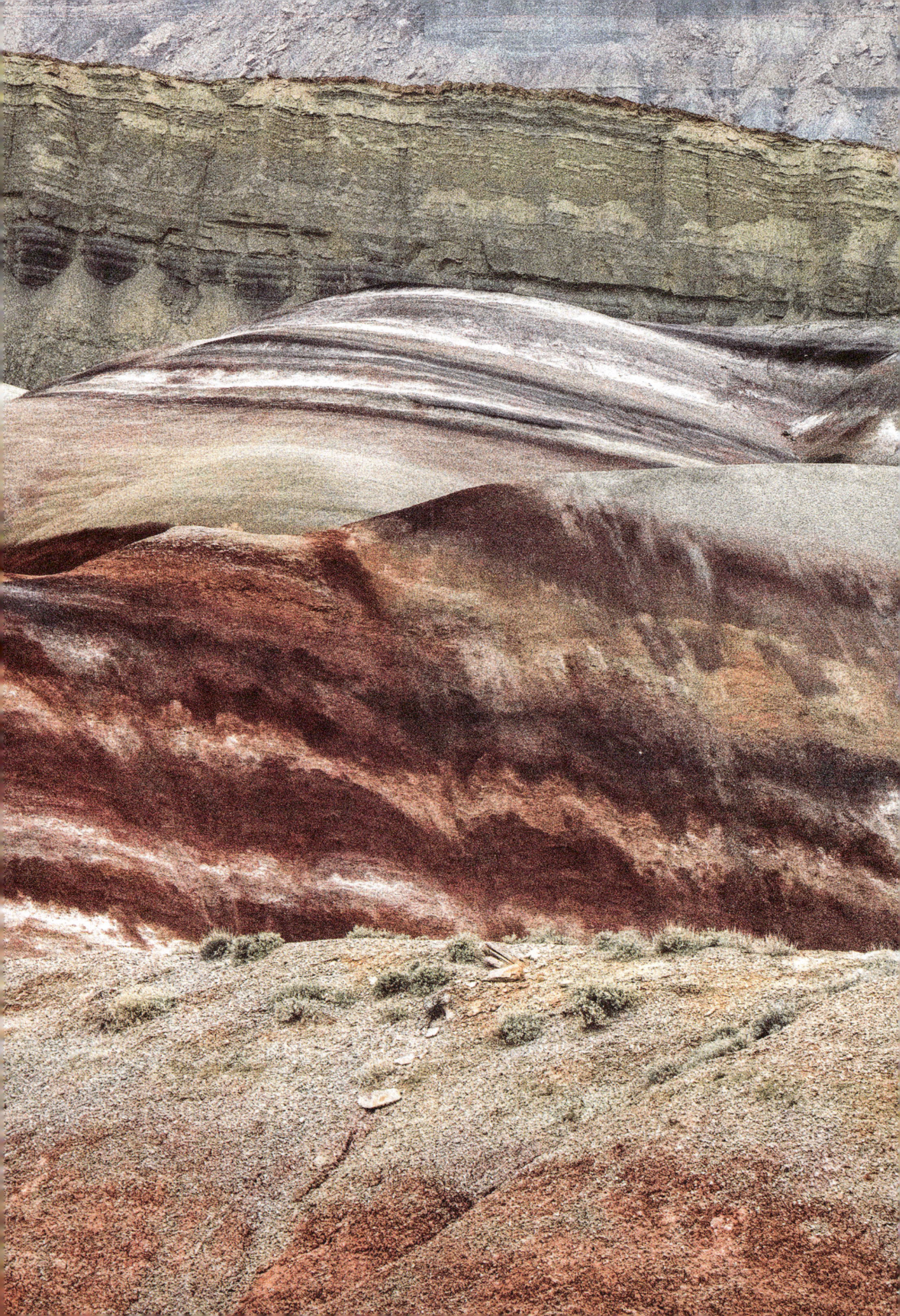

Earth & iron

In my experience, iron and oxygen plus something inexpressible is all that defines ochre. With a piece of ochre, we hold earth in our hands. And not just loose dirt—Earth.

Earth's most abundant elements, by mass, are nearly identical to ochre's makeup:

1. IRON, Fe (32.1%), concentrated largely in the planetary core or heart

2. OXYGEN, O (30.1%) in protective atmosphere, and bound in rocks and our own bodies

3. SILICON, Si (15.1%) in land, soil, sand, crystals, crust, and mantle

The estimated chemical composition of Earth overall is roughly $Fe(Mg)SiO_3$ (which is a slightly green ochre). Iron (Fe), oxygen (O), and land (Si): Think of these as elemental seeds—the main ingredients in ochre.[10]

Fe, the chemical symbol for iron, is central to ochre, our planet, and the entire universe. Ochres feel timeless and can be anywhere from a few seconds old (fresh rust) to three billion years old (banded iron formations) to who really knows—but astrophysicist Freeman Dyson theorized that in the whole wide, cold universe, "all matter eventually decays to iron."[11]

Iron is a fascinating element on its own terms, and is considered unequivocally key in the interplay and circulation of Earth's creatures and "disproportionately abundant" in the universe. So what's the big deal? Iron—one of the most stable known elements—occupies a rare cosmic threshold between energy *releasing* and energy *consuming* elements: a singular, crucial interface.

In stars, for example, lighter elements fuse with heavier ones to generate energy over billions of years, until they begin to fuse with the twenty-sixth element, iron.[12] Once iron tries to fuse, it acts like the straw that breaks the camel's back: Instead of creating energy, iron initiates a heavy metal catalyst. A star will almost immediately start to collapse on itself once iron is introduced, creating an inner shock wave that explodes the stellar body into a bright supernovae. Sounds bad, right? Yet all complex forms of matter, every weird element in the universe that comes after iron (up to element 118 on the periodic chart thus far) *would not exist* without the nuclear breakdown made by iron's self-destructing power![13] Stellar nurseries are full of iron dust that clumps together to form future planet cores. Thus, iron in space could be considered a skeleton key of galactic creation cycles. Eerily, iron meteorites could be some other star's collapsed, broken-apart heart.

So I imagine ochre (iron earths, iron oxides) as a shared creative portal between the cosmos, Earth's heart, and my tiny life.

If I look closely within my own little stardust body, I, too, find that I am full of microscopic, ochre-rich mineral deposits. Magnetic black ore gathers in my brain. White earth dust in my bones. Iron shepherds oxygen through my veins via my blood—a direct line to red ochre and hematite, or "the stone that bleeds." When I metabolize, breathe, or move, it is thanks to iron's long bond with oxygen.[14] When I run hard and breathe too deeply, I'll probably taste bloody rust on my tongue. We contain inescapable intimacy with earth and dust. We *are* made up of earth. As traditional Italian pigment artist Carlo Romiti muses, "there are landscapes that have long been digested, which live inside of you."

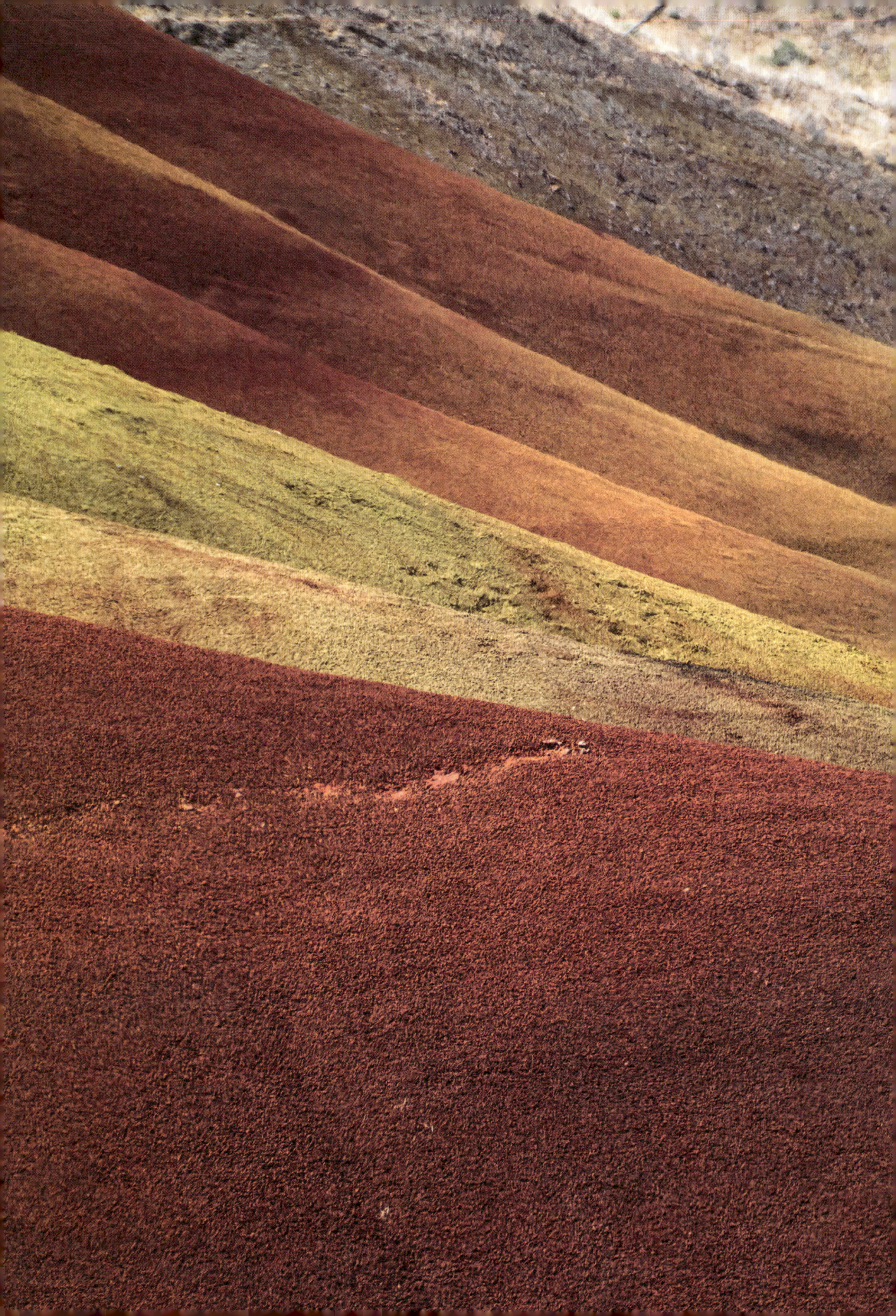

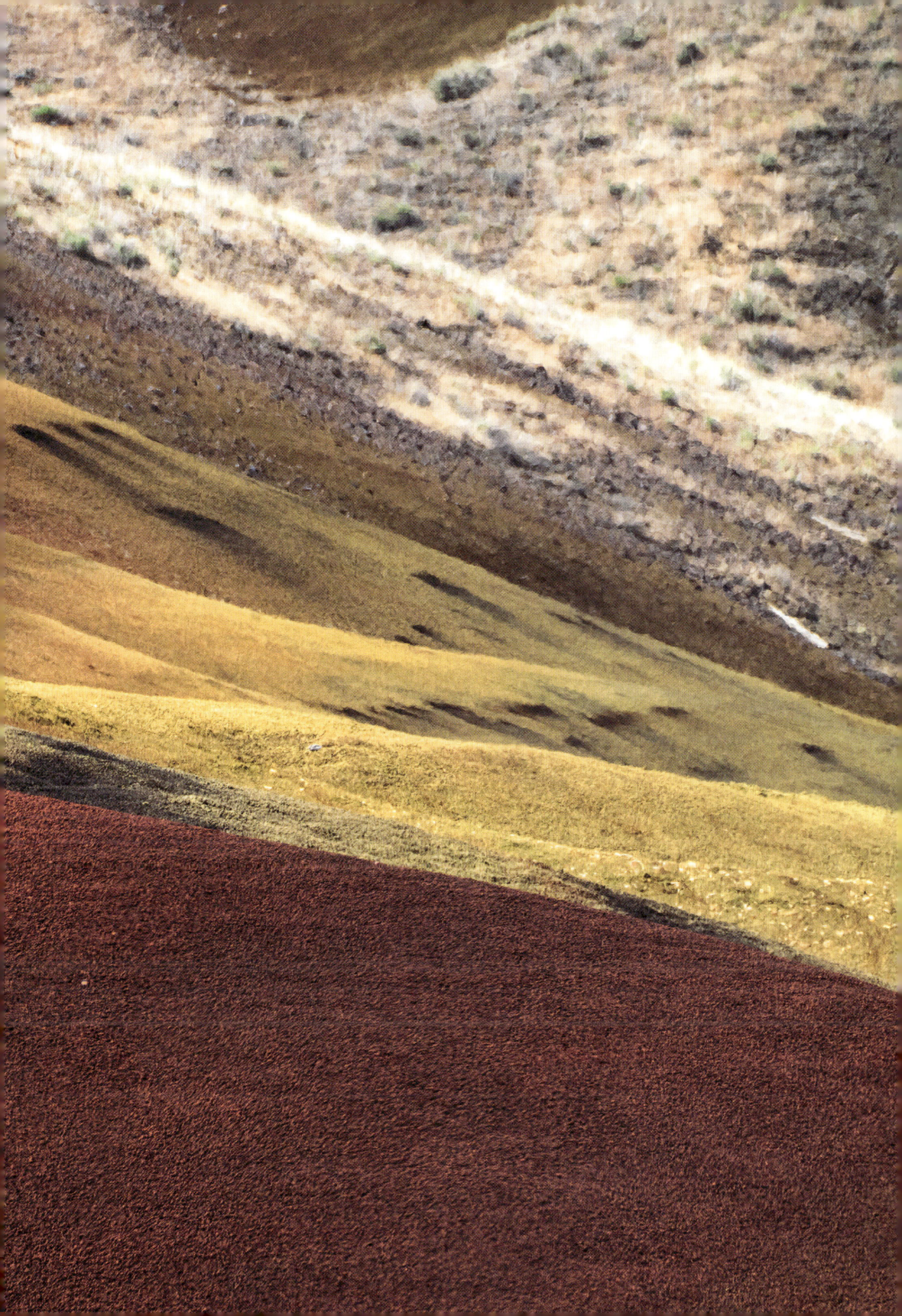

A very brief history of humans & ochre

Just how long have people made natural paint from rocks? Humanity's ochre practices developed over the last 500,000 years[15] in Africa and the last 250,000 in Europe. According to the archaeological record, our ancestral human (and other now-extinct relatives, like Neanderthals and Denisovans) big brains emerged at least 300,000 to 200,000 years ago, alongside red ochre procurement. Neanderthals, for example, collected and used red ochre-rich liquid at Maastricht-Belvédère in the Netherlands approximately 250,000 to 200,000 years ago.[16] The development of "modern" human cognition more clearly emerged around 120,000 years ago, coinciding with an increase in collection and processing of ochre and earth pigments, and utilizing them in diverse ways when applied to body, hair, hands, shelter, shell, hide, sinew, ground, wood, vessel, field, flint, bone, grave, and cave. This advance in ochre use in the archeological record appears to align with the emergence of symbolic thinking, and a segue to durable, written language. Modern humans exist in an ancient lineage of ochre pigment gathering, processing, and use—without which we likely would not behave or think or, most importantly, imagine or create how we currently do.[17]

As early humans shapeshifted their bodies and minds, larger amounts of worked ochre material indicate an expanding and meaningful influence (over eight thousand ochre pieces were found in South Africa's Blombos Cave [70,000 to 100,000 years ago] and 5,000 in Sibudu cave [37,000 to 77,000 years ago][18]). Pieces of metallic red rock were purposefully engraved. Pieces of ochre were also rubbed into hand-shaped, heart-sized paint stones. Sometimes they were powdered and mixed with tree or plant resin and water, stored in abalone or oyster shells or special sacs. These were formed, presumably, in something akin to prehistoric workshops, where humans purposefully developed recipes for long-lasting and portable paints, facilitating the spread of human culture across the land.

With red and yellow and burnt black paint lines, more complex images and marks developed. Nearly all ancient painted or drawn pictographs—found on every continent except Antarctica—rely on ochre, starting around 75,000 years ago (though these dates are always rather controversial), with the oldest so far found in South Africa (of engraved lines), Indonesia (of jungle buffaloes and wild pigs), and Iberia (of handprints).

Our ancestors seem to *need* ochre to metabolize their existence. With ochre (and, to be fair, a lot of manganese and charcoal, too), ancestral humans bore witness to ecosystems, marked secret spots, expressed themselves, and displayed an expansive reach. At first glance, the primal process of rubbing, scraping, and spraying walls with pigment feels similar to other apex predators who mark their territory and erotically signal in a similar process with urine, scat, and fur.

Thousands upon thousands of early ochre images depict flourishing animals, seasonal herd encounters, constellations[19] and starry skies moving overhead, seasonal events, and of course, fellow people—their handprints, activities, and habits. I love how some drawings are positioned in caves at points of echo, making what appears to be a static image more alive, even responsive when sung, whistled, or whispered to. Similarly, flickering firelight helps activate the whole scene, as if in an early cinema. This gets especially fun when encountering infamous strange hybrid beings with radiating luminous heads—are they shamans? spirit? some kind of talk-show host? gods? aliens? half-animal/half-human (therianthropes), badass boss people?

In academic communities, "ochre use" (the historical when, why, how of it) is commonly split into distinct categories:

- **personal expression and display (body paint, marking color on shell, bones, beads, etc.)**

- **functional, domestic and medicinal use (sunscreen, insect repellent, hide-tanning)**

- symbolic and communal ritual or ceremonial use

- funerary practices

- and if none of that is quite right, it is often labeled as art or "just" graffiti.

Color historian Michel Pastoureau muses that early humans were just "drawing attention to oneself," and for that reason, were particularly interested in adornment and display. Red ochre in particular has been shown to influence physiological responses—heart rate goes up, skin flushes, warmth rises, etc.—arousing and empowering those who wear and are near it (even those who are blind!).

I see the art of applying ochre, across these various categories, as an act of anointing with iron dust. Why? I cannot really say. Maybe the reason can be found in the handprints found on rock walls and caves. There are hundreds of these places worldwide. Ocherous handprints left behind on stone leave evidence of a sensual, intimate place of contact. But *why*? I agree with Diane Ackerman, who says "hands are messengers of emotion." To me, the potency and attraction of ochre indicates an empathic power: an ability to, say, merge with that "something inexpressible."

In my experience, ochres open deeply sacred, and sometimes dangerous, thresholds. Because ochre initiates a psychic negotiation between places, beings, and feelings we can't always grasp, understand, change, we can use it to draw nearer to unknown places. Disparate worlds bleed together with ochre. I touch a rock and, perhaps, Earth touches me back.

I promise it is not as weird or uncanny as it sounds. Interspecies interaction can take huge creative leaps and material energy. Think of all the equipment (e.g., microscopes and telescopes) we invented to try to relate to microbial viruses and cells, or alien planets and galaxies. Or what of all the books and prayers and rituals to connect with felt but unseen beings?

I like that ochre helps us grasp and imagine worlds beyond our own immediacy. For example, I could paint that big scary animal smaller, and make it less frightening to me. Or I could see yesterday's eclipse again tomorrow in my painted recollection of that moment. Through our use of ochre and paint, not only do we draw on others and other places and display and communicate that, but we can enable a part of our memory to live outside ourselves. We can share our inner, secret worlds. In this capacity, it seems safe to speculate that people learn to inhabit themselves differently, and perhaps inhabit others ("put yourself in their shoes"), through art and the materials associated with it, which may in turn lead to an ability to subjugate, manipulate, or control others—or dominate and destroy them, or invite greater curiosity, understanding, and empathy. The creative power of art and images goes every which way. Human heritage gives us a creative ability to traverse deep time and space, and yet each person is ultimately destined to settle and sink further and further into the earth eventually. Life, art, belongs to Earth.

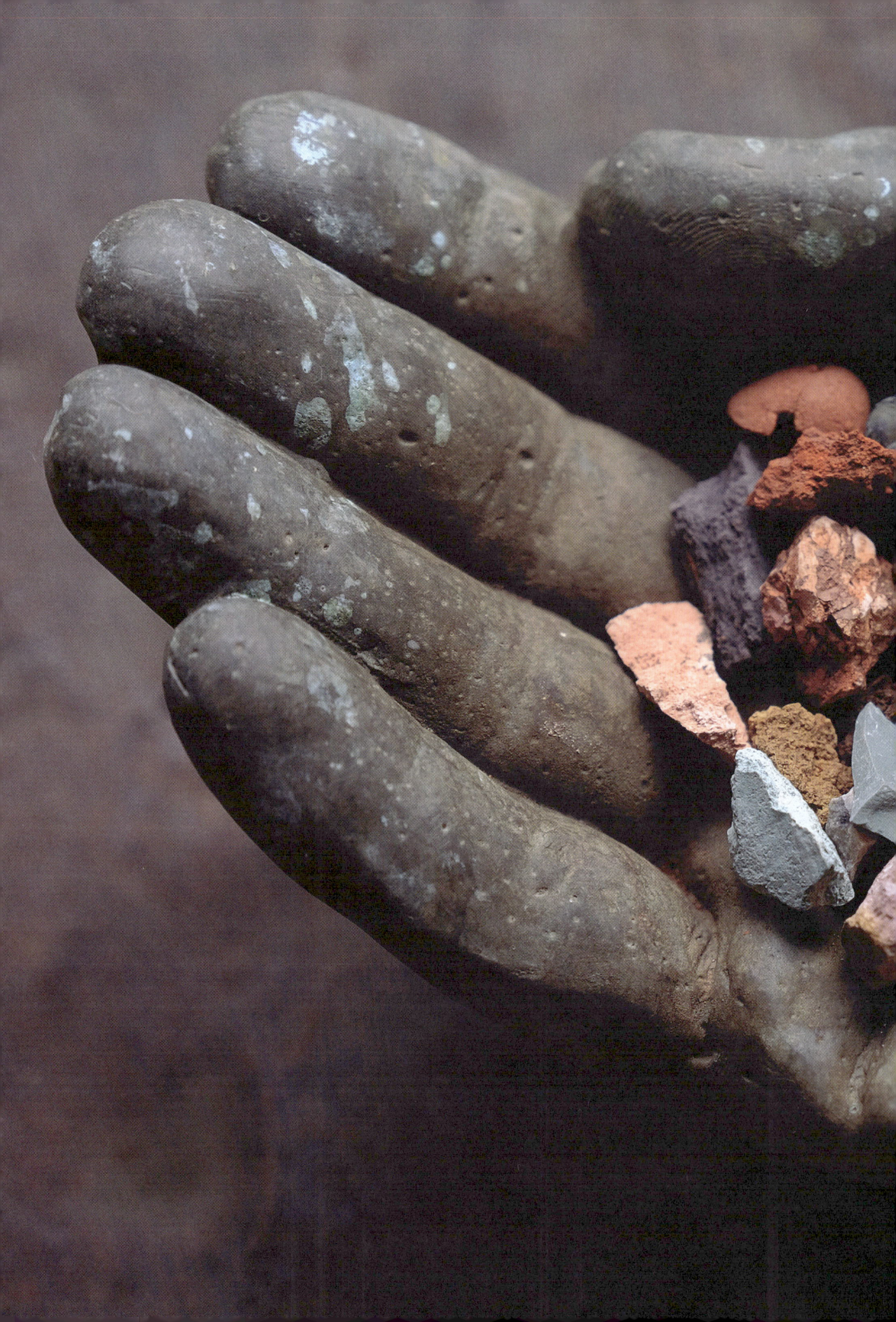

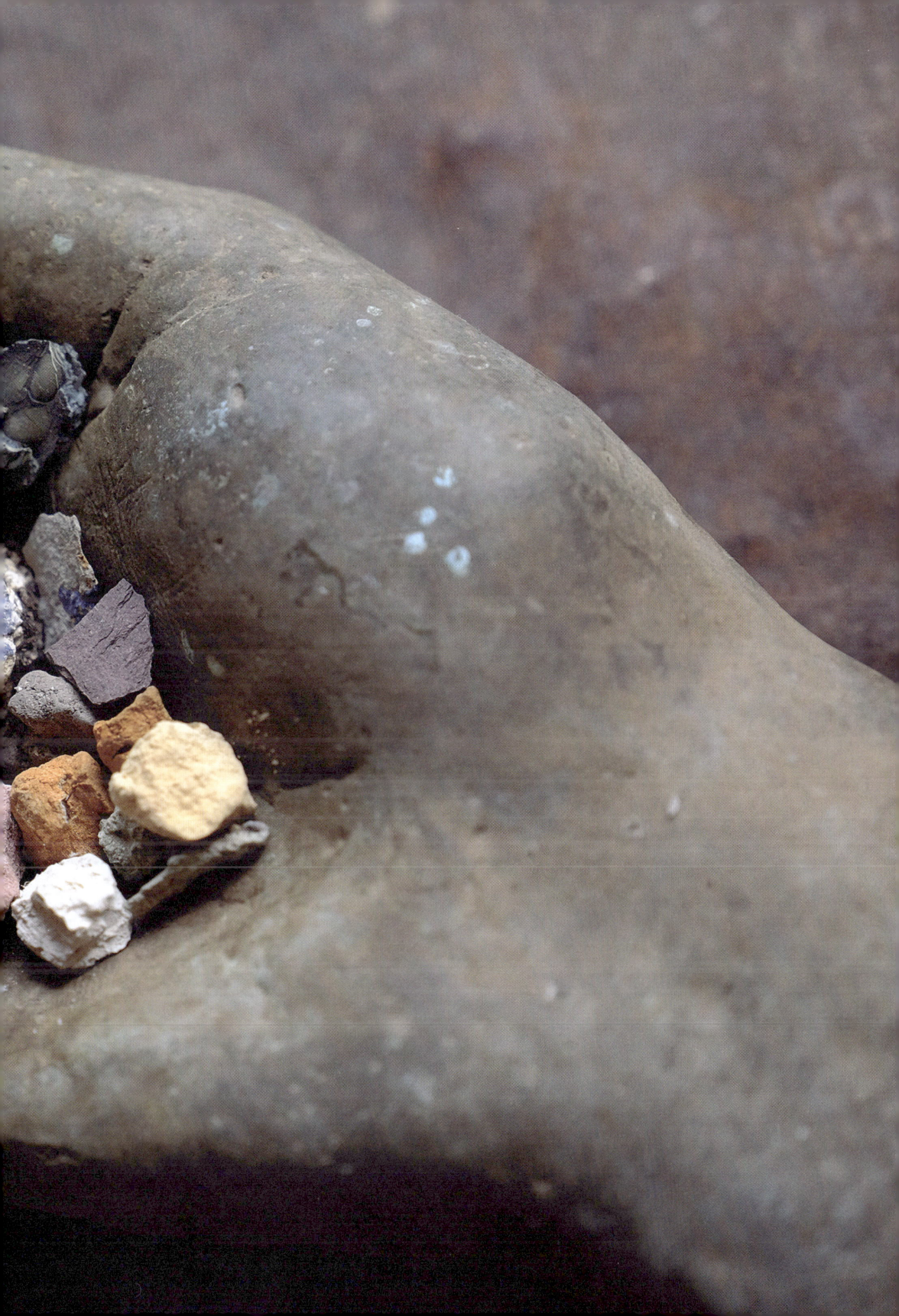

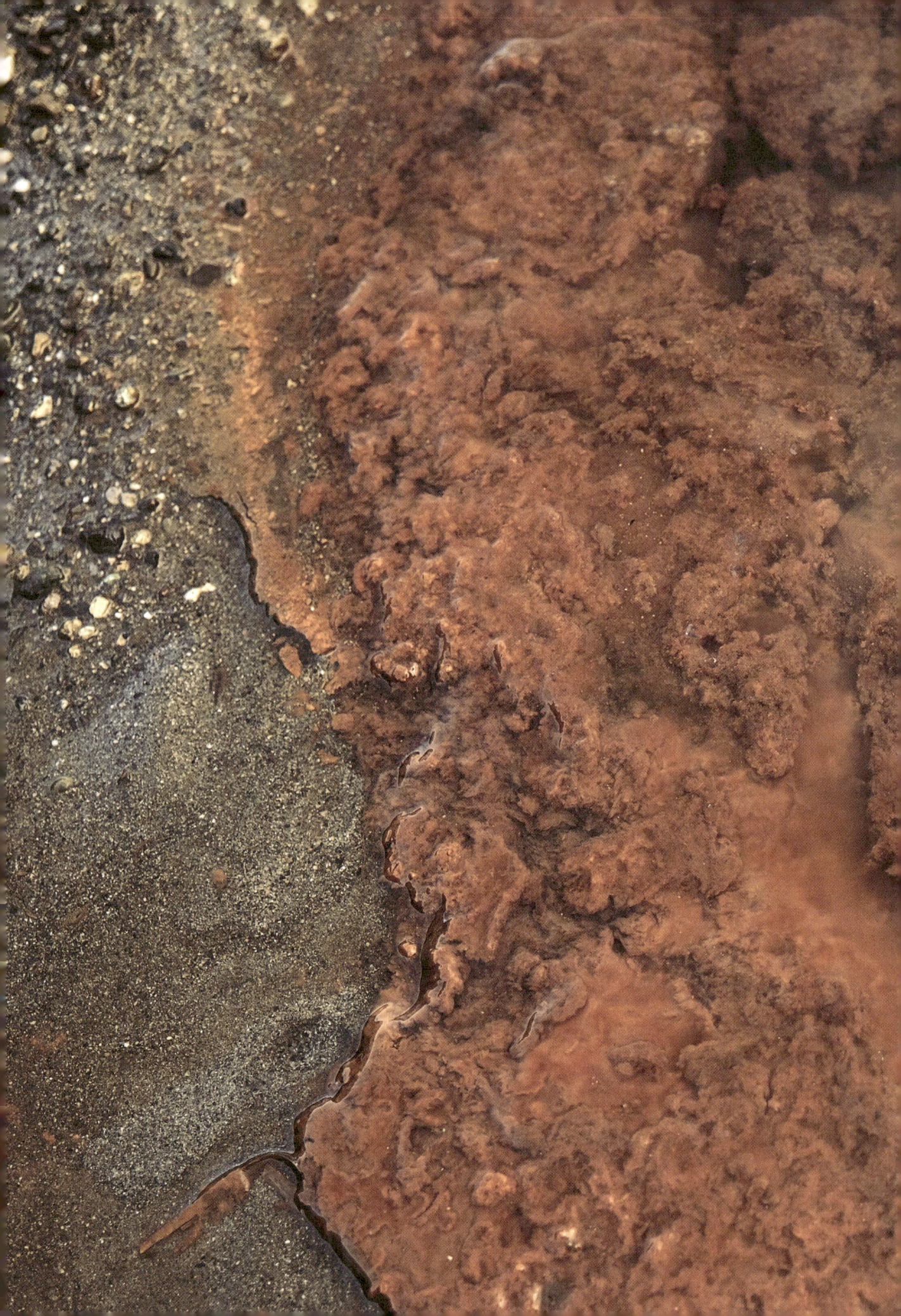

BIOGENIC OCHRE

BIOGENIC IRON, OCHRE SCUM

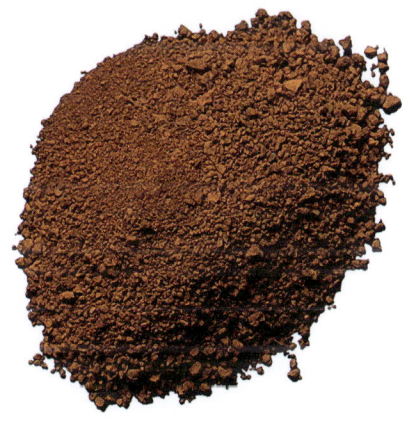

FERRIHYDRITE OF IRON-OXIDIZING BACTERIA ▽

Etym. from *bio*, life, and *genic*, producing, "produced by life"

FeOB, Fh, $Fe_5O_8H \cdot H_2O$

PIGMENT SPECTRUM

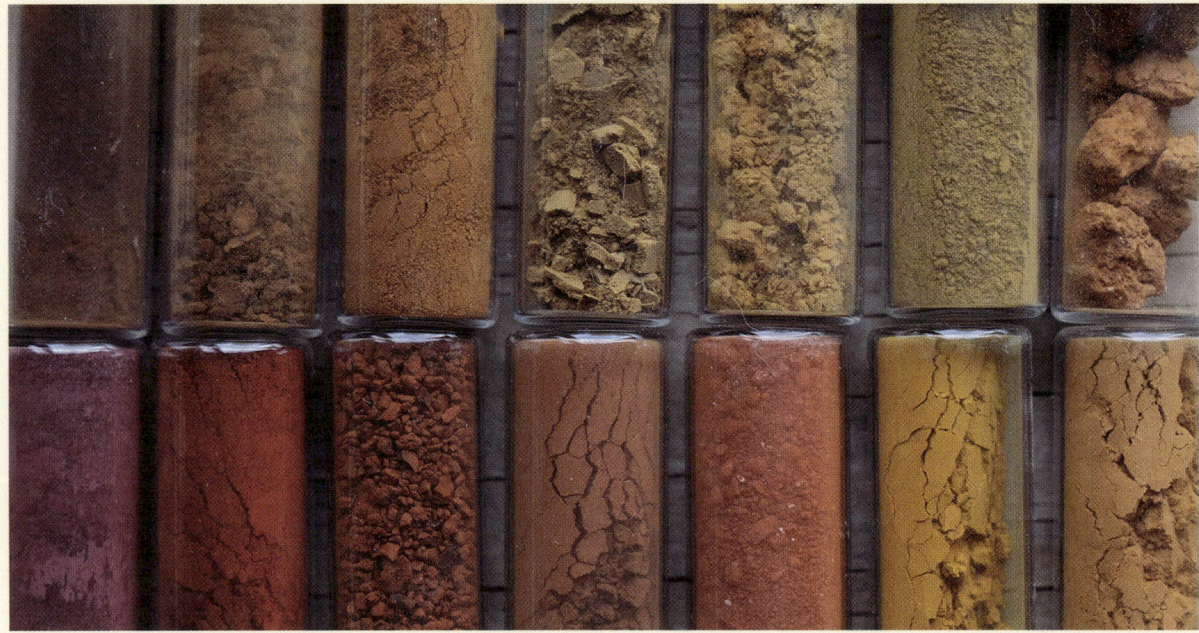

MATERIAL DESCRIPTION

Commonly observed as primordial ooze, yellowish scum, mud, or muck near warm springs, biogenic ochre emerges from microbial organisms found in water. So ochre begins already alive. Interconnected thick mats, flocs, and delicate veils of ochre form as "waste" from microbes as they basically eat iron and excrete it as iron hydroxide (FeOH). Specifically, ferrophagic (iron-eater) bacteria cells seek dissolved iron and oxidize it into essential energy, while they grow a wild tail or flowing rock-record of their metabolic activity, secreted in filaments, hollow sheaths, ribbons, and twisted DNA-like stalks of the FeOH, which clump together to form prolific cathedrals in their wake. From a human's perspective, it just looks like wispy clouds or filthy orange water.

Dissolved, or free, iron is found in freshwater rivers, swamps, creeks, and saltwater oceans, primarily at deep vents on the seafloor, hydrothermal springs on the surface of Earth, or near iron-rich weathering rock (often iron sulfides, pyrite, and lignite) and acidic ponds and streams (near coal, copper, and sulfide mines). Why do they form here? These places, sometimes called natural paintpots, create unique diurnal, thermal, and acidic pH conditions in the water that allow for the seasonal thriving of chemolithotrophic proteobacteria families, such as *Gallionella ferruginea* (in spring), *Leptothrix ochracea* (year-round), *Ferriphaselus spp.*, *A. Ferrooxidans*, *Mariprofundus ferrooxydans*,[1] that are chiefly responsible for ochre creation (notice how their names honor iron, *ferr-* in Latin). What appears as iridescent sheen on top of slimy waters is actually a collection of thriving bacterial organisms. These ochre creations, delicate as a fluffy pillow or thick as mud, are unstable and very easily disturbed, and can dissolve completely *or* solidify into stable, harder iron ore and then mineralize into a prismatic spectrum of more mature iron oxide family members (goethite, hematite, magnetite). Thus, biogenic sludge may be honorably referred to as "the mother of ochres" or "they who give life."

NAMES

goo, slime, mother of ochre, living mud, living ochre, biotic ochre, microbial ochre, biogenic iron oxide, orange scum, fat of the earthe, *sphragis*, acid-mine drainage, oxidized dissolved iron, iron water, natural acid rock drainage, chalybeate, yellow boy, irn-bru "iron brew," járn-brák "iron slick," ferr-ochre, fairie ochre, red ochre, yellow ochre

HABITAT

deep springs — wetlands — iron seeps — aquatic habitats — old mines, especially coal — pyritic watercourses — acidic pH liquids — hydrothermal vents — volcanically active seamounts — crustal spreading centers — black smoker vent fields — hot springs — holy springs — where iron-loving microbial communities live — waterways with a fluctuating or low water table — waterways after wildfire — agricultural runoff drains — creek beds — bog iron drainage — marshes — swamps — origin places — creation stories — wet and free places

CRYSTALLOGRAPHIC SYSTEM

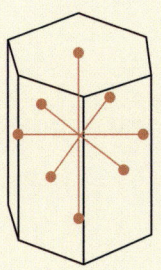

HEXAGONAL
UNSTABLE

CHROMA

ORANGE	UNEASY
REDDISH	TEEMING
YELLOWISH	RAINBOW SHEEN
VOMITY	SHIMMERING
BLOODY	WARM
FATTY	

BIOGENIC OCHRE — Made by Life

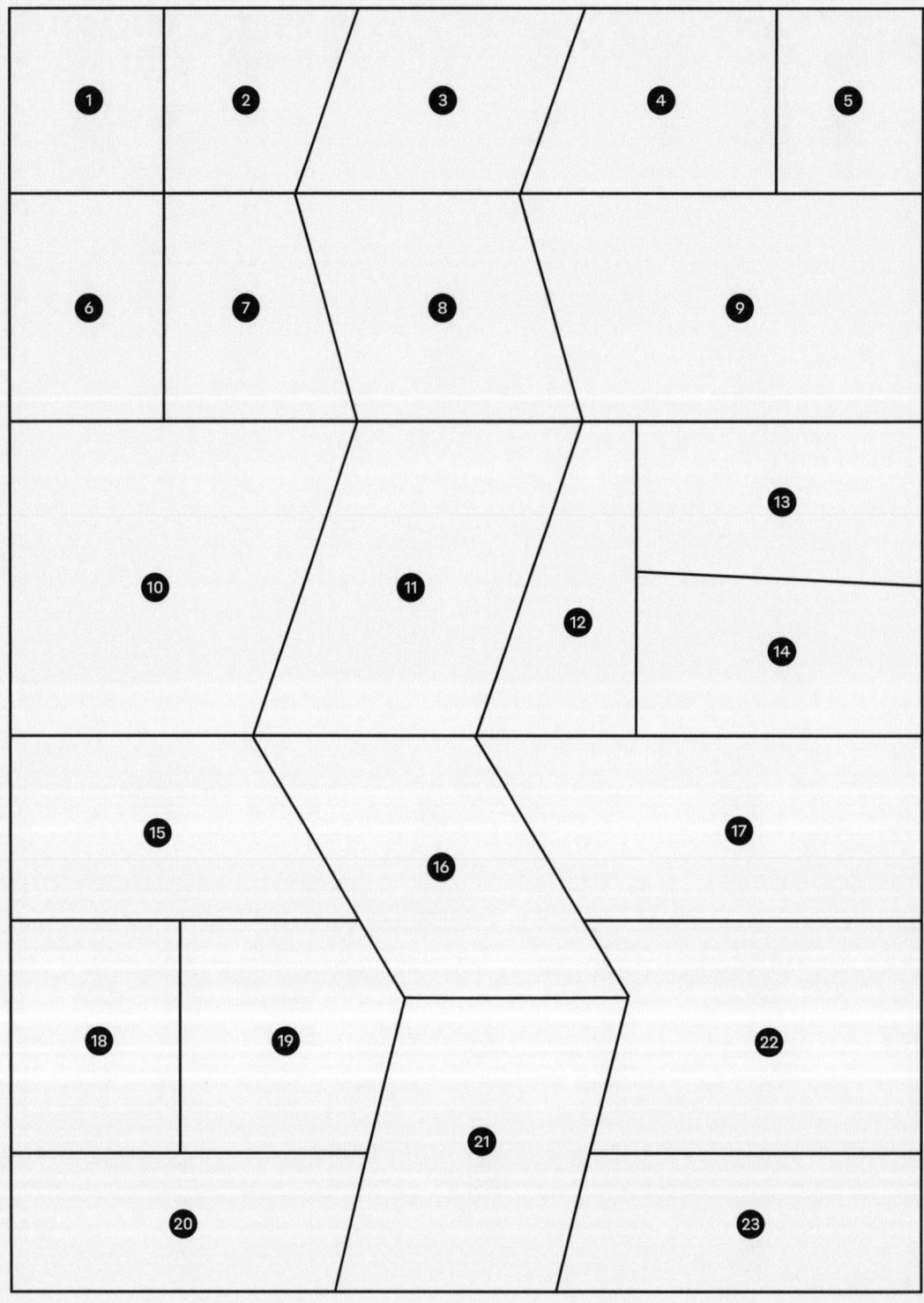

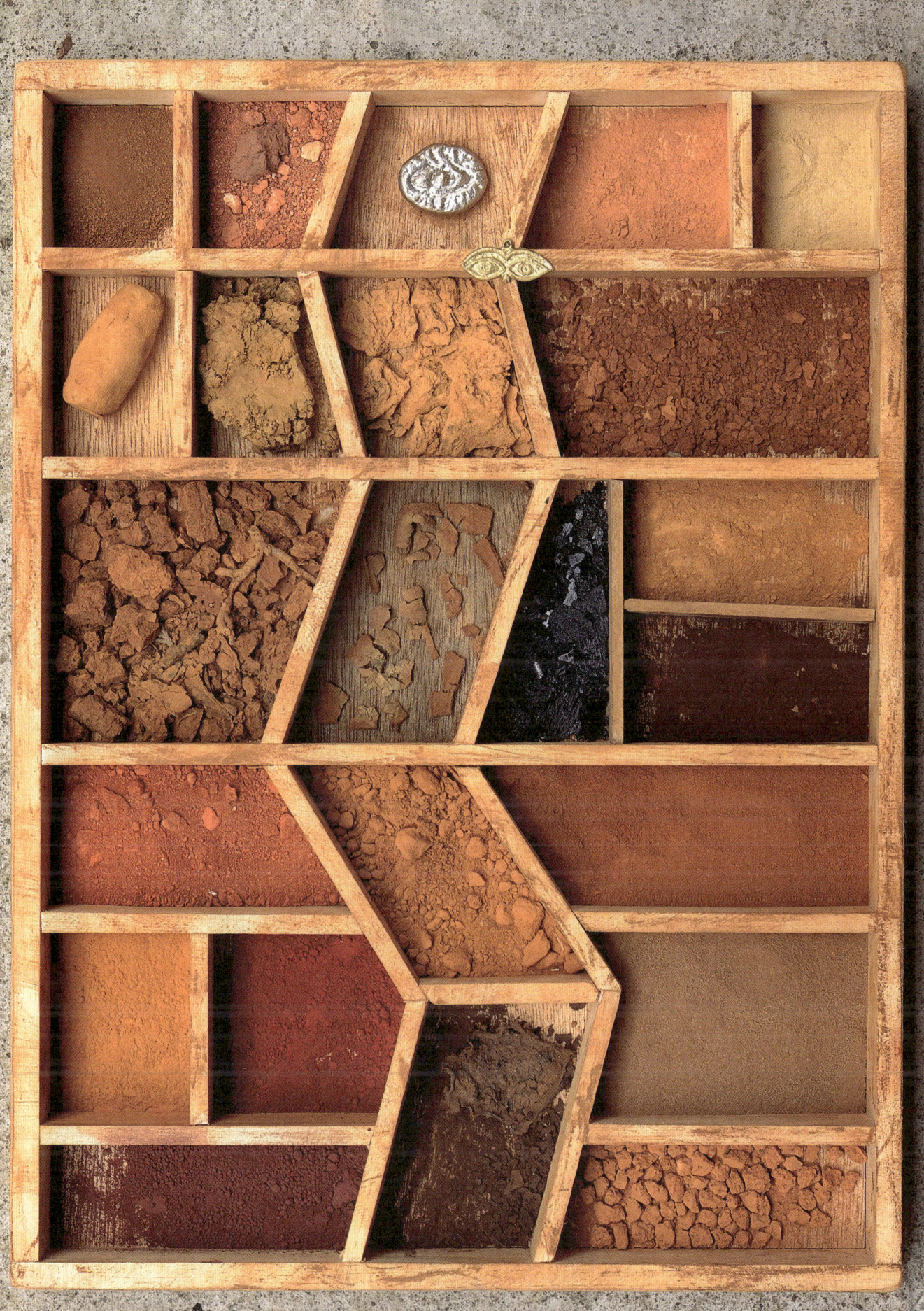

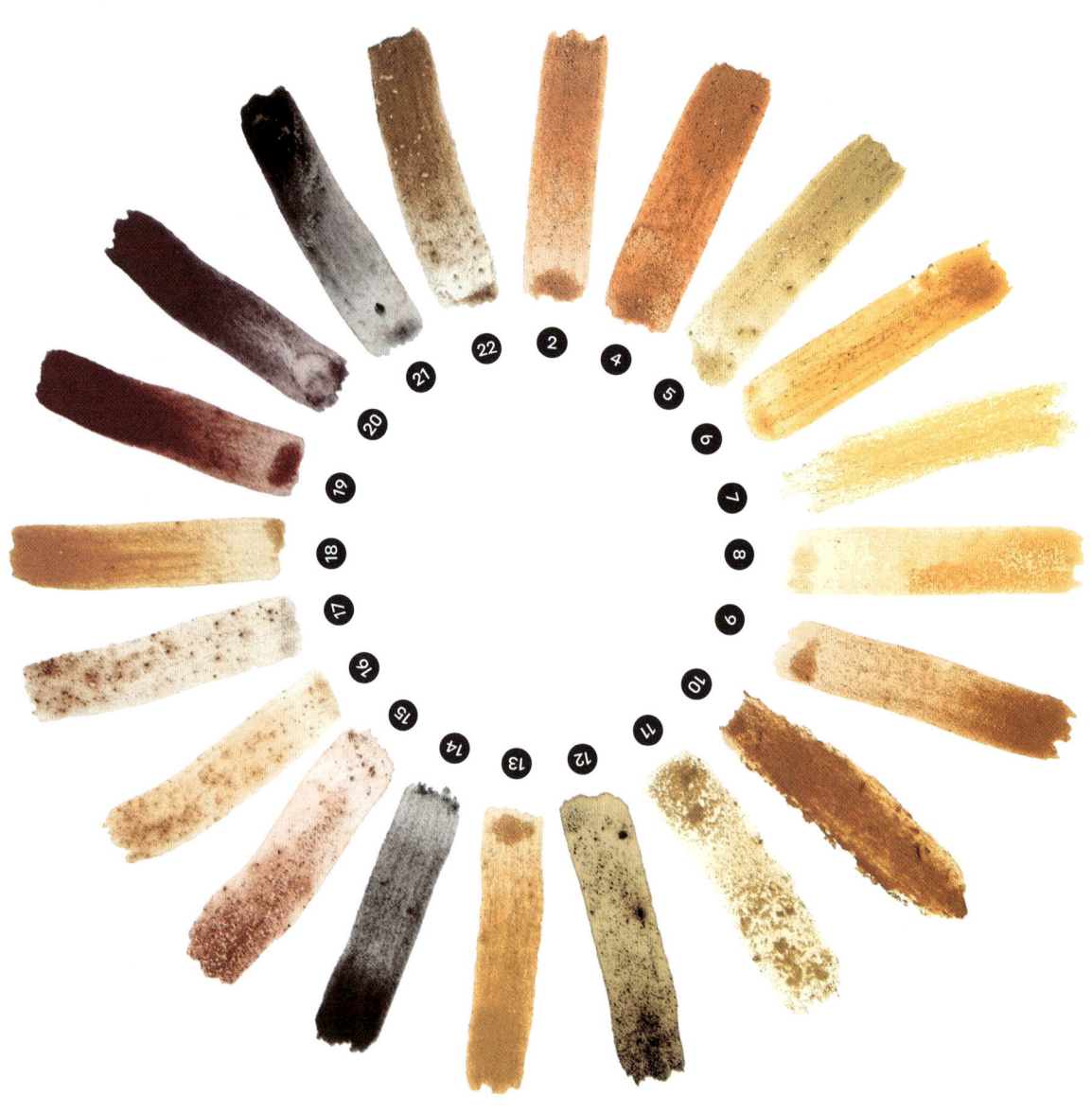

*1. *Pangborn in the Spring.*
Thick biogenic ochre forms
around an electrical ground
in rural ditches around the
bog iron field within biking
distance of my home near the
Canadian border, ancestral
Nooksack land.

2. *Iron Brew.* A gift from Icelandic
volcanic streams. Gratitude to
Dylan Young.

*3. *Return to the Eye of the
Beholder.* Magic metal amulet
made by Despina Papadeas,
a descendant of Lemnian
iron workers.

4. *Hephaistos Amphigúeis.*
Lemnos, Greece.

5. *Hephaistos Klutotékhnēs.*
Lemnos, Greece.

6. *Muck Rolled into a Crayon.*
Gathered on the coast with
friends in daze.

7. *I have no memory of where
this is from, but it smells like
goat's blood.*

8. *Mother of Ochres.* From the
swamps where I came from.

9. *Mother of Ochres.* Quietly in
fire. So soft, as if baby powder.

10. *Biogenic Ochre.* Born fresh
and plentiful in the roadside
ditches around the Nooksack
and Sumas Rivers after the
floods of 2022.

11. *Microbial Cloud.*

12. *Microbes Cooked in Lemon.*
Unfiltered biogenic ochre with
lemon juice added, cooked on
high in an oven. Very reflective
and almost creepy black.

13. *Microbe Mecca.*

14. *Microbe Ceremony.*

15. *Falun, Acid Mine Drainage.*
Ochre from copper mine
acid drainage ponds, used
for house and barn paint in
Falun, Sweden.

16. *Vulture Mud.* Gathered from
possible bird bathing pools in
the alpine high country.

17,22. *Acid Mine Drainage.* Used
by research scientists to help
remediate toxic algae ponds.
From two mining overflow
pools in Pennsylvania.

18. *Life.* Gathered from acid mine
drainage pollution by artist
John Sabraw and team in
Ohio. The extracting of acid
mine drainage ochre for paint
pigment simultaneously cleans
and harmonizes the water
stream, remediating the toxic
pollution.

19. *Hot Life.* Same as **18** and
heated in a kiln over 1,200°F
(650°C) to transform
yellow ferrihydrite into red
iron oxide.

20. *Super Hot Life.* Same as **18** and
19 and heated in a kiln over
1,500°F (850°C).

21. *Biogenic Magnetite.* Muck
gathered and roasted from
around Kenmore Elementary
near north Lake Washington,
traditional Sts'ahp-absh,
Duwamish, Sammamish, and
Coast Salish land.

*23. *Biotic field.* Gathered for paint
pigment, from Colombia,
South America, by dear
Catalina Christensen and her
mother, and in connection
with a spiritual leader of the
Muisca peoples, Suaga Gua
Ingativa Neusa.

*Swatch not pictured, see raw material/
color on page 35*

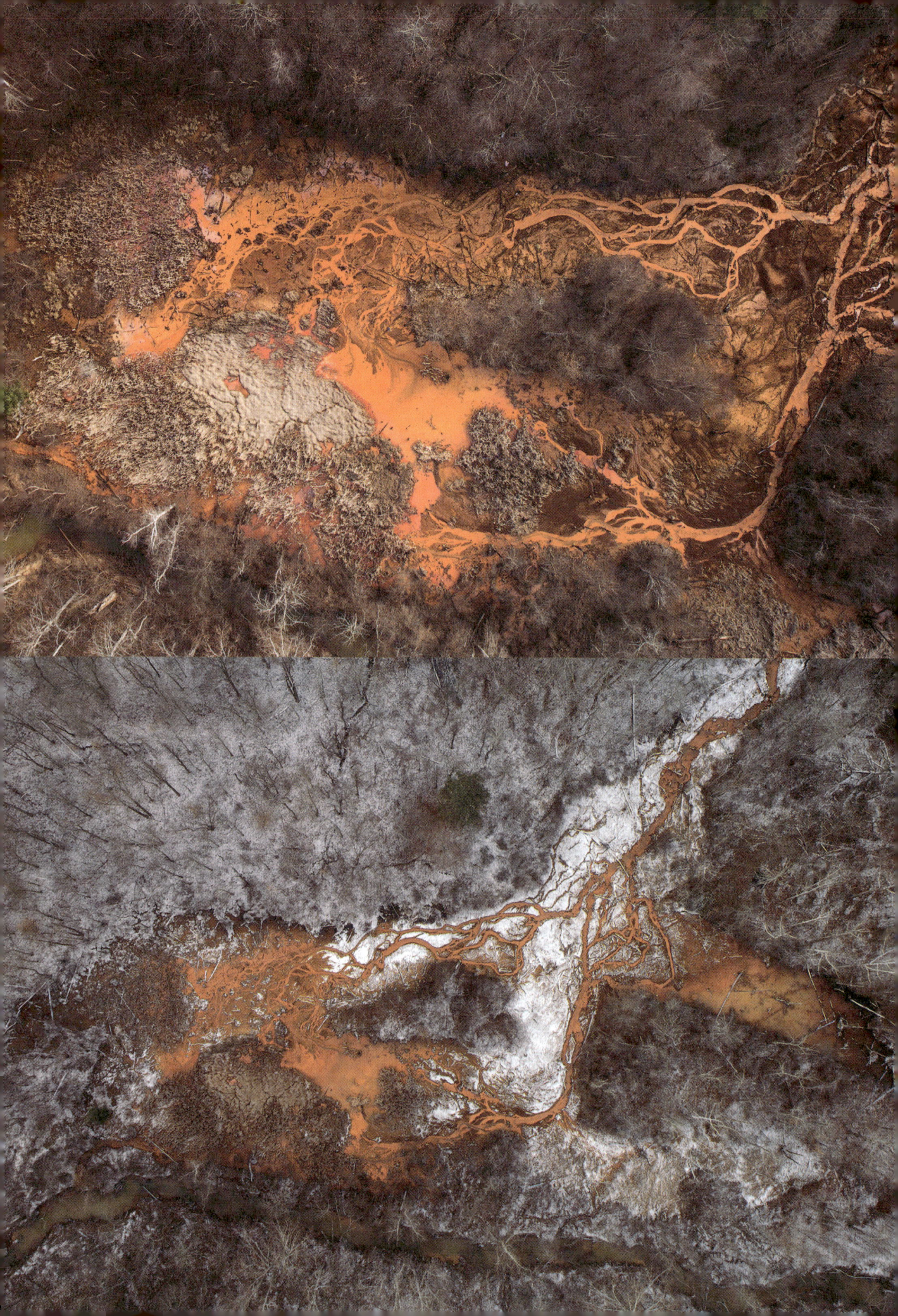

Vulture Mud

MICROBES—THE TINIEST AMONG US—are iron scavengers. They eat iron like we eat food. For bacteria, iron's an ideal power supply on which their metabolism, biodiversity, and, perhaps even communication, depends.[2] To digest iron, they alter "free" ferrous Fe(II) iron ions to a "bound" ferric Fe(III) state—an elemental change of valence and potential—essentially gaining energy during the switch. Metabolism, generally across all living beings, converts matter into growth and strength.

Iron-eater microbes in fact *secrete* ochre. As they metabolize, they weave out beautiful wings of iron hydroxide, converting ions into momentum. In *The Writing of Stones*, Robert Caillois describes the process brilliantly:

> Life appears: a complex dampness, destined to
> an intricate future and charged with secret virtues,
> capable of challenge and creation. A kind of precar-
> ious slime, of surface mildew, in which a ferment
> is already working. A turbulent, spasmodic sap,
> a presage and expectation of a new way of being,
> breaking with mineral perpetuity and boldly

exchanging it for the doubtful privilege of being able to tremble, decay, and multiply.

Spasmodic sap! The stuff from which diverse life appears and multiplies (to sum up a bunch of novel science). Which is why we casually call this outcome *biogenic*, aka life-born. When I see primordial iron brew seeping out of a ditch, spring, or sidewalk, I feel like falling to my knees, as if in the presence of the greatest gods. Maybe that sounds ridiculous, but why not? Aren't I witnessing a four-billion-year-old lineage of life creation in front of my eyes? No story, no propaganda, just pure unadulterated nature—our very oldest ancestors *still right here and here and here*, pumping out iron and oxygen, without which life as we know it would cease to exist.

Moving on from microbes to animals, I observe large, tusked mammals (elephants, rhinos), and certain birds (sandhill crane, rock ptarmigan), are fellow devotees of ocherous mud, a mud they intentionally rub on themselves and bathe and cover themselves in. To me, the most poignant homage to microbial ochre shows itself in the lifestyle of bearded vultures (*Gypaetus barbatus*), also known as lord of the skies, ossifrage ("bone-breaker"), lämmergeier ("lamb vulture") homā (هما in Persian; "spirit of paradise"), or བྱ་རྒོད། (bya rgod; the sacred bird of Tibet) and found throughout cold high mountain regions of South Africa, the Arabian Peninsula, the Caucasus, the Pyrenees, the Greek islands, the Altai Republic, the Alps and others.

Bearded vultures *love* to take ochre baths. Almost no other vulture does that. They fly into thick shallow puddles or streams, plunge their beaks down into the mud, and cover their throat, heart, chest, and leg feathers in reddish, dawn-colored muck. As devoted parents, couples take turn intentionally brushing wet ochre on eggs, and later cover their hatchlings' vulnerable white down with it, naturally rust-dyeing their coats with a kind protection.[3] Very little is understood about why they do this—some say it's a sexual signal or mating display, some say it has antibacterial or functional purposes (to prevent feather wear, or reduce smell of rot).

Did we learn to use ochre from vultures? Or how to protect our vulnerable babies? Some scientific researchers, who study the possible biomimetic behavior (skills mimicked from other life-forms) of early humans, think we did. They speculate that vultures and humans were often scrounging around food sources together. Early humans could have easily observed the long-standing and evolved behavior of vultures: breaking bones for marrow and arriving to the feast decked out in red ochre.

I imagine that, if I was alive back then, and knew bearded vultures would lead to potential food sources (delish marrow! fresh or dried meat!), and maybe even longed to mimic their skyward behavior, eventually I might try to do what they do. Like rub myself in ochre. If I cover myself in ochre mud, it would do something to me (could I grow wings!?). Watching vultures (and their five million years of successful evolution), and then rubbing reddish ochre on my face and throat, along with fresh marrow, I might come to realize it becomes a UV-protectant sunscreen, like a fatty sun lotion, and antibacterial sanitizer, masking foul odors and insulating my body against cold winter air.[4]

Did bearded vultures also teach us how to paint visions with feathers? Or apply makeup? How to use stagnant pools of rusty color? Our oldest paint *is* wet ochre, or "art water," as my toddler niece brilliantly put it. Our early brushes and pens were made of feather, fur, bone, twigs, moss—the stuff of nest and bird.

Curiously, in newer studies of Paleoindian hunter-gatherer paint technology in mountainous North America,[5] anthropologists revealed that red paint used in some rock art was sourced from local biogenic ochre pools. Not only was the bacterial ochre mud harvested, it was also heated to specific temperatures ($1,200°F/650°C$ to $1,500°F/850°C$) in hearths, which transformed the hue into an even more vivid, stable red or a "highly thermo-stable paint," according to Brandi MacDonald. Biogenic paint was then used to create the long-lasting pictographs we still see today.

Elsewhere, it is well documented that Ancient Egyptian and Near East queens and goddesses wore vulture regalia and crowns, to represent and honor divine wings, and would paint protective "red rouge"—ochre and bone marrow—on their lips and hair. For those women, vultures held prestigious mythic status as oracles, "able to see any event in the world."[6]

I could speculate, elaborate, or even just stop here. Massive wild birds maybe taught us some transcendent beauty tricks like a million years ago—so what? Haven't we adapted these skills and evolved beyond playing in the mud and putting on makeup? Maybe. Maybe not.

Unlike us, bearded vultures don't kill any creatures. They are the only vertebrate on Earth that exclusively eat bones of the *already* dead. Literally they break open cold bones and get energy from shard and marrow. They metabolize pure bone for energy. Gentle and sane, they digest lion, lamb, human all the same (but they refuse to eat birds—won't touch the bones of their own kind). For them, fresh bones contain 108 percent as much energy as fresh meat.[7]

Once swallowed, their wild acid stomach (pH of <1) completely softens and dissolves the bones in twenty-four hours. Bearded vultures' "peculiar, macabre diet"[8] is so extreme they also seek the last of freeze-dried skeletons, stragglers that don't ever fully decompose in wild, permafrost heights. Long leftover bones—full of parasites from all the gnawing by hyena, crow, larvae, and other carrion eaters—are easier to break and swallow, and vultures fly lighter while digesting.[9] In this way, they transform old dry death into soaring and soaring.

With a wingspan over nine feet, more than the length of any human, "the stomachs of these birds become a living coffin flying in immense space," writes Van Cam Hai, a Vietnamese poet, in conversation with the celestial burial master Lama Rigpaba, a Tibetan Buddhist in Lhasa.[10] Celestial burial masters are trained ritualists who observe and work very intimately with vultures to perform a sacred ceremony called sky burial. A sky burial is a more than ten-thousand-year-old Vajrayana Buddhist funerary practice where vultures are fed the dead.[11] Families carry their loved ones tied on their backs far distances, and they are delivered to a spiritual master, ritually cut up, and offered to diverse kinds of vultures—vultures who are willing, after all that, to metabolize a person's most hidden, hard parts. Human flesh and bones are transformed into energy to fly, and are compassionately "buried" in the vast sky. In a spiritual up current, the end renews the beginning. A soul "melting into the body of the rainbow."[12]

There's an immense empathic logic to this practice. Purity, even. "There is someone in the wind," as surrealist André Breton says. Perhaps to a Western mind, sky burial could sound unfamiliar, gross, even hellish. But I see empathy: a song of continuity worthy of our contemplation, prayers, profound flights of imagination, and, frankly, fierce protection.

And I see ochre. Imagine a frozen femur bone on the ground. In my vision, a vulture swoops down and grabs hold of the bone. I see fresh ochre pigment from their chest and face rubbing onto the bone, coloring it redder and redder, as it's ground in their beak and then moves down through their throat and warm innards. Soon the vulture is airborne, with the bone inside its gut, flying over a deep valley. Flowing through a meandering passage, a harsh acidic waterway, the ochre-covered bone shard attracts all kinds of excited digestive company: very active, very

infectious, thriving kinds of fusobacteria and clostridia (which live happily in a vulture's gut, but inside a human, these bacteria can cause several serious illnesses). These creatures begin to crawl all over the bone, eating it, and dissolving each sharp shard edge, edges of leg memory, a leg that ran and jumped and walked for years and years. The bone disintegrates and transforms into a different substance, a kind of energetic ether.

Elementally, these shifts between matter and spirit are akin to alchemy. Alchemy is an ancient tradition whose mission was to change matter and transform nature through specific human-directed operations (i.e., chemistry, art, psychology, prayer). In fact, some alchemical vessels (known as alembics or containers) for doing experimental work, especially for the processes *digestion* and *circulation*, were named "Vultures."[13] The Vulture vessel helped digest and circulate energy. I can see why vultures were an ideal inspiration for such important processes: Think of a gigantic bird dressed in red ochre, eating the long-dead and defying gravity by flying up an enormous mountainside. That's a very powerful representation of alchemy's aim.

It's important to note that if an alchemist were to choose the wrong experimental vessel or glassware, all bets were off. The same is true in chemistry today—you need the proper container. If you were to choose a different apparatus (like the "Pelican" or "Serpent") for processes that are meant to take place only in the Vulture, you run the risk of serious explosion, injury, chaos, and even death. Choosing the wrong vessel can make things go seriously awry.

I think it is fair to say that humans have now turned the whole planet into an experimental vessel with devastating results. Vultures and biogenic ochre are very good life-forms to help explore this problem, I think. While I praised biogenic ochre as a wonderful and protective support for vultures, I must admit that the murky material has come to signal something far more deadly and dangerous today.

Around the world, biogenic ochre goes by another name: toxic acid mine drainage (see images on page 38). Bio-ochre shows up not only in sacred springs, at the origins of life, to protect vulture babies

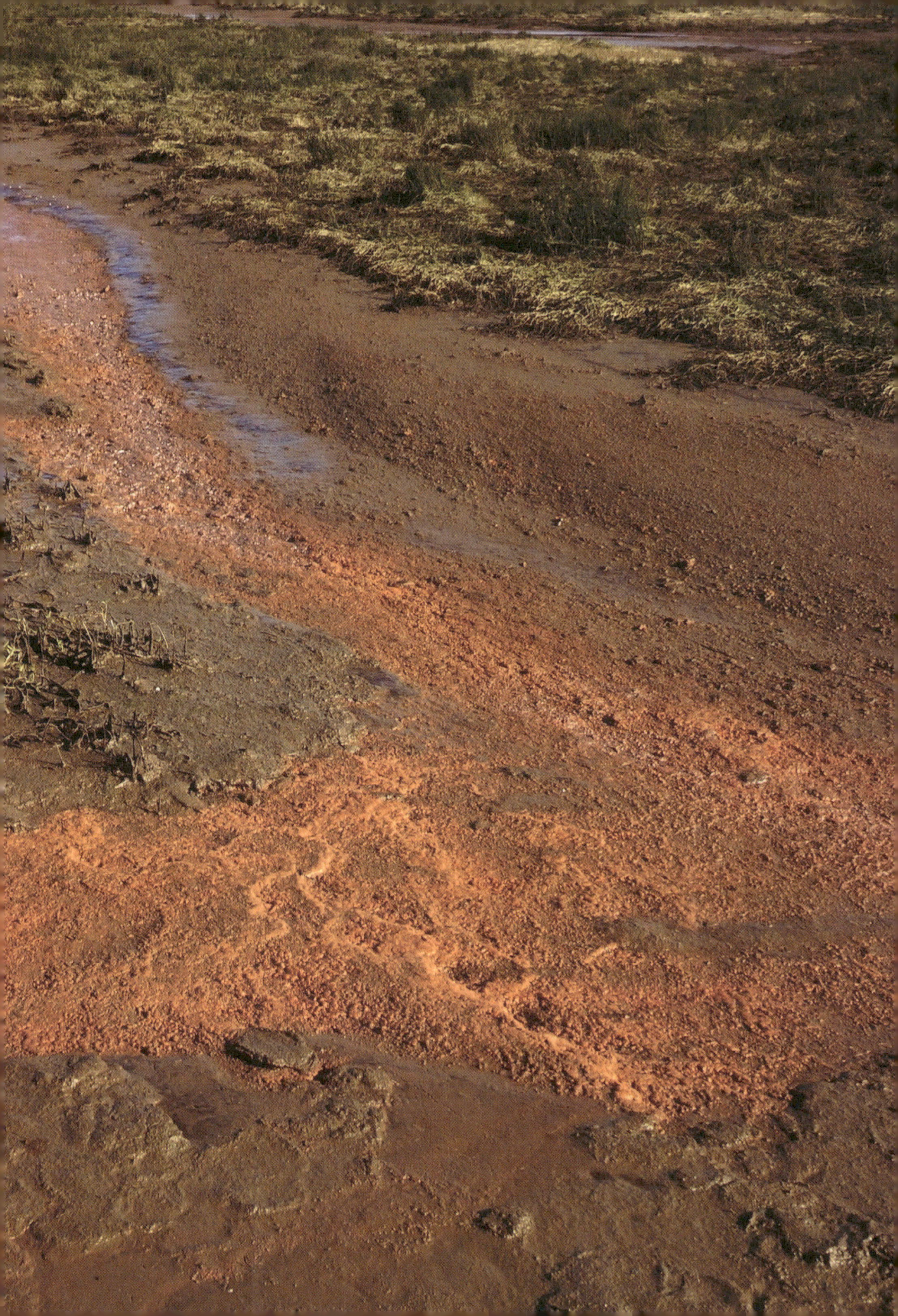

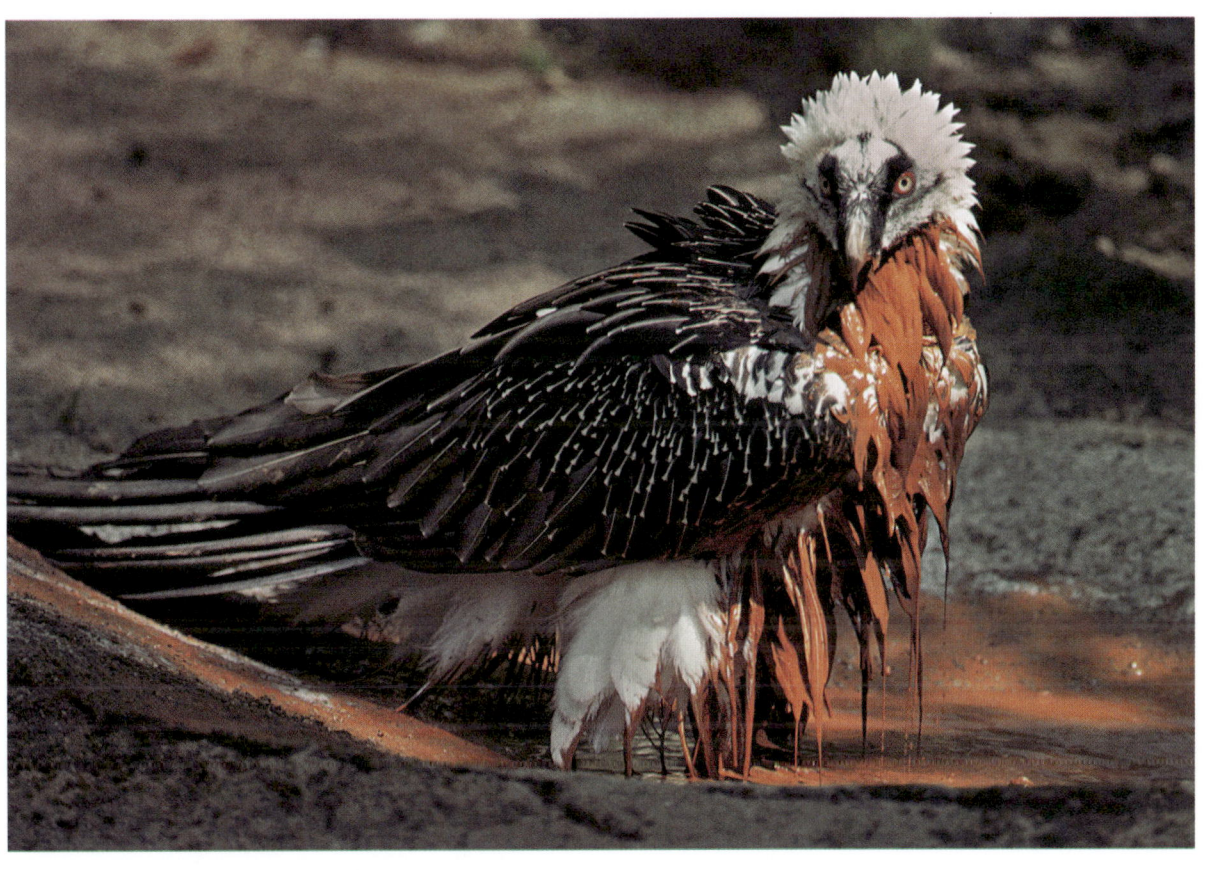

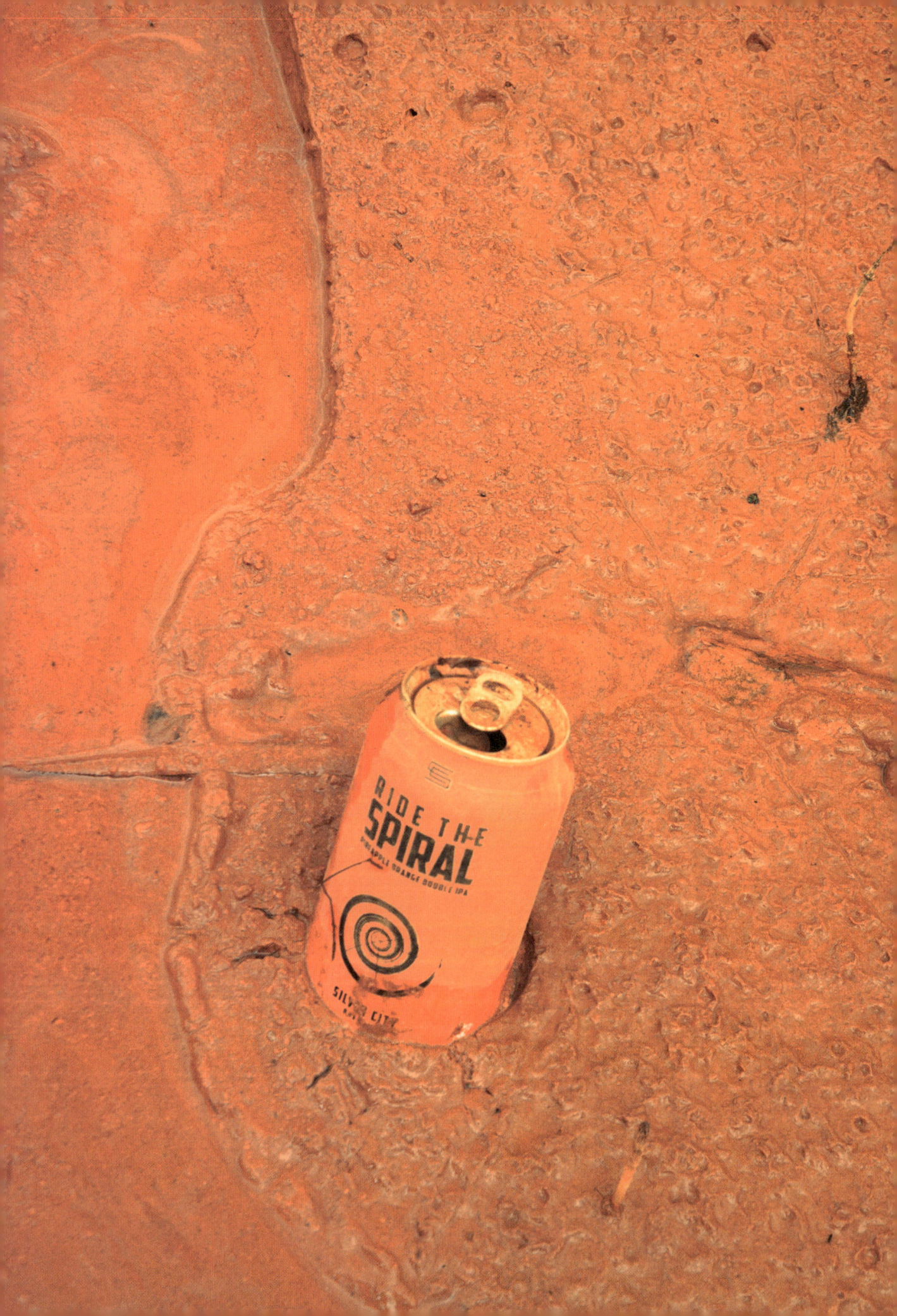

and spirits in shimmering rainbow flights; it also proliferates in very acidic rivers, around abandoned quarries, and poisonous pools created around mines of fossil fuels, in particular (coal, oil shales, anthracite, bitumen, peat, etc.). Biogenic ochre is considered pollution. Why is this generous, powerful ochre emerging in such toxic waters?

If you were to ask about the chemical or scientific specifics, technically it makes sense. Industrial mining creates excess rock piles called *overmined tailings*. Often these rocks include pyrite (iron sulfide), which degrades and erodes into industrial waste streams, aka rivers and waterways. Water breaks down the rocks. And rock breaks and unbinds sulfur and iron. Sulfur makes water more acidic (and stinky!). Acidic water further breaks down elemental structures and "frees" iron (and worse, many other metals, like arsenic). Bound, unavailable iron becomes accessible. And, as I said before, microbes fight to feast on fresh and free iron.

Clear streams fill with fatty orange curdles built from iridescent microbial feasts. The ochre itself isn't toxic. Ochre is a *natural symptom* and a *response to* sulfuric water with excess free iron. Mine waste rivers basically turn to battery acid. No creatures can actually enjoy or drink from these streams, so animals may die of thirst, plants will wilt, fish suffocate, people puke and get very sick. In rural Colorado, acid mine drainage from coal mining killed most biological life in a seventeen-mile radius along the Alamosa River.[14]

Acid mine waste pollution in waterways rose exponentially in the past twenty years (think of bloodied, red polluted rivers caught on news camera in Spain, Brazil, South Africa, India, Siberia, and of course, the United States[15]), causing undrinkable, undigestible water throughout the globe.

And peaceful bearded vultures? Their populations are declining upward of 95 percent in many areas, nearing extinction.

Are they dying because they bathe in super toxic water? Perhaps getting sick because they rub acidic ochre all over themselves? Causing a kind of burning injury or digestive failure? Nope. In recent vulture breeding programs, where birds are raised in protective captivity to help repopulate them, the rules *require* iron-rich mud baths as a basic prescription for successful survival. Vultures actually remove the ochre from the streams—effectively helping remediate them.[16] "Toxic" ochre is not what is killing the celestial bird of paradise. So what does?

Painkillers. Tiny little pills kill vultures. Human-developed anti-inflammatory drugs like ibuprofen, aspirin, diclofenac, and most other NSAIDs (non-steroidal anti-inflammatory drugs)—laboratory-synthesized chemical compounds—are consumed by people to ease hurt and fed to domestic animals (cows, pigs, horses, dogs, sheep, etc.) so they can live longer in often uncomfortable conditions. Their pain goes on (until it doesn't) and depending where and when animals die, vultures arrive to do their powerful alchemical work dissolving bones into spirit.

As drugged-up mammal bodies get digested inside the vulture's alembic system, painkillers act as a homogenizing poison, annihilating the compassionate microbial communities inside a vulture's precious and rare intestinal ecosystem. Instead of majestic rainbow flight, vultures' necks droop over, heavy, lethargic, as if forced into a subservient bow. Most die within two days. Our suffering gets displaced. Their symbiotic power, and death-defying achievements, lost. Or, at least, alchemically transmuted.

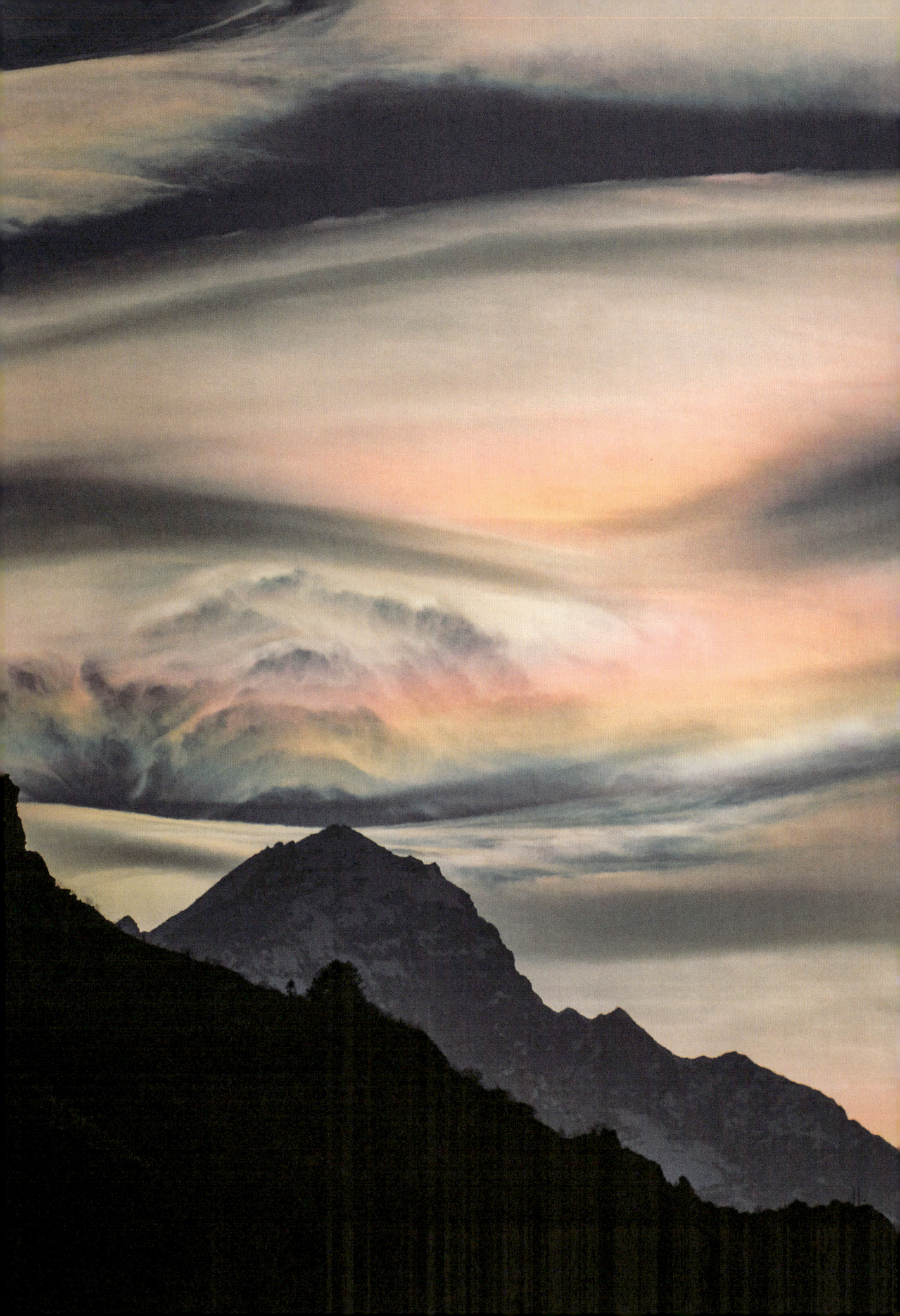

THE COSMIC ORACLE

trans. from Ugaritic cuneiform tablet, part of The Baal Cycle,[17] *ca. 1200 BCE*

————————

Salt in the earth sacrifice

Ochre in the dust harmony

Pour peace into the liver of earth

Offer bliss into the glory of field

tick tock tick

to life force

let's run quick

for i have a word to give to you

matter i deliver to you

a word of tree

and a whisper of stone

moaning of heaven with earth

and of deep with stars

i understand blessbolts even the heavens do not know

a word unknown to humans

which masses on earth cannot comprehend

come

i will reveal it to you

in my holy holy mountain

in the sanctuary

in the high peak of my heritage

in the beautiful place

innermost inland of power

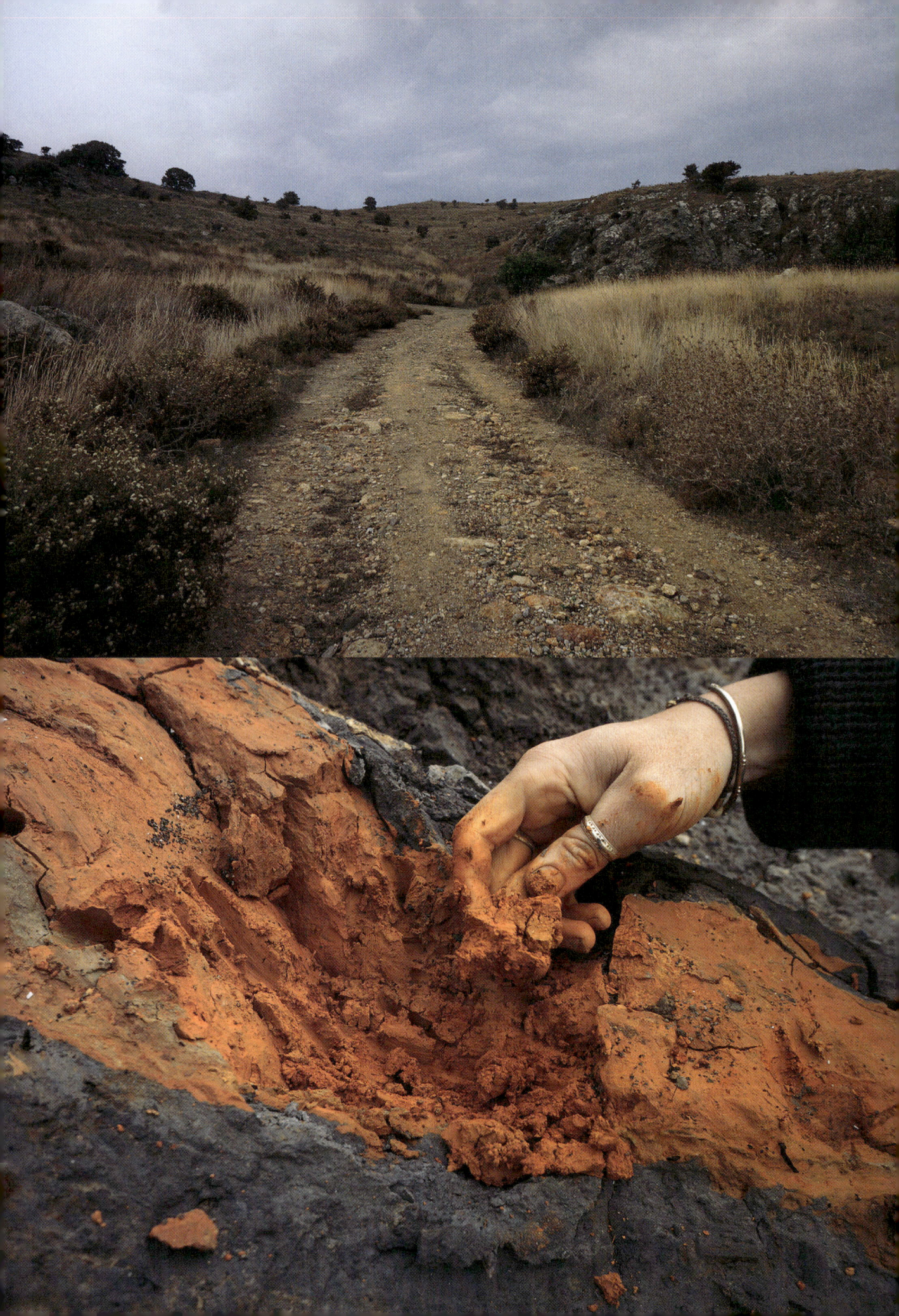

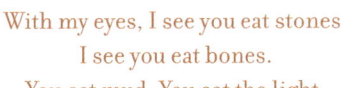

Terra Sigillata

I TRAVELED FROM MY RURAL CABIN in Northwest Washington to an island far away in the northern Aegean Sea, hoping to meet and commune with the ancient Greek god of iron, Hephaistos (Ηφαιστος).

Hephaistos was born around the seventh century BCE, conceived by goddess Hera from her own body. According to mythological tales, he emerged deformed, "lame, ugly, and red." Too grotesque, Hera hurled him down from "the sacred threshold of the gods"[18] onto Lemnos (a Greek island a few miles offshore from Turkey) or, in other stories, into the sea. It took him all day to fall, spinning like a wheel through space.[19]

In *mukos*, the "innermost place" or "secret place" or "women's part of the house," he gains forge skills:[20] the art of matter, metal, iron, light, and fire, and *technê* (of *tec*tonic, *techno*logy, *techn*ique). Hephaistos, or Vulcan (volcano), the god of iron and art in the West, becomes a tortured, filthy crafts-man able to manifest mystical armor for all the other gods. He blacksmiths cloaks, shields, nets, chains, thrones, and lifelike automatons, "metal objects imbued with life."[21] Every solid thing becomes animated, magical.

He even mixes earth with water and pounds from the ground Pandora—the first mortal, or "the all-giving" being who can become anything.[22] He labors and labors for living works of art. Some say he's the midwife to nature and even Earth's own birth.[23] As a god of Earth's soul, and hot inner forge (that core of iron!), he is considered a descendant or parallel mythic figure to celestial iron smiths farther south in West Africa, who feature prominently in Dogon, Bambara, and several other ethnic groups' religions. The Yoruba people's well-known god of iron, Ogun (or in Haitian Vodou, Ogou), is also considered the "First Smith and true son of the Supreme God," as historian Mircae Eliade notes in his research on early metallurgy and ritual.[24]

Hephaistos, like other gods in the ancient world, was known by several names or epithets. These lyrical nicknames signal unique attributes and behav-iors of the god—they are teachings and subtext. The act of knowing a god's several identities is important, revealing specificity and details: not unlike realizing that "ochre" could come in many different mineral forms and goes by several distinct names. Hephaistos has at least twenty-six epithets signifying his capacities and powers, and I will mention a few of the most significant as I go along.

Sadly, nowadays in the West, people tend to think of celestial iron smiths, namely Hephaistos, as clunky stereotypes, mythic bygones and remnants, chained underground and simply there to operate the rusty steampunk bellows and furnaces that delivered the Iron Age. Of course, I think he's far more fascinating and noble. Why? Hephaistos does all his magic in an unusual place (for a god, anyway): *on Earth!*

Hephaistos Ἀμφιγύεις (*Amphigúeis*) "of both ways" or "lame one" works exactly where he was outcast, on Lemnos. What else do we know about Lemnos? Lemnos is an old island made of a Miocene volcano (5 to 20 million years old). And it is the beloved home of a "race of man-slaughtering women," as the Ancient Greek lyric poet Pindar says. (Because they killed all the men on the island!). These ancient women inspired the phrase "Lemnian deed"—an idiomatic legal phrase that translates to "greatest of all transgressions imaginable."[25]

And Lemnos is the key birthplace of modern Western pharmaceutical drugs.

Those criminal ladies? Their land gave rise to widely distributed, little round colored pills. And not just any old placebo: These highly effective, antibacterial healing clays were known throughout the ancient world as Lemnian sealed earth, or *terra sigillata*.[26]

While forged steel is made from burning iron oxide with extreme heat from coal, Lemnian sealed earth was made of moist iron oxide clay or scum. Soft metal. According to several historical accounts, women priestesses (and *only* women) ceremonially gathered and processed this "fat clayey earth" and formed it into round cakes "about the bigness of ones thumb," dried in the "shade of the sun,"[27] on only one single day each year.

Iron processed *without* a forge, by priestesses. No fire required. Not really a place where Hephaistos, "god of the forge," is typically thought to belong. But is he here? Theophrastus (4th century

BCE), Dioscorides (1st century CE), Pliny (2nd century CE), Galen (2nd century CE),[28] Agricola (1450), Ghistele (1485),[29] Belon (1588), and Covel (1677) and dozens of others spanning over two thousand years all visited Lemnos and documented various stages in the seasonal production process of medicinal "sealed earth":

> extract ochre clay from a jumping and overflowing spring, pool or vein
>
> bless with grain
>
> rinse or wash with water
>
> sort and create different quality grades and colors
>
> dry in the shade

These early eyewitnesses saw the clay being transported by donkey cart[30] to a nearby cultic, red-floored temple of Hephaistos and shaped into small coins and impressed with image and words (these would change depending on the century), becoming official Lemnian *sphragis*, sealed earth. The term *sphragis* came to be known as a literary device, meaning "a copyright or trademark," used to signify and identify the symbolic stamp (most often depicting mountains, moons, stars, keys, goats, eagles, grain, place names) of an authentic origin and maker. Once sealed, the ochre transformed into a powerful medicine, *pharmakon* in Greek, or as locals called it, "sacred earth."

What makes earth sacred, exactly?

Probably the body of a god.

Ancients mention "the place of its [Lemnian earth] extraction was synonymous with the place where Hephaistos is supposed to have fallen."[31] That location is very specific to northern Lemnos, in the few Miocene volcanic outcrops. The healing iron clay literally forms where Hephaistos touches down, in that "secret place." His story lands. He's not stone-cold ore. He's an entire therapeutic landscape, a deep time metamorphic presence: Hephaistos Πολύμητις (*Polúmētis*) "of several wisdoms," in the earth.

I want to mention that visions of a buried spirit transforming into mineral iron were deep-seated in the ancient world, and, curiously, are affirmed today by scientific inquiry. "They bury him beneath a mountain; his body changes to iron," a Greek proverb hints. "The connection between metals and the body of a God may likewise be detected in Egyptian traditions," Mircea Eliade notes, writing, "iron was 'bones of Set' and haematite 'bones of Horus.'"[32] Recent archeological digs have found mineralized bones of big extinct animals, discovered in huge numbers, at the cultic centers of Set in Egypt. Even the mummified preservation of dinosaur flesh, too, is due to iron's power.[33]

When I went to the historical extraction place of Lemnian earth, up in the rocky hills overlooking a fishing village, I found a lot of weathered volcanic rock, a literal Vulcan-Hephaistos body. "In the area of the extraction of the Lemnian earth there are yellow slightly altered . . . rare relicts of fresh volcanic rock . . . and Miocene tuffs," as archeologists[34] note. I saw ancient pyroclastic minerals—from *pyro*, fire, and *clastic*, broken into pieces. I remember a faraway chanting in my mind, as I recalled the Orphic Hymn to Hephaistos (ca. 6th century BCE)[35]:

Powerful and strong-spirited Hephaistos,
unwearying fire

that shines in the gleam of flames,
god bringing light

to mortals, mighty-handed and eternal artisan.

Worker, cosmic part and blameless element,

highest of all, all-eating,
all-taming and all-haunting,

ether, sun, stars, moon and pure light;

for it is a part of Hephaistos
all these reveal to mortals.

All homes, all cities and all nations are yours,

and, O mighty giver of many blessings,
you dwell in human bodies.

Hear me, lord, as I summon you to this
holy libation,

that you may always come, gentle,
to make work a joy.

End the savage rage of untiring fire

since, through you, nature itself
burns in our bodies.

I could imagine a burning-hot flame, perhaps akin to Hephaistos's hearth heart, now broken up into pieces and spread across the ground. I tried to envision—as an ancient priestess perhaps once would—that at this northern rocky outcrop, I *am* in the presence of iron fire, bones, blood, and the fat of an earth-bound, iron-knowing Hephaistos. Geomorphologically, it's true: When volcanic formations decompose, they weather and alter, and veins of iron oxides do remain, spreading out across the rock face. At the outcrop base, a hidden hydrothermal spring is a perfect home for thick ochre-producing microbial organisms feeding off sulfuric water and free iron, and I wonder, how could I *not* imagine the molten ooze, a sign of godly bellows below? Greeks have a technical name for this viscous stuff: ἰχώρ, *ichor*, ethereal fluid or blood of the gods and immortals. Ichor, ochre—how alike they are!

When entering a place with mythic blood, I often return to the gods' names for clues on how to handle and approach them properly. So far, I've only mentioned three of Hephaistos's: *Vulcan* (volcano), *Amphigúeis* (of both ways), and *Polúmētis* (of several wisdoms). I turn now to Κυλλοποδίων (*Kullopodíōn*), "the halting." The Greek root *kulobos* indicates "something broken, curtailed" or halted, held, and was used to describe joint deformities of the body.[36]

So let's consider what Lemnian pills healed.

Lemnian earths cured the plague. They stopped the effects of poison (snake bites, in particular), dysentery, and inflammation of the eye, and were antibacterial against *Staphylococcus aureus* and related pathogens, could absorb toxins, and generally, on a deeper level, brought a "strengthening of the heart."[37] Pharmaceutically, the island's medicinal earth proves to be alexipharmic (functioning as an antidote to poison), an astringent (drawing together or contracting the tissues), bioactive against deadly pathogens (such as *Pseudomonas aeruginosa*), and desiccative (absorbing or drying).

Basically, it halts disease.

Lemnian earth, made from a god's body, actually enacts Hephaestian-embodied qualities. His whole mythic identity contained in a little earthen pill, stomached and somatically dwelling in our bodies, in return, reanimates us, reviving our life force.

And proof of efficacy may lay in the sheer fact that Lemnian seal ritual extraction and international distribution was sustained for thousands of years. How is that possible? As Christine Downing points out in *The Goddess*, Hera's child (Hephaistos) was born without a father and obviously deformed, as "the parthenogenetic mother is the pathogenic mother as well." Parthenogenetic power—the ability to give birth without anyone else, or in secret—turned pathological. And of course, the pathological knows pathology (i.e., disease) best. Which is to say, Hephaistos's birthright gave him special power and knowledge about pain and disease, and with that, an ability to relate to those wounds in others. Or, at least, offer them relief in the form of his pills.

In the ancient world, drugs were called *pharmakon*—where we get the words *pharmacy*, *pharmaceutical*, *pharmacist*. Physicians' early apothecaries all point back to Hephaistos's sealed earth as the model for common medicines as we know them: They are small circles, stamped with proof of maker, and effective, distributed medicine. As with drugs today, there are generics, dilutes, fakes, and spin-offs.[38]

Authentic Lemnian earths—let me rephrase: microbially rich iron clay ritually extracted from old volcanic veins and pus of Hephaistos's body, harvested sustainably only once a year, mixed with sacrificial grain,[39] ceremonially stamped and empowered on the red-floored temple of Hephaistos, by chanting offspring of murderous women—were a *uniquely* effective pharmaceutical.[40]

Not just because Hephaistos intimately knows pathology. Or because he employed those women who were capable of killing. He's sacred earth. As ingested earth, he falls and dissolves all day into our innermost place. Our plague, that venom, our inflammation . . . the god who knows the belly and womb of the forge surely knows what it takes to tend to a painful inner flame.

Gods heal in these strange ways. And medicine and poison are paradoxical opposites, part of the same spectrum. We usually utilize this strange logic to talk about drug safety or psychedelic dosage. Take too much, or the wrong mix, and you'll drop dead or lose your mind. Do it just right and border on the ethereal. Early pharmacopeias of Europe and the Mediterranean account for this logic, sometimes labeling generic medicines and stamped earths with "P S QUIBIUS," an abbreviation of the Latin phrase *Pilulae sine quibus esse nolo*, "Pills without which I would be nothing."

But the original meaning of *pharmakon*, "sacred earth," takes on an even richer multiplicity and harkens back to my favorite of all of Hephaistos's nicknames: Κλυτοτέχνης *Klutotékhnēs*, "renowned artist" (due to his endless capacity to make). What are the various meanings of the term *pharmakon*? I'll list them:

medicine	ointment
poison	perfume
remedy	talisman
cure	amulet
pigment	soul-altering substance
artist's color	
paint	sacrifice
spell	scapegoat
intoxicant	idea
cosmetic	concept

And these are held all together, in ochre—alive between two worlds (gods—earth; color—toxin; iron—life). In part, I think that paradox is what unconsciously motivated me to seek his ichor spring on Lemnos a few years ago. I only found the road to the spring because I recognized the hills from a black-and-white photograph I saw of the place in an obscure academic research article. Today there's a church building obscuring what's left of the outcrop. As Lemnian earth extraction devolved over the years, Greek Orthodox clergy took over the land.

I drove up the small road (top image, page 48) as far as possible, got out, and walked to a long cement trough. Was it part of the ancient priestess' spring? Or a watering hole for animals? Still a modern ritual extraction site? I smelled violets in the wind; there was a large solo fig tree, a raptor feather. Further tucked into the rock body, behind sharp grasses and medicinal weeds (*helxine* in Greek), I saw a metallic mirror covering the spring hole. Maybe an entrance to Hephaistos's secret place. I prayed: *Hear me, lord, as I summon you, I pray that you may always come, gentle!* I lifted up the mirror. *Cosmic part and blameless element, highest of all, all-eating, all-taming, and all-haunting, cosmic ether!*

Gazing below into the spring, I saw the water high and murky. My heart sank. There was no inflamed microbial wonderland of pus or fat. No ochre, no sacred earth.

Where is the ethereal presence of Hephaistos? Did he leave our world? How do gods go extinct? Perhaps his iron wound is bled out, microbial communities long gone. Or maybe the gods and microbes are all dormant, hibernating for the season or the long haul?

As I was overwhelmed with a thousand racing thoughts, I felt a warm sensation on my finger and realized I had cut my hand—maybe on the grass blade, or on the mirror. There was a small bit of blood. Something I read flashes in my mind: "When the god is offered symbols of blood, of cutting, or of tears, the ritualists provide them with the means to gain control over their wound, and to overcome it."[41] *O powerful and strong-spirited Hephaistos! I pray to you, I wish for your safe return!*

A subtle sound of bird wings passed overhead. I think maybe a partridge or a dove, but in my heart eye I imagine it is the ghost of my dear celestial vulture friends, who nested throughout the Greek islands. What a strange place to be. I'm staring at an empty, clear, hole. A hole where anti-inflammatory pills claim their origin.

Vultures, Hephaistos, *pharmakon*, and me: all devotees of fatty sacred earth. And not just any sacred earth—very ambiguous iron-rich *mukos*, "secret places." I realize now it is maybe not ochre or vultures or land or people or even myself that I seek to commune with and protect, but perhaps something more like those unseen sacred thresholds. Could ochres be a type of compassionate and healing fascia between realms? An uncanny point of contact and connection between the dead and the living?

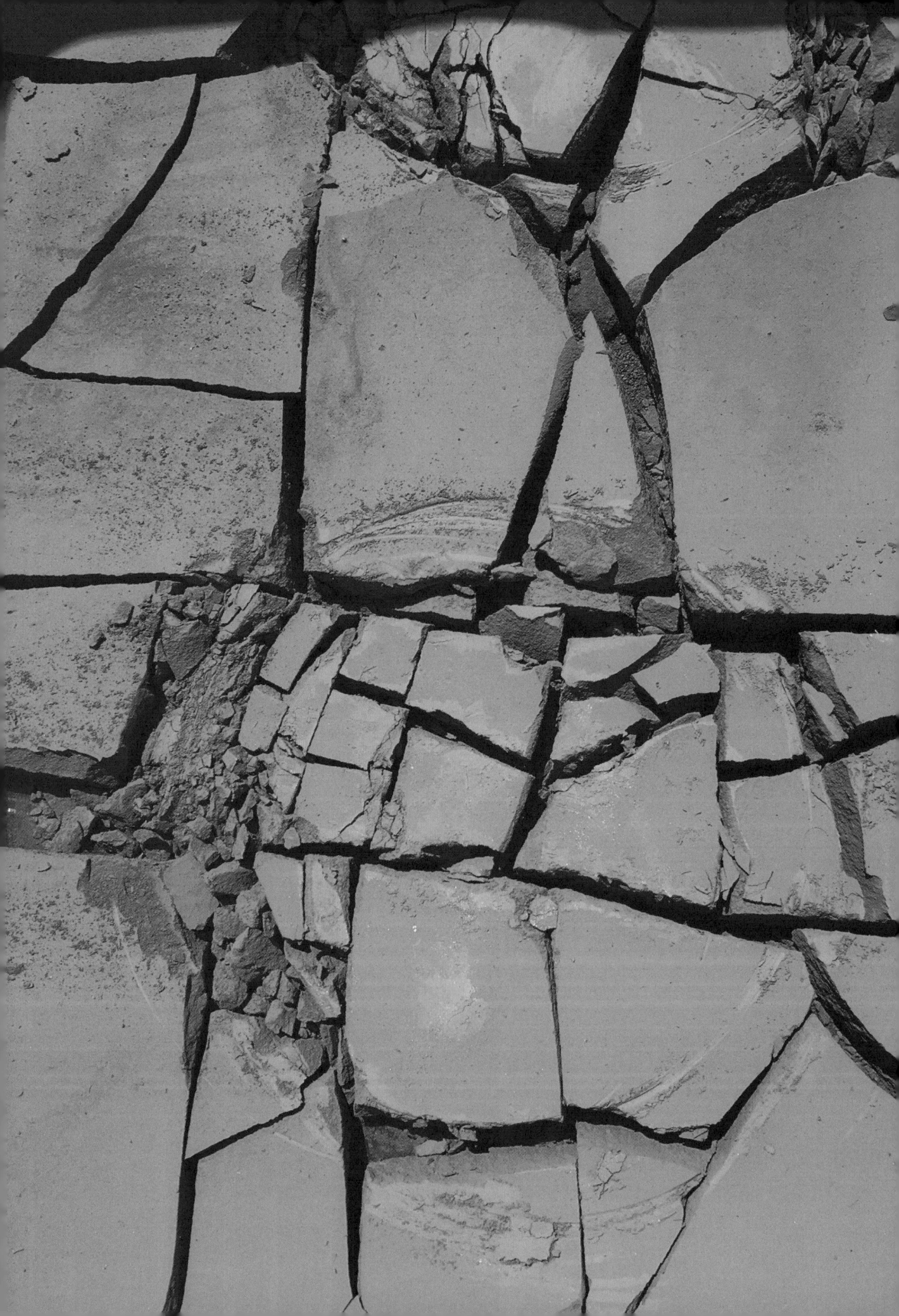

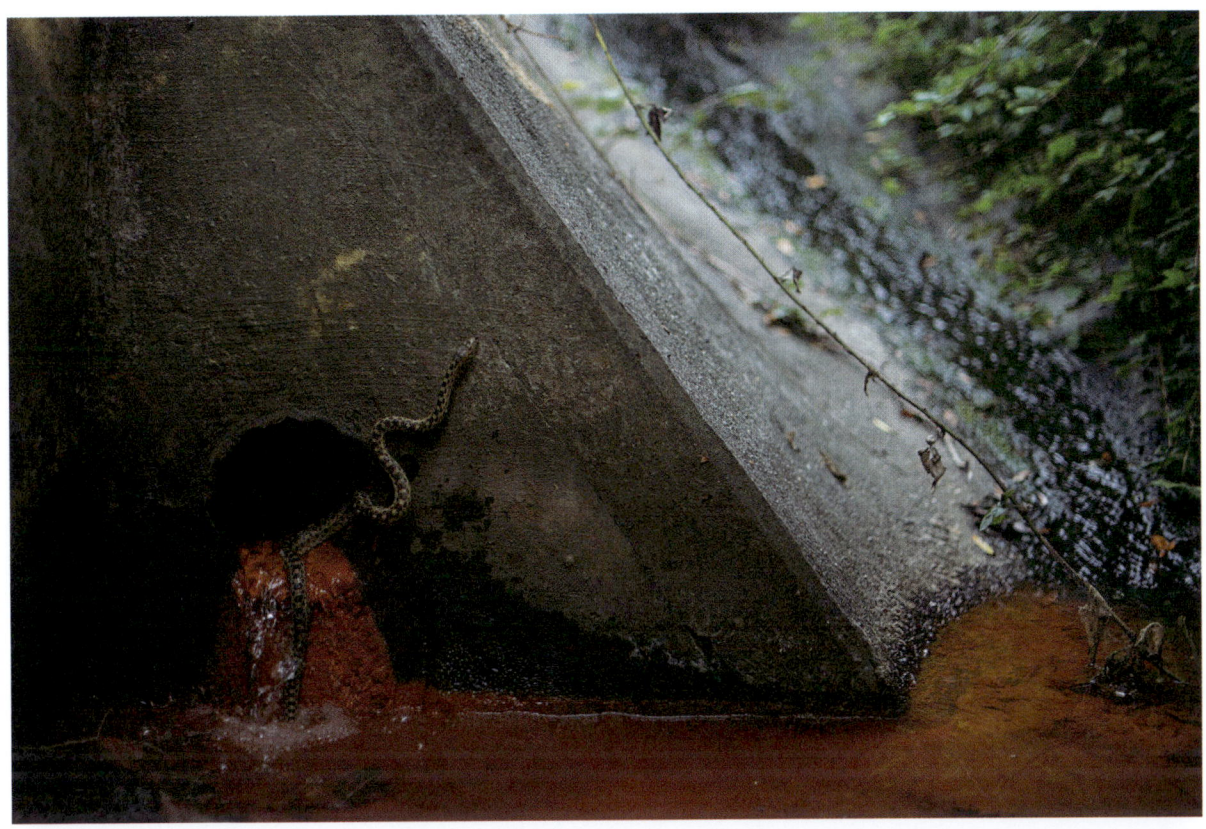

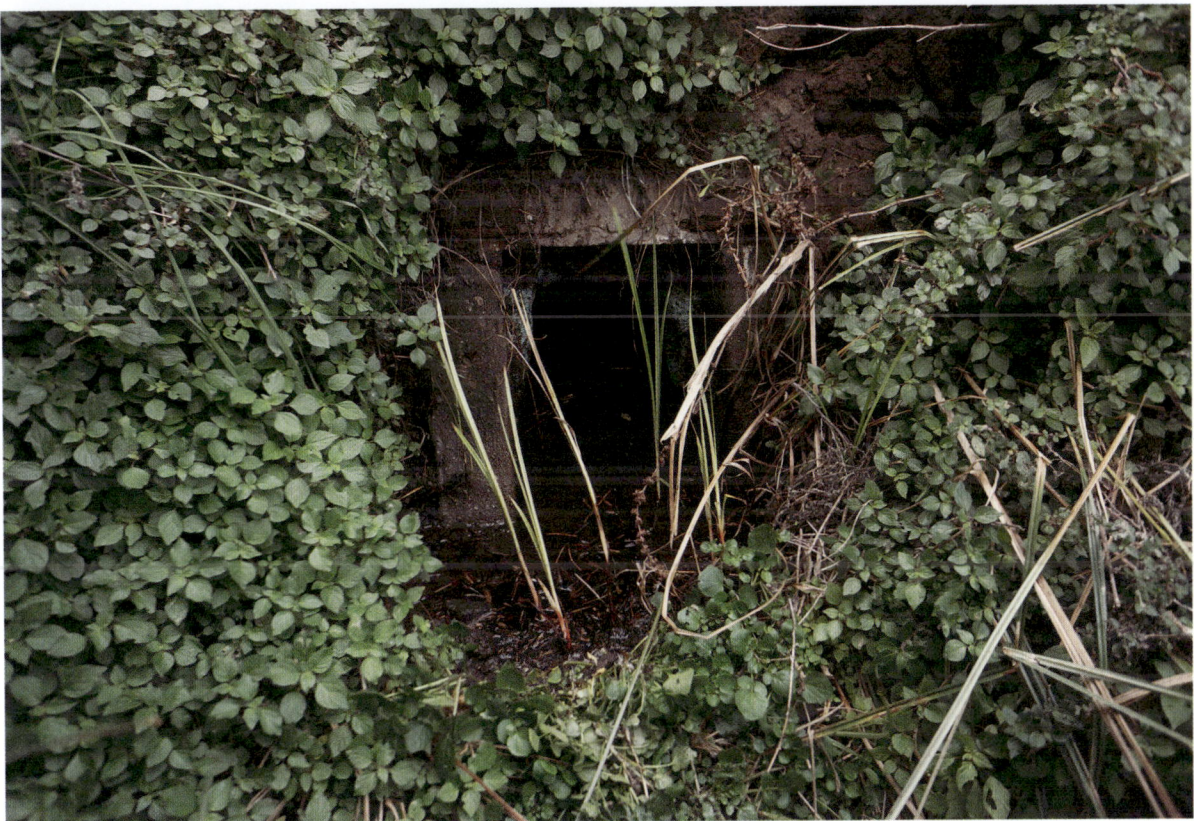

A NOTE ON PLACE

the land sounds you
given a motioning over uneven and broached and breached
land farms you
touches you

—ABRAHAM SMITH, *DESTRUCTION OF MAN*

There's a cliff. I lift a clump of earth to my cheeks, to my lips. Slowly I rub the gray little chunk between my middle finger and thumb. Gentle. I move my fingers up to my ear to listen—scratchy sand, silt? Or smooth and soundless? Clay. But feels chalky. I taste a tiny bit and chew. I get on my knees—what smells? Hints of moon snail skirts, driftwood smoke, distant runoff, clam juice, horsetail root, some dog, some sail, mugwort, yarrow, oak moss, sparrow nest. Down the shore an erratic boulder tells of extinct ice floes. Carved initials and crude images in friable sandstone indicate people: letters hearts signs people smiley faces hearts names scribbles dicks slashes initials of people and our manic habit to leave a trace for what? I don't truly know. This embodied pulse feels intimate though. I feel like I'm in the middle of a world. I'm in a secret someplace. Land inhabits me. My body seems to go forward? Backward? Into something humming, beached in deep time, an ecstatic fullness of multisensory, multivalent, polyvocal, emotional heft. And there's a shore. Perhaps to talk to a stone first I must become one.

WHERE I LOOK FOR PIGMENT

in my own body — in my own home — in my dreams — in my visions — during meditation — in my own memories — the ocean — in stories — near volcanoes — and all waterways — and in fluctuating water tables with seasonal gradients — in foreshore and on sea-cliffs — in extinct bodies of water — in the bodies of gods — in caves — where erosion has occurred — in exposures — where there is tectonic tension — where an explosion has occurred — after an eruption — after an upheaval — where there are rifts — in holes — in animal trails — where there are flow blockages — at uprooted tree bases — in tunnels — mines, audits, tailings — landslides — construction sites, namely profile, footing, or foundational work — digs — sidewalks — roadsides — suburban yards — alleys — museums — soil and mineral maps — google or satellite maps — old photos — shops of all kinds — junkyards — broken buildings — the flood-washed — graveyards — burials — ruins — where are you from — why are you from there

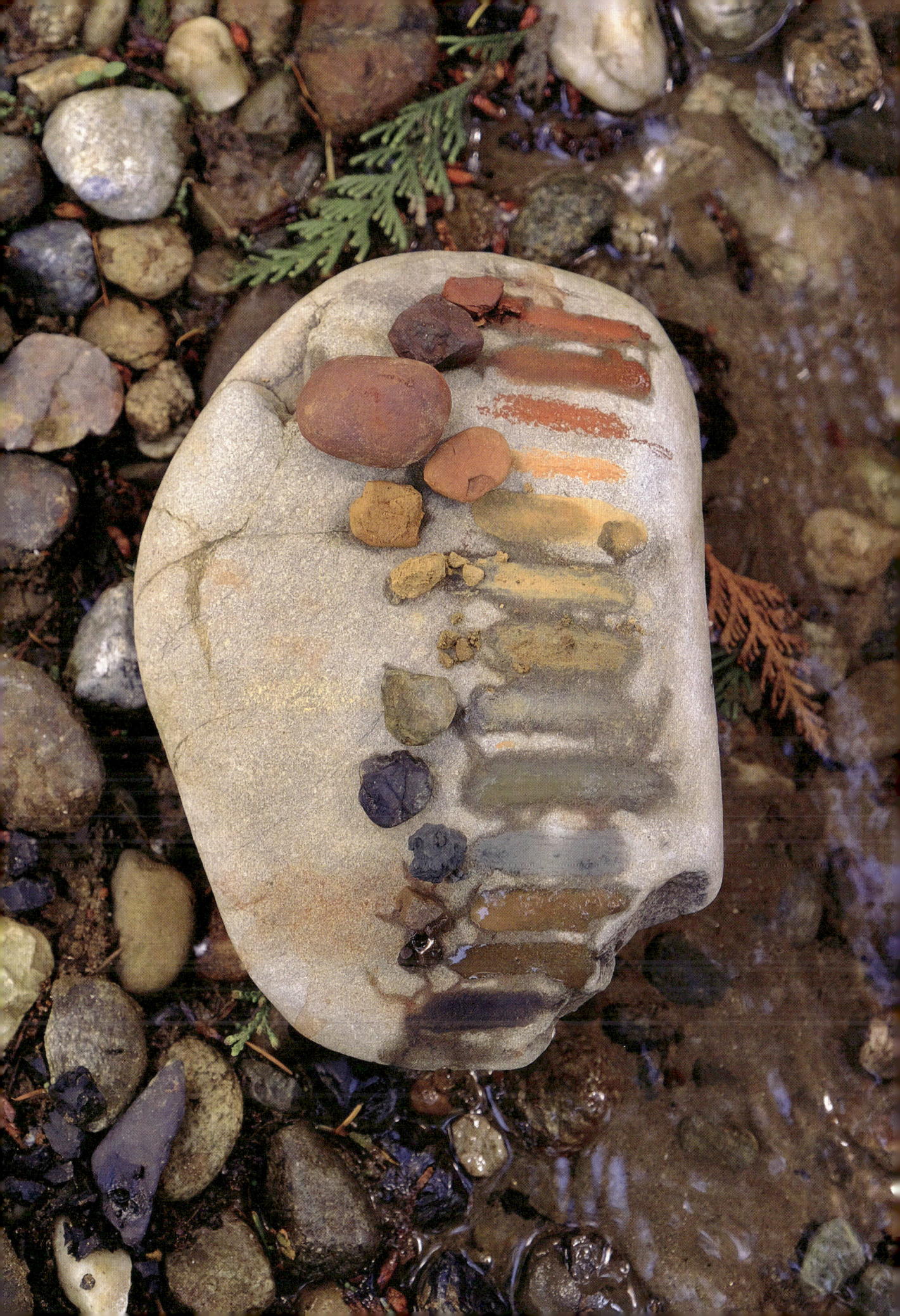

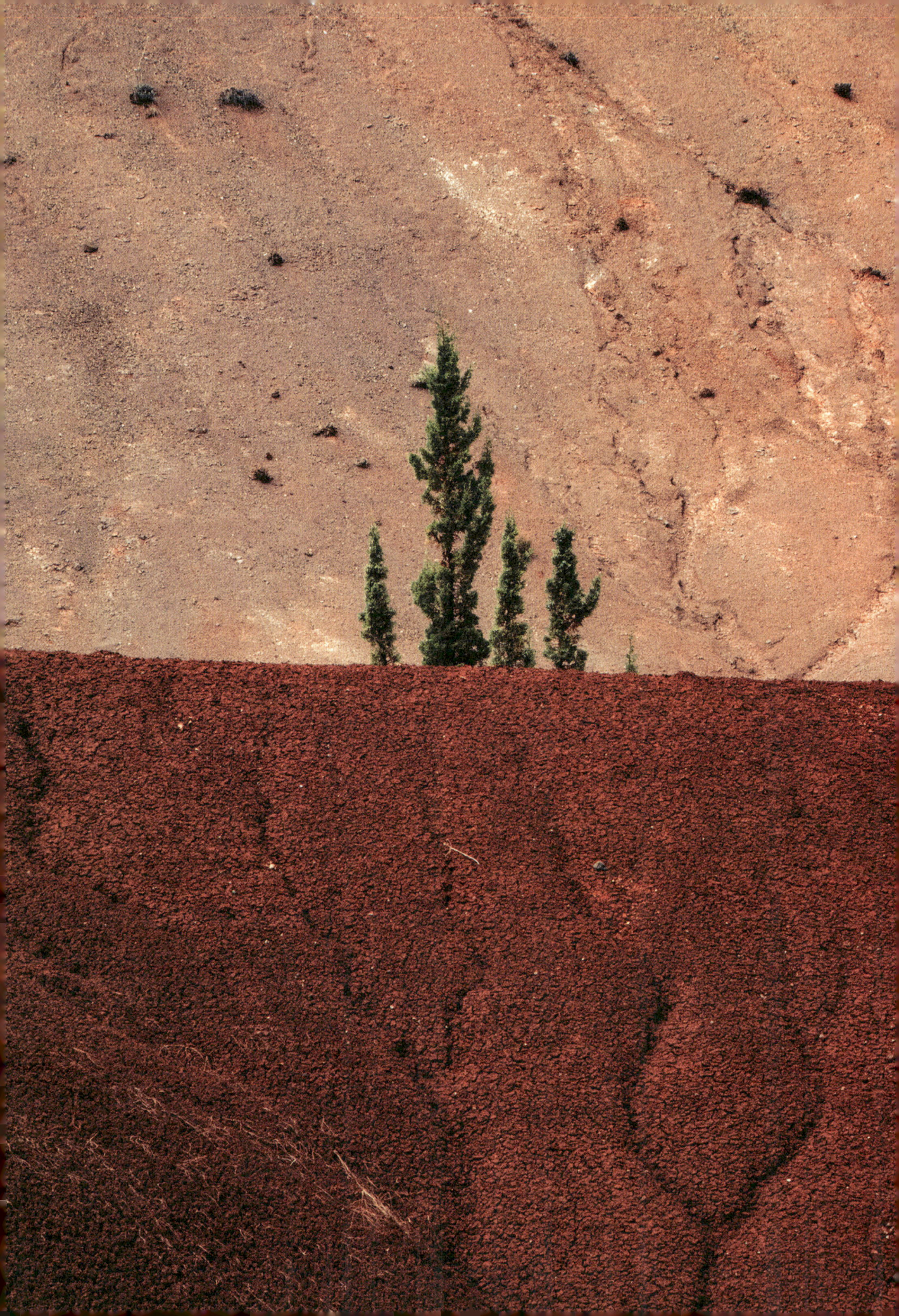

RED OCHRE

HEMATITE

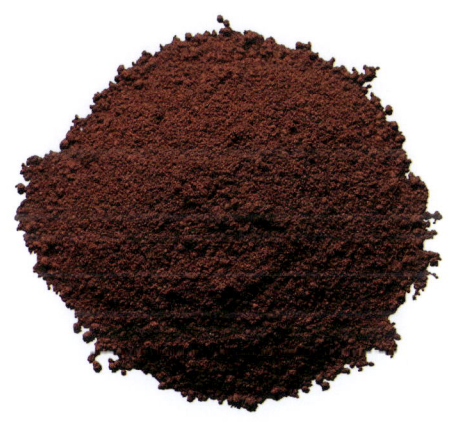

IRON OXIDE ⚹

Etym. from *haima*, blood and *lithos*, stone, "stone that bleeds"

$\alpha\text{-}Fe_2O_3,\ Fe_2O_3$

PIGMENT SPECTRUM

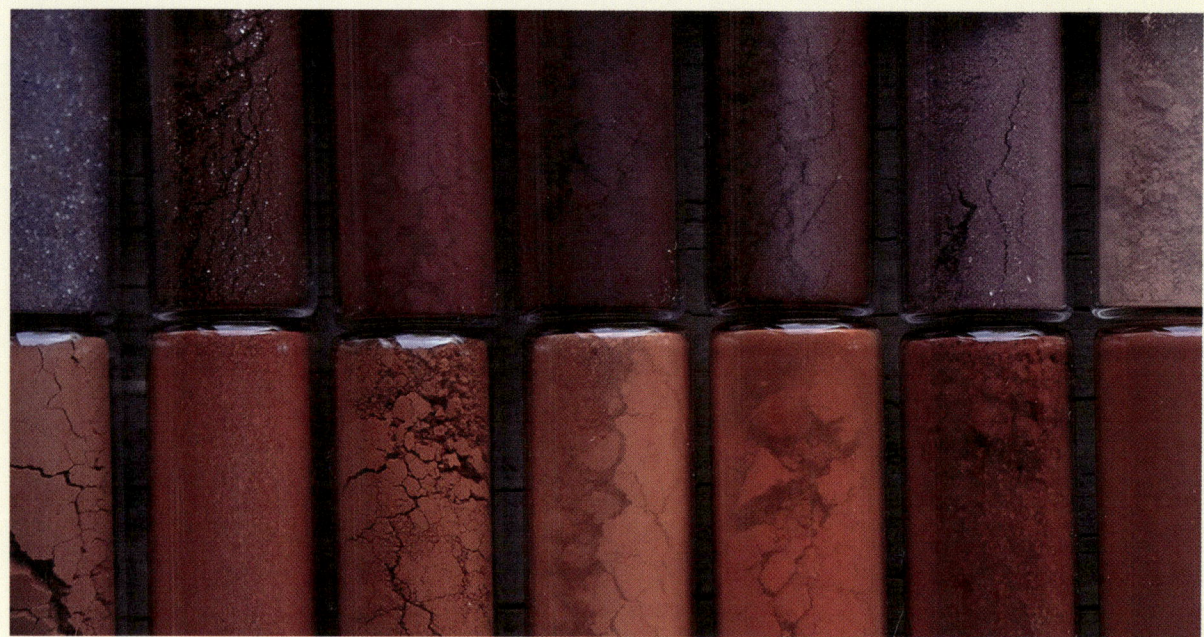

MATERIAL DESCRIPTION

WIDESPREAD ON EARTH, MARS, and the moon's nearside poles, hematite is a primary part of global iron cycles that shapeshift easily between life, air, water, land, aeolian vapor, solar wind, and deep flowing planetary mantle and core. Hematite forms in and penetrates a few key environments: the deep wet (ancient marine-based, or banded-iron formation), rising hot (metamorphic or hydrothermal), superficial soil or weathered rock (sedimentary clays, mudstone, siltstone, shale),[1] thus we find red rock rusting and weathering out of interbasaltic beds, volcanic zones, tectonic fissures, microbacteria, and ancient oceans, even meteorites. Softer, friable hematite soils and clays form faster in "tropical climates, and indicate past warmer climates in the geologic record," according to pigment geochemist Ruth Siddall.[2]

Famously called "the stone that bleeds" and found in veins, hematite tastes, smells, and looks like mammal blood. We sweetly bleed red like rocks do. In many old stories, the presence of red soil may indicate mythic blood, either of a god, ancestor, or animal spirit. On Mars, "freaky little hematite balls"[3] indicate conditions for extraterrestrial life. The oldest hematite on Earth are called banded iron formations, or BIFs, and were formed over three billion years ago as the result of a complex microbial process that also built the oxygen-rich atmosphere that we thrive in.

Hematite does not always appear naturally blood red to the eye, however. Iron oxides often look deceivingly dull, even black, gray, bluish, purple, gunmetal silver, or shiny metallic. Deeper red hues emerge when hematite is well-loved. Ground into finer and finer dust, rich blood red is easy to see, especially suspended in water. When processed in coarser particle sizes, a shady purple, metallic, or bluish-gray tone may dominate. It comes in various forms, some of which may mirror or appear as starry sparkles (particularly the botryoidal, specular, or micaceous forms). In fact, early mirrors were sometimes made by simply polishing shiny hematite! No matter what form, when hematite is struck on hard ground, a rust-red mark is left behind. These stones are sometimes used in divination as the "truth stone" or "true word stone" by prophets and high priests of the Ancient Near East[4] due to hematite's deeply reflective and lifelike powers.

NAMES

bloodstone, ferric oxide, red iron oxide, iron sesquioxide, sky iron, red ochre, haematite, specular or micaceous hematite, specularite or "mirror-like," kidney ore, raindrops of metal, blueberries, "freaky little balls," mirror stone, stone of truth, crocus martis, martite, red bole, rubricae, rust

Note: There are many other terms for red ochre that are specific to individual cultures and languages. I have only listed main alchemical terms used in my culture.

HABITAT

volcanos—ancient dried-up oceans—body or blood of a god or ancient animal—geothermal and hydrothermal vents—interbasaltic beds—caves—uplift—extinct human dwelling realms—human shelter—weathering of basalt—places on high—sandstone—sentimental sediment—seashores—runoff—shallow waterways—burial grounds—wastelands—steel landfills—car graveyards—railroad tracks—ceremonial holes—innermost insides—yawning hollow—heart of Earth—hearth of hearths

CRYSTALLOGRAPHIC SYSTEM

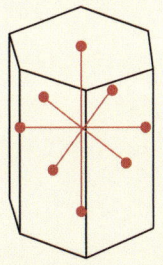

HEXAGONAL
RHOMBOHEDRAL

OTHER COMMON RED OCHRE MINERALS

SIDERITE, "IRON" $FeCO_3$

LATERITE, "BRICK" HEMATITE

LOOK-ALIKES

Cinnabar or Vermillion (Mercury Sulfide)

Minium or Red Lead (Lead Tetroxide)

Realgar (Arsenic Sulfide)

CHROMA

MIRROR	MENSTRUATION
METALLIC	RUST
BLUE-GRAY	RUST'S RUST
IRON GRAY	ROUGE
DULL STEEL	UNASSUMING
INVISIBLE	MOUNTAIN-ALIVE
REFLECTIVE	CORE
BLUE	WHAT I CARE ABOUT
GUNMETAL	WHAT I LEAVE BEHIND
RED	
BLOOD	

RED OCHRE — The Stone That Bleeds

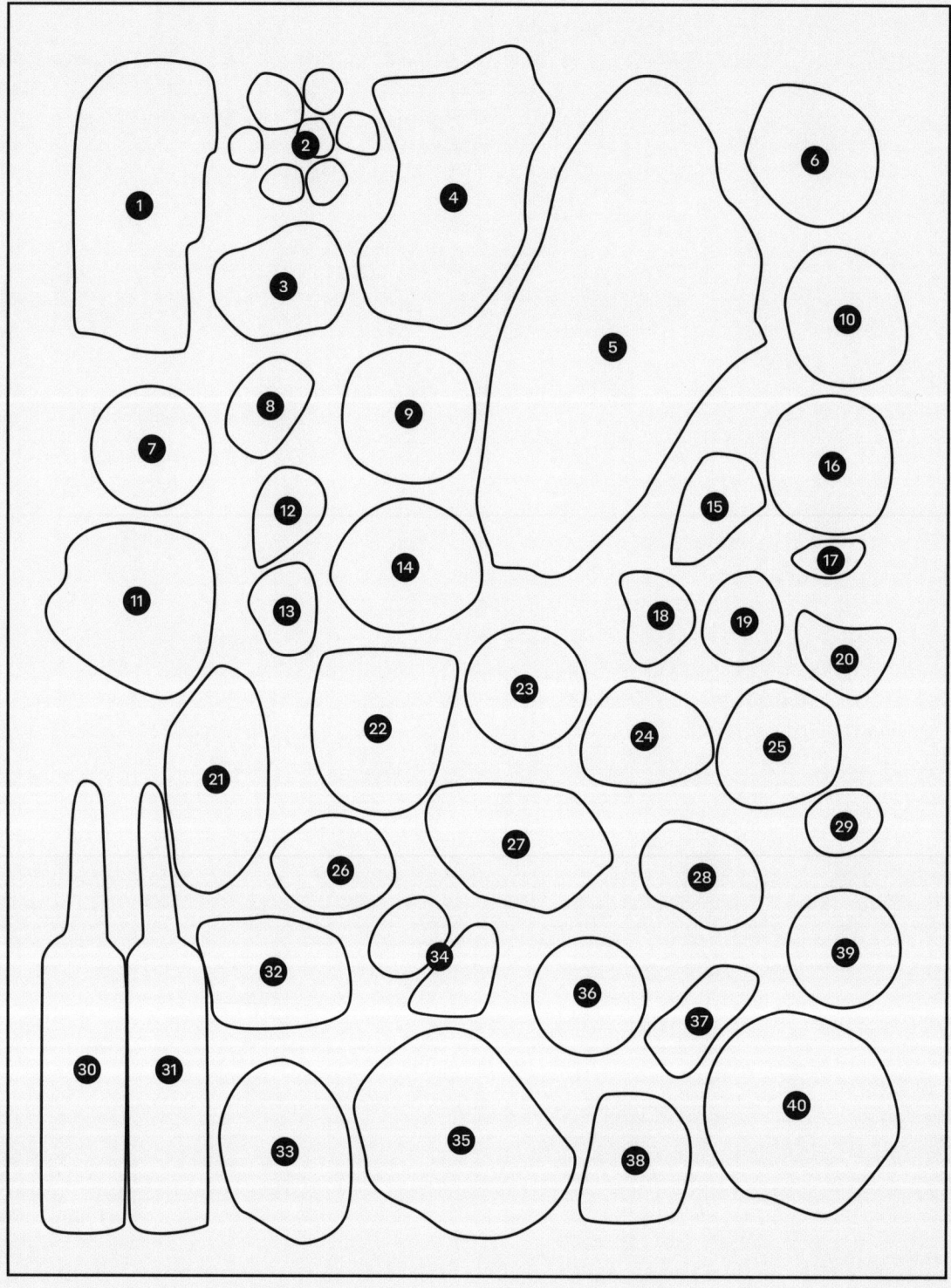

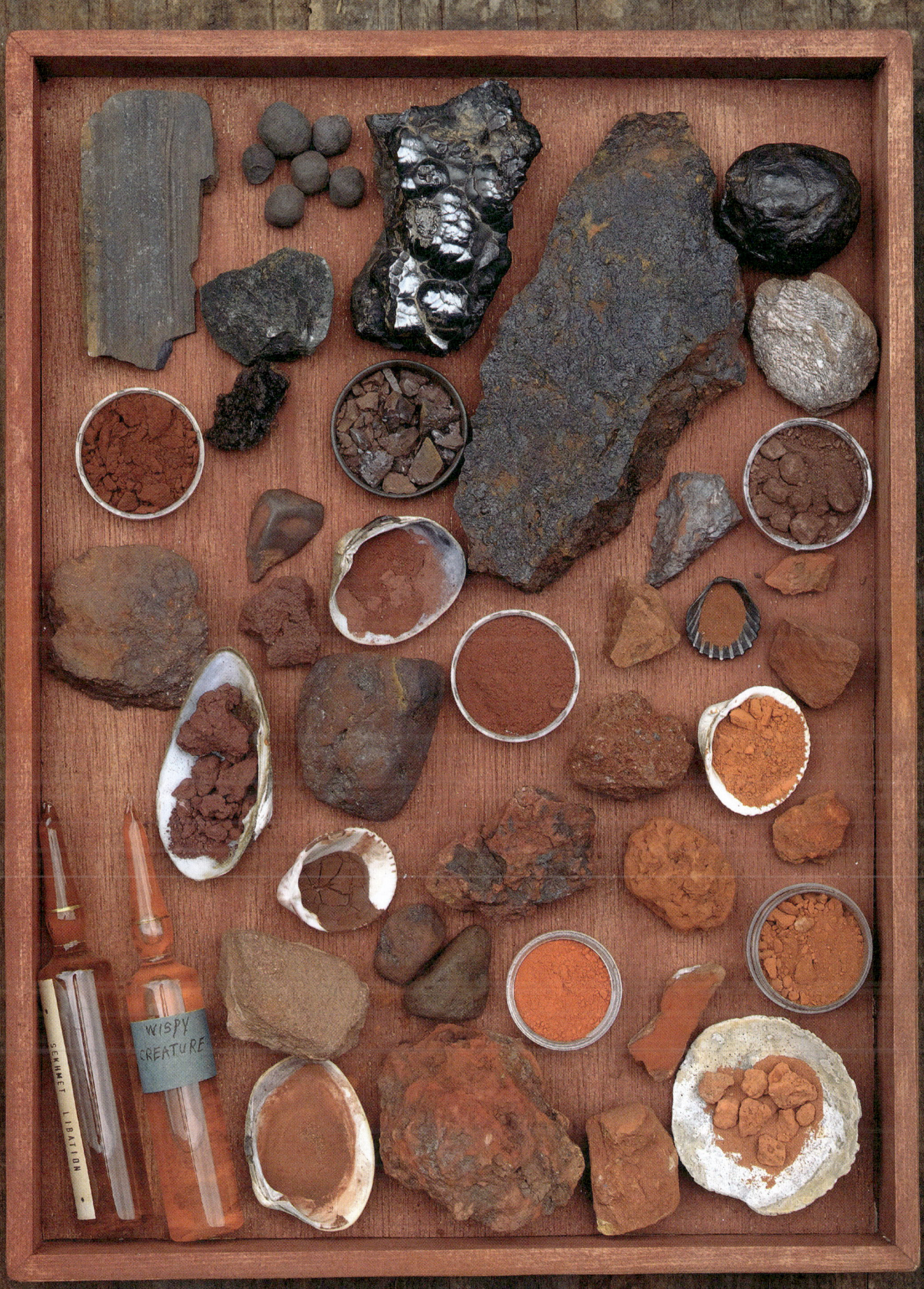

SEKHMET LIBATION

WISPY CREATURE

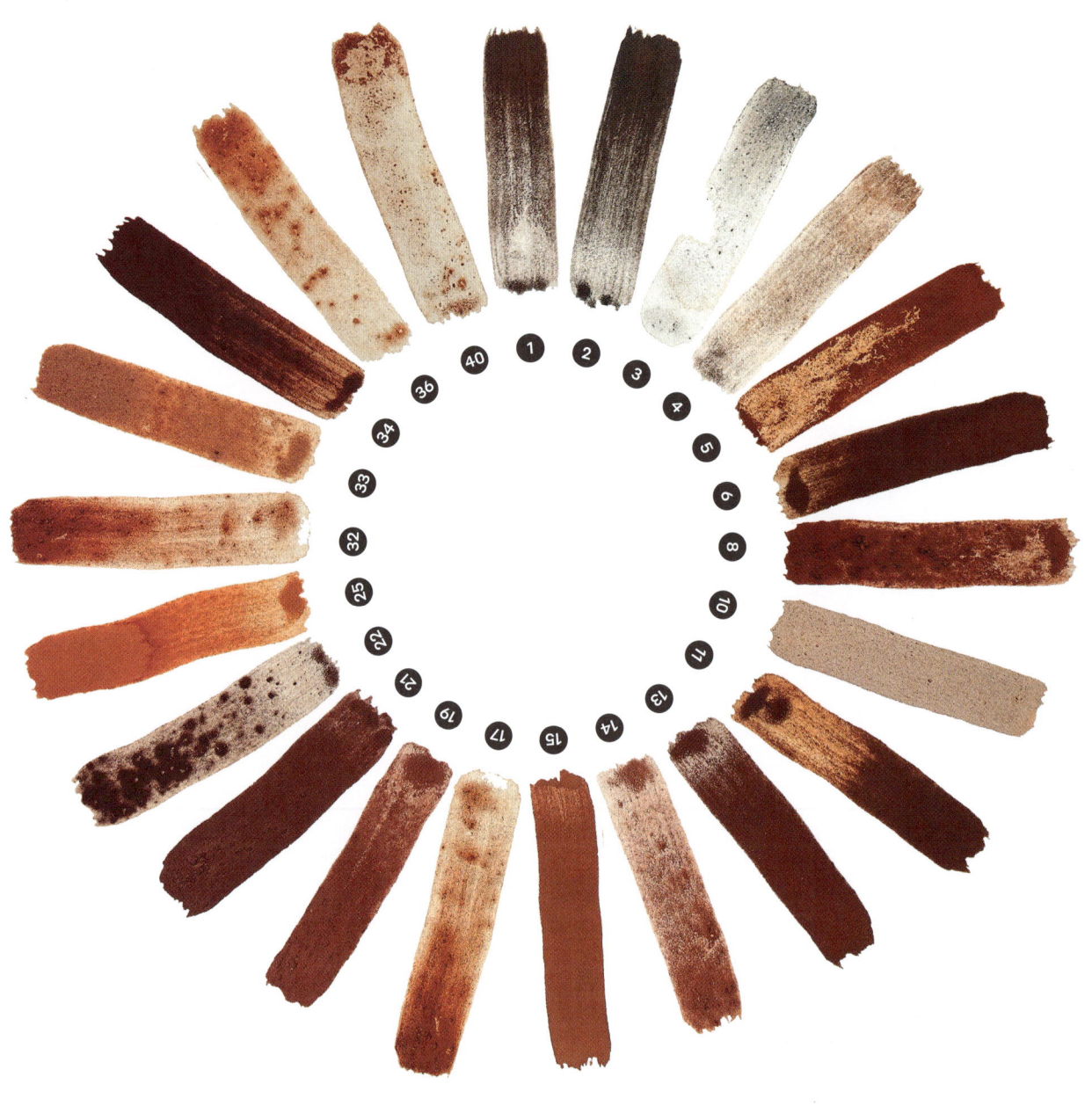

1. *The Great Breathing*. Three-billion-year-old Banded Iron Formation (BIF) from the pivotal times when oxygen became central to the atmosphere. Modern-day Michigan, ancestral Anishinabewaki.

2. *Taconite*. Mined iron ore of "lesser quality" that is then processed into pellets for use in steel-making. Carried by railroad from Iron Ranges of Michigan and Wisconsin, and shipped throughout the world. Gathered from near Bob Dylan's hometown, thanks to Kim B.

3. *Sparkly Hematite*, or so I was told. Near Banded Iron Formations of the Iron Ranges.

4,*9. *Kidney Stone*. Extracted deep in caves in Morocco and that is all they'll say.

5. *Red?*

6. *Fear Red*. From a place of mystery.

*7,8. *Unknown Prospect*. Hematite soil of north Utah, gift of Elpitha.

10,*16. *Buffalo Blood*. Micaceous hematite bedrock from the "Good Fortune Schist" formation, ca. Precambrian age. Gathered by archeologists from a Paleoindian habitation dwelling and butchery pit area, considered by some to be the "oldest hematite quarry in North America," 13,000 to 11,000 years ago, Wyoming.

11. *I Yearn Mountain*.

*12. *Sibudu Cave*. Ochre rubbed by early humans, 70,000 years or more ago. Gift by way of archeological steward and curator of the Wits University Origin Centre in Johannesburg, South Africa, Dr. Tammy Hodgskiss.

13. *Clearwell Cave*. Crunchy, sparkly hematite, ca. 300 million years ago. Mined from Clearwell Cave, a contemporary ochre mining cave complex in use over 4,500 years, in the Wye Valley, England.

14. *Supergene Ore*. Red earth from the island of Hormuz, Iran. A generous gift from the brilliant British Iranian painter Hana Shahnavaz.

15. *Underground Red*. Heavy. Bone-breaking. Silver-gray hematite mined commercially in Zaragoza, Spain, for red iron oxide pigment.

17. *Superfund Mine Iron*. Hematite iron ore build-up on excavated gold and copper ore at a Superfund site with stunning views of the Methow Valley, Washington State.

*18. *We Don't Know This*.

19. *Nourishment for Those Far Away from Their Homes*.

*20. *Southwest Red Ochre*. Take a 4×4 up a very long rock road, but not if it will rain soon. Go until you reach the Saguaros. Circumambulate by foot at night under the right moon before entering. Eat baby food for sugar. Ask for nothing, leave with who you are.

21. *Forest Ochre*. A wild galactic red ochre that makes the forest floor appear pink through a camera lens. Beautiful on skin. Found in the overgrown spoils of an old iron operation near the border of England and Wales, by swan wing'd Caro Ross and flock.

Swatch not pictured, see raw material/color on page 63

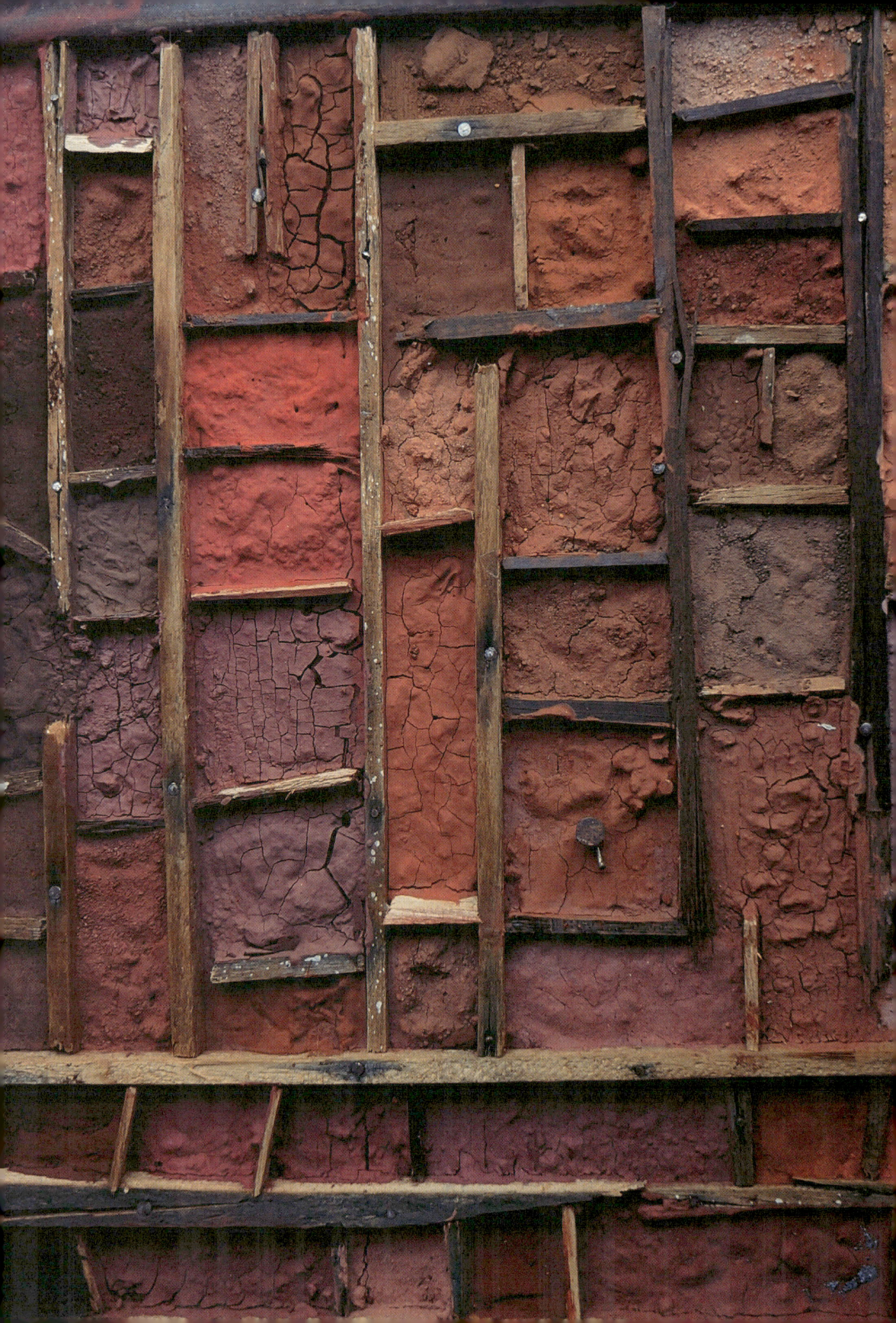

22. *Irondale Ore*. One of my most beloved teacher ochres that escaped becoming steel. Originally was anonymous iron ore shipped to Irondale, Washington, to be transformed into steel. Likely brought by boat from China or gathered from local bog deposits in the Chimacum Valley, Washington, on traditional Chimacum, S'Klallam, and Coast Salish land.

*23. *Red Ochre from a Dream Cave Place*.

*24. *Red ochre from the Tourist Beach Volcanic Fish Place*.

25. *Luberon Ochre*. Formed around 200 million years ago in ancient ocean deposits and now are the traditional Luberon mines in the Ochre Massif region of France.

*26. *Barouk Chestnut*. Red soil gathered by Joumana Medlej's mother in the Barouk area of Lebanon, whose name shares the root word *baraka*, meaning "blessings."

*27. *Hematite*.

*28. *Not Red*.

*29. *You Will Know Me Only When the Time Ripens Again*.

*30. *Sekhmet Libation*. Red ochre liquid made according to ancient recipes for fake blood, used to pacify raging goddesses, especially Egyptian goddess Sekhmet and Ugaritic goddess Anat. Blessings to Thomas Little.

*31. *Wispy Creature*. Hibernating biotic red ochre ink, as alchemized by Thomas Little.

32. *Desert Red*. Cemented red sand from the vast deserts of jabal al-mazmar, "Mountain of the Plague" in Wadi Rum, Jordan. Gifted as an omen, not long before COVID-19 pandemic.

33. *Red Several*. A ritual blend of red ochres for descent.

34. *Home Red*. Takes many washings to reveal rich, greasy red. Gorgeous and well-hidden hematite (with magnetite particles in the matrix) that washes down from a creek bed deposit, on timber land up the road where I live. Traditional Nooksack land.

*35. *South Red Ochre*.

36. *Ercolano Red Ochre*. Loves to spread out to the edges when put in water. A beautiful red ochre laced with dolomite from San Giovanni Ilarione Veneto, Italy, I believe.

*37. *Red Ochre Guarded by Blue Butterflies and Rattlesnakes*. With gratitude to Zev.

*38. *Land Core*. Red earth from a soil core taken by soil scientist and pigment researcher Zeide Nogueira Furtado (of GeoColors Brasil) in São Paolo, Brazil.

*39. *Sick in the Center of Their Being*. Healing iron topsoil from a recovered Superfund mine site in the United States.

40. *Crater Core*. Red earth from the core of a volcanic crater. You'll remember the rest later.

Swatch not pictured, see raw material/ color on page 63

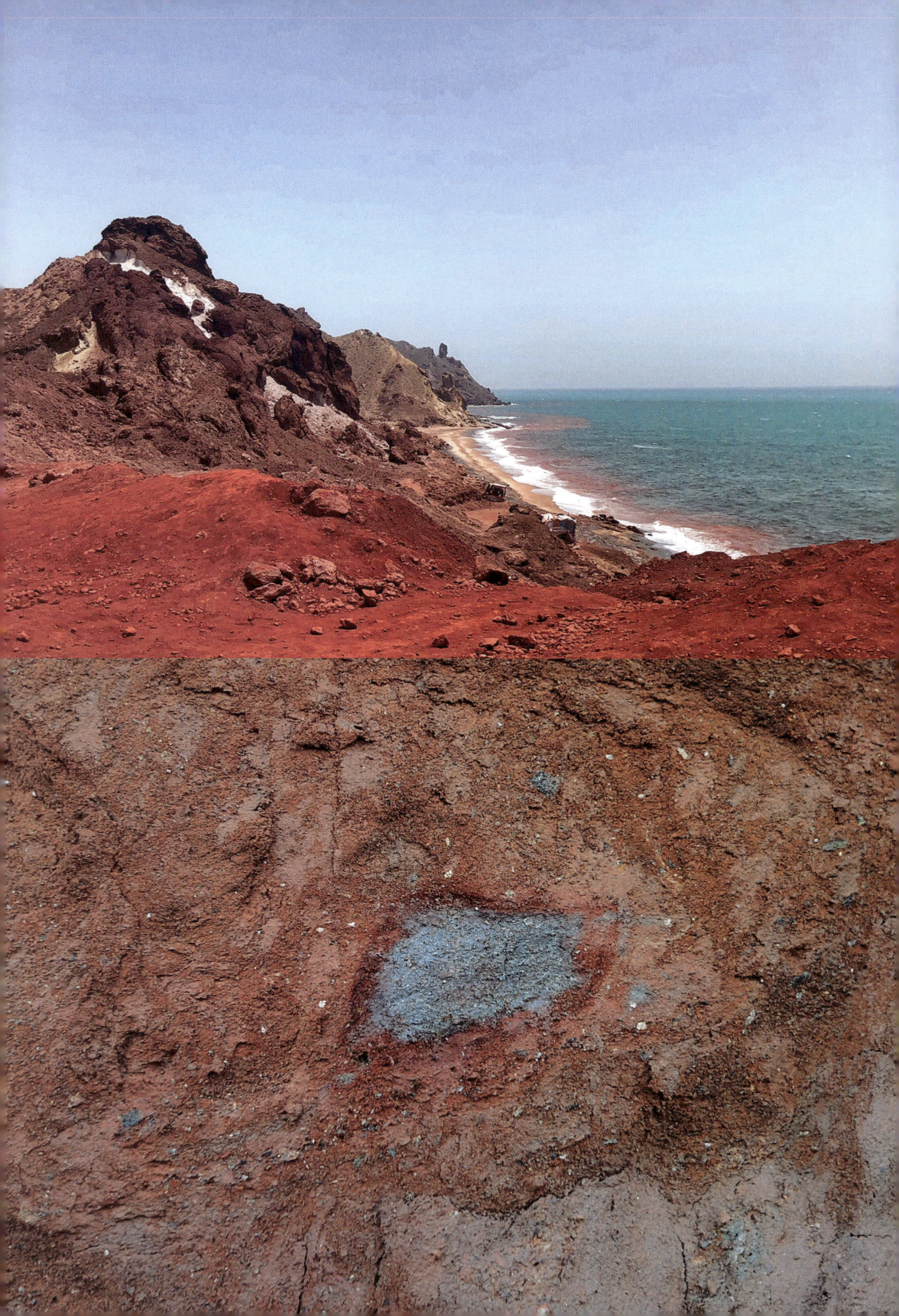

Supergene Ore

DEEP RED SOIL—more than 500 million years old—on Iran's Hormuz Island is called supergene ore. The ocean tides resemble menstruation on a larger scale, red wave after red wave. "This island is a part of our body. We are not just born here, but it's *in our blood*," remarks local boat captain Abdullah Gederi.[5] *In our blood* recalls the early Semitic word for the first human, *'dm* or Adam, from adama, earth or soil, often translated more broadly as humankind, or as red earth, common clay, clod of red earth, or simply . . . red ochre.

The red mineral color itself comes from the elemental presence of hematite, an iron oxide whose name comes from "stone that bleeds." *Human*, or *homo*, is derived from *humus*, meaning "living soil," earth. Human being bleeding soil.

So then were does life force come from? Something shared between human and rock? Iron? Oxygen? Supergene ore? Spirit?

Earth?

Captain Abdullah makes a traditional dish (suragh) of little ocean fish fermented and cured in ochre and salt over several days. He takes into his body the old dust of bleeding rock. Somewhere else, far away, this red island is considered by Western scholars to possibly be the historical origin of the Garden of Eden.

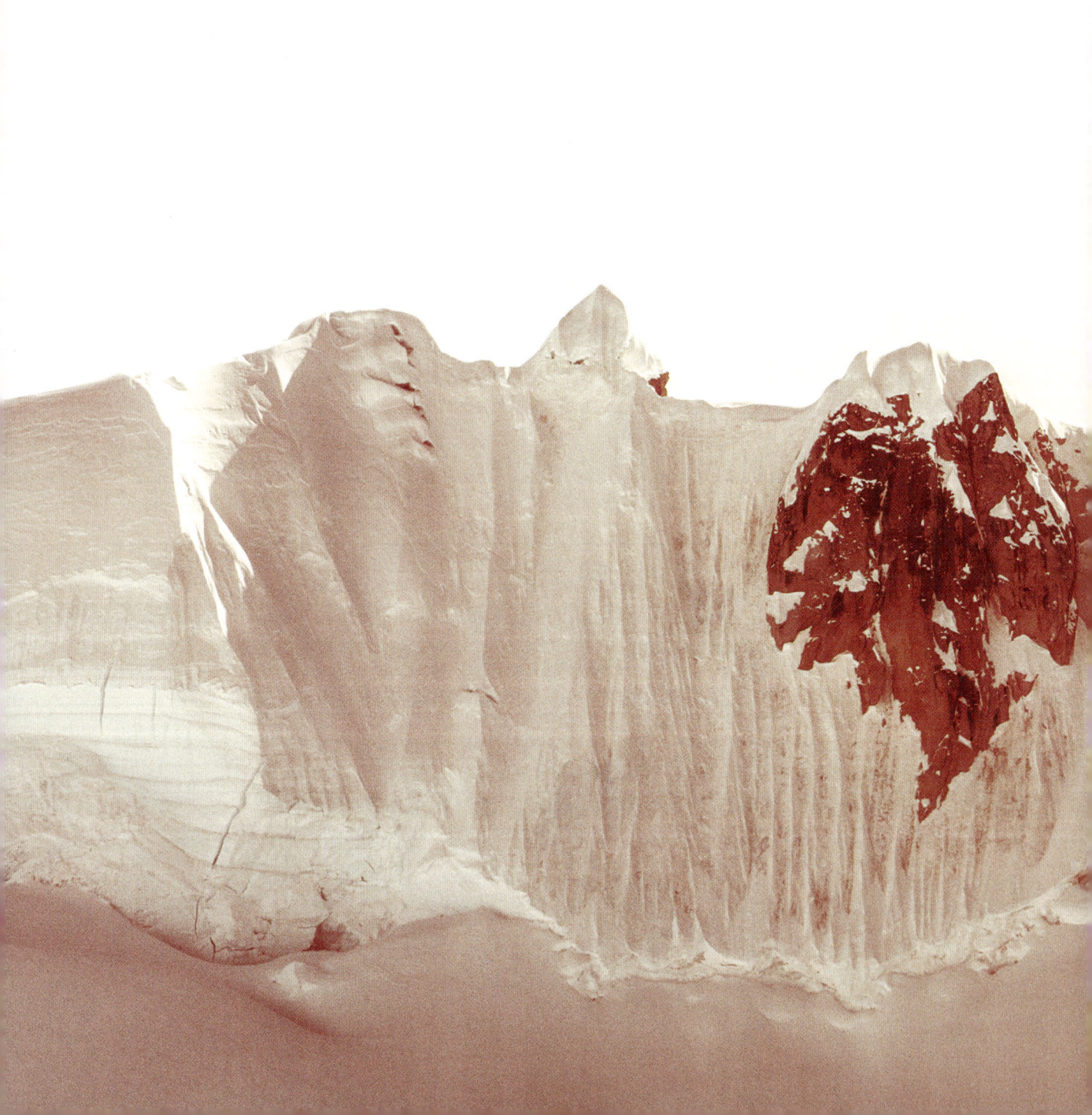

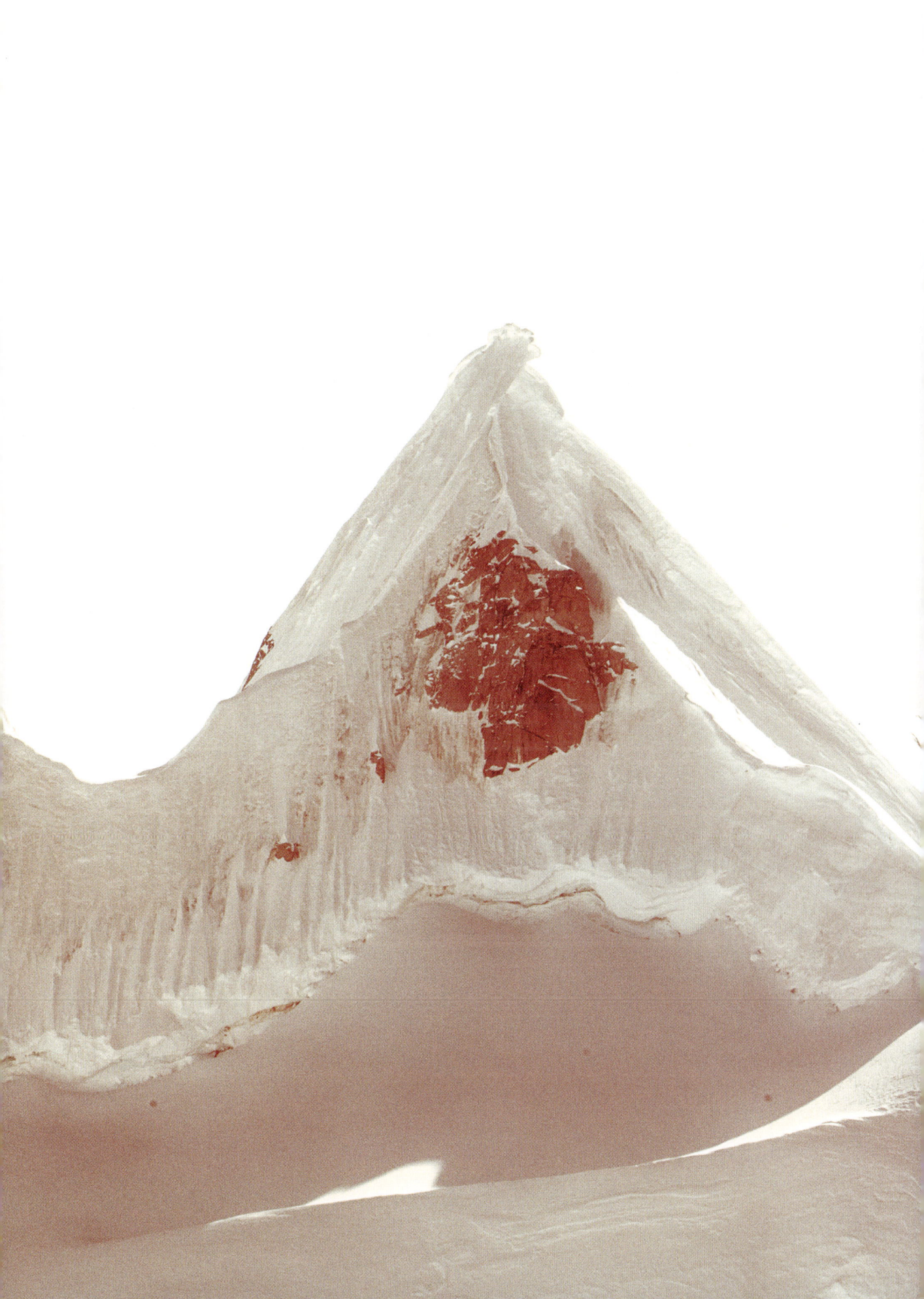

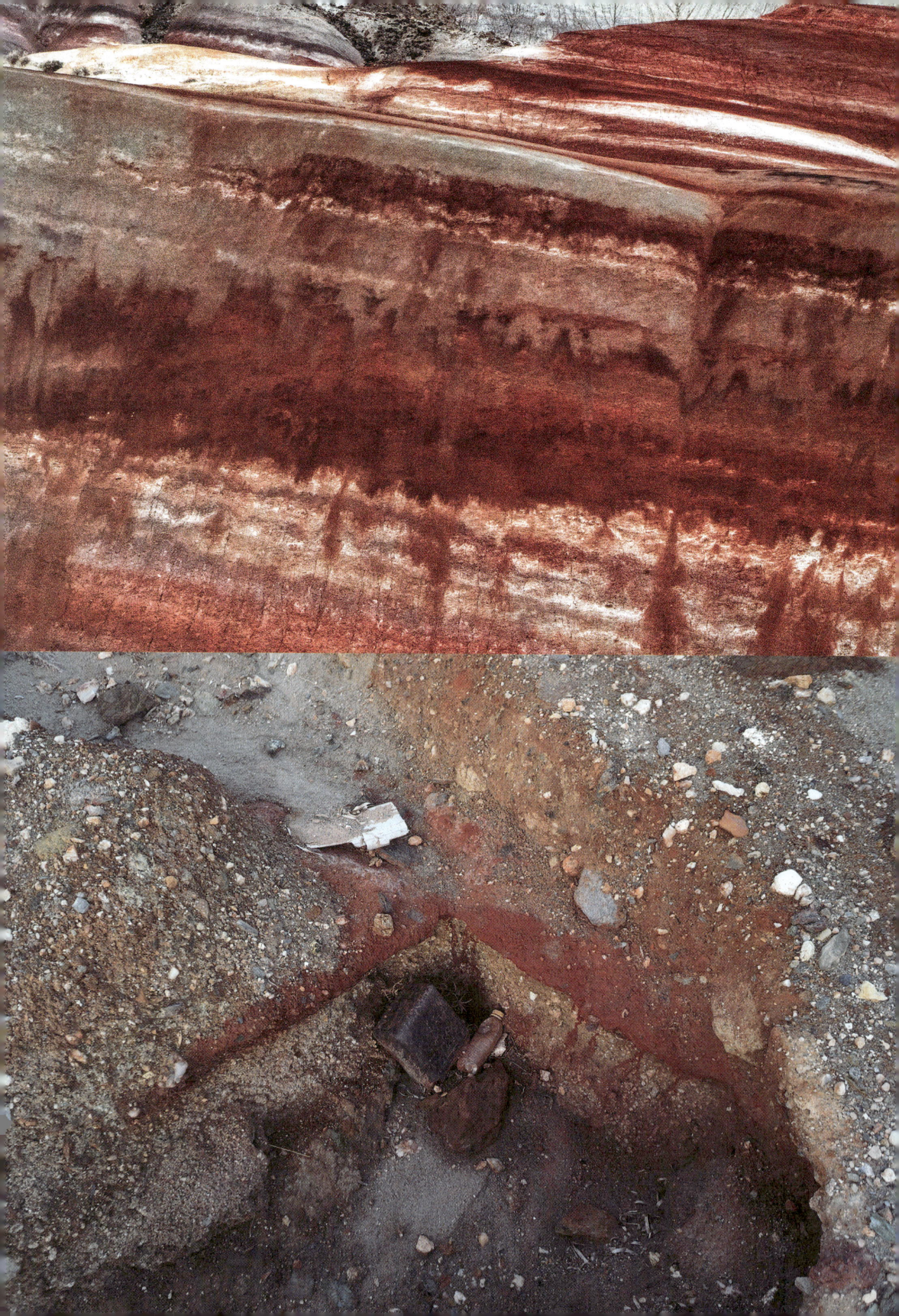

dream began red then slipped out of the vat and ran

—ANNE CARSON, *AUTOBIOGRAPHY OF RED*

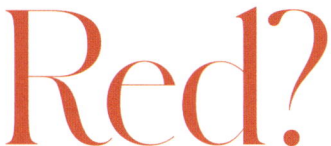

ground where the flower grows turns deep red
this ground that keeps turning deep red ground.

—JOHN TAGGART, "THE ROTHKO CHAPEL POEM"

In a way it's inside-out, red.

—ANISH KAPOOR

Red?

BEWARE OF THE POPULAR MYSTIQUE around red, especially red ochre. Red shouts prehistorical blood, aggression, heat, eros, sex, bonds, power, war, "the brains of men in a red fog."[6] Or the sisterhood of menstrual blood, the mania of wombs, vulva caves, the moon's red code. Red persists through the ages as dangerous, hot, big.

Red is known as the color currency of nature, or as Tennyson writes, "Nature, red in tooth and claw." Mammals and especially carnivores and predators, like us, still seek red in ripe strawberries, red lips, and raw meat. Our mouths act rather red-handed. And we find it difficult to escape the fleeting, fickle, brutal impulses of nature—we kill, breed, bleed out, annihilate ourselves over and over. Red ochre conjures power and desire, but nowadays natural red hues are overruled by technicolor attractions of neon reddest red. But that is not the main story I know or wish to tell here.

We're descendants of red earths—we are not separate from them.[7] Talking about red ochre is very personal and intimate to me. And I may not be able to do justice here, even with the help of many friends and their voices.

I used to know very little about red ochre (and am joyfully aware how little I may continue to know). My initiation began over several years of heartbreak and trials. I vividly remember when I met red ochre first: in a dream with *no red* in it. In my journal I wrote this:

I'm on an old path.

There's nothing remarkable about it.

Two turkey vultures fly above.

I descend down a dusty road.

I'm looking for something.

On one side of the path I pass by many uninteresting stones.

Get to the end and must go back because apparently I missed what I came for.

I see graffiti tagging on an old mine shack.

Go back up road.

Reach down and pick up stones from earlier.

The rock looks and feels metallic blue gray but . . . somehow . . . it's not?

I wake up haunted by a longing, and a voice in the back of my mind saying,

Because they really ochre.

Despite the mysteriousness of the dream, I didn't think much of it and wasn't sure what to make of it. Not until many weeks later did I realize the dream was probably a geologic ochre place in the "real" world. I remember the day I found the path (or, perhaps more accurately, it found me). I'd been looking on Google Earth for somewhere that might have old mining shacks. Sometimes I get these geographic intuitions that certain places on maps might want a visitor. Sort of like a child tugging at my hand, inviting me to look down at a shimmering, fleeting insect.

Honoring that feeling, I drove to where there seemed (on the map anyway) to be old mining stuff similar to my dream and began aimlessly walking around toward a dusty hidden trail. As soon as I walked onto the path, two vultures flew overhead, just as in the dream. Suddenly I felt I'd shifted through a portal, as if in an alternate dimension, full of déjà vu. Everything unraveled along the dream path. I walked by metallic rocks and didn't see them at first. I saw old mining graffiti. Dream turned reality. Eventually, after shocked hours and days and months of going back and back and back to this place, I came to realize the blue-gray metallic stones were not really gray inside, but would turn into intense red dust. *Ochre.*

Red ochre is not ordinary red. As in the dream, many red ochres are *not red at all.* Blood ochre stones can appear steely, gray, silver, blue, aubergine, lilac, golden, black, sparkly, brown, dusky pink, and even faded green on the outside before they warm up, break open, and blossom to reveal red pollen. They look like different colors when painted on human skin than they do on paper or stone or wood. And red ochre is so very often *invisible,* as Frank Andrew, a Yupik elder, says: "Some won't see it, even if they go to get it. They say that it is only visible to some people. Red ochre is strange because it disappears." In alchemy, they say red is not red but the *color colorum* ("color of colors").

Many other pigment folks I've encountered over the years have their own personal origin stories of seeing, of truly catching a glimpse of red earth's splendor for the first time. How were they first led to ochre? Almost no one speaks proudly of remote wilderness experiences. Rather they tend to speak of a dream, their grandmothers, a spirit, child, or a work of art. Transformative shared experience. Life affirmations where Earth reveals a *secret red joy,* a "red gentleness, very red like an open wound," as Buddhist Chögyam Trungpa put it.[8] It is a profound honor to hear personal tales of silent and tender discoveries of red ochre—a red guilt, the red of shame, "red made deeper by black."[9] A forgotten red, a wet red, a cold red. The peripheral red pulse felt in holding a small young hand. Pleasure red of a warm embrace. Plain old red heart red. Home red.

Certainly red pigment contains creative power. But not only in the typical "art" sense. As it turns out, profound abilities of red to initiate psychophysical recovery are well attested to. In human studies, red's been "shown to have special significance for humans because of the range of physiological effects it induces, including changes in the heart rate and brain activity. . . . When brain lesions impair colour vision, red is the most resistant to loss and the quickest to recover."[10] When shown red light wavelengths under certain conditions, male blood testosterone increases,[11] and mitochondria in human eyeballs age more slowly.[12] Red alerts, orients, and helps us recover.

Consider astronauts living on space stations, separated from Earth altogether, where they exist in constant freefall. This floating feeling can sometimes make them sick. The cure? According to a NASA study, eating reddish earthen clay was one possible way to counter the effects of weightlessness (and bone loss).[13] Red matter from Earth tethers an astronaut's body to their home, even when they are far, far away. There's something quite spiritual in this to me, akin to how red ochre painted in the margins of religious texts (like the Koran) indicate places where one is supposed to bow, to drop one's forehead down directly on the earth.

Red earth doesn't only ground us during space flight or spiritual ascent. Across deep time and space and into rituals of today, red ochre is used to help bury the dead and reconnect a soul with the heart of Earth. Paleolithic graves show plentiful red iron oxide that was offered with dead bodies—placed carefully as talismanic chunks or as powder sprinkled generously over the head and heart (especially of children).

As if offering a return of one life force to another? Earth to earth? Dust to dust? In more recent times, ochres are used in tombs (to paint "earth spells," as I described in my dedication), graveyards, and other burial contexts. As chromatic journalist Adam Rogers puts it, so many of humanity's earliest written texts were "full of references to iron oxides, the ochres. In fact, they were some of the first pigments humans ever wrote words about, in Assyrian cuneiform and Egyptian hieroglyphics. They're all over Egyptian tombs, Assyrian ruins, Greek ruins, Roman ruins. It's a literal testament to how important they were."[14]

In Chinese, red hematite is called 代赭石 ("dài zhè shí"), "red stone that perpetuates the generations," and a "descending stone" that helps return souls to the earth. Red hematite is considered the heaviest yin stone (female and rest principle) and stone of the pericardium (heart muscle).[15] In an ancient burial site in Xiaohe Cemetery, China, dated to around 4,000 years ago, archeologists found women buried with several talismanic objects, including red paint sticks—a kind of lipstick or rouge stick—that was found to be red ochre mixed with *a fat that is only found in the muscular center of a cow's heart.*[16] Some red lipsticks made in China today are strikingly similar, advertised as providing "plump pillow lips." They're still made with hematite or iron oxide and enhanced with animal collagen. This magical act of combining heart with heart recalls for me, also, the Atsugewi Native Americans in Northern California, who, as historian Paul D. Campbell notes, "prized red ochre above all other colors and they kept it even in a deer pericardium, that's the sac that holds the heart that pumps the blood."[17]

I think about how in death, human bodies, too, circulate through the Earth. It's almost as if we become the blood of Earth, as our elemental structures' sediment, and get moved through a million- or billion-year geologic pumping system fueled by the planetary heart. In that image, it makes sense that red ochre shows up as heartfelt, serious support to prepare a body to re-enter the bloodstream of Earth's soil. I wonder if this is why I am often asked for red pigment by women who need to bury their miscarried children, to dye death shrouds, and by others to support ceremonies or to give strength to the terminally ill. I remember sending red pigment to a woman dying of cancer, who told me she felt a great power in their bundles and that even before she unwrapped them, "My fingers started to tingle . . . my whole arm got goosebumps."

The art of celebrating burial and/or the dying process is not easy. A fellow earth alchemist brontë velez, a Black-Latinx activist who co-founded Lead to Life, an organization that "bridges racial and environmental justice through ceremony and art practice," shared with me a traumatic, heart-wrenching story that describes this experience of feeling earth returning us to earth.

In brontë's heritage, there are Mexican/Indigenous curandera (healer) traditions that describe a folk illness called susto, when the soul or spirit becomes detached from the body. The remedy given by a curandera is to literally bury the afflicted in cold earth to effect reconnection. "This was significant for me," brontë shared, "because my grandmother on my maternal side, a Black woman, had been buried alive in a very traumatic situation. I found this out through being buried in an art project that my mom saw footage of and later told me the story." In recounting their own experience of being buried alive as an art project (on the Bespoken Bones podcast), brontë describes emerging out of the ground, opening their eyes, and seeing the tree canopy through a "dirt filter over my eyes."[18] Being able to "see through soil" was described as a kind of powerful subterranean lens: being buried in the earth, and re-emerging, alchemized and brought healing to their ancestral lineage, through soil, their body, performance, dance and ceremony. One of the most profound things I learned was that this kind of intensive, intimate earth art could be considered a part of "the *pleasure* of giving back . . . giving back power to those who are in right relationship with place," as brontë puts it.

This teaching moves deeply with me. I remember the early days of my ochre training, when nothing about earth pigment felt right at all. My soul was lost, my heart broken, everything seemed shallow, sad, painful and depressing. I felt very alone and didn't know where to go, who to talk to, or how to listen. I only knew how to do obscure, introverted archive research. I would scour the library stacks for books with anecdotes on ochre or pigment. I read a lot of burial descriptions, transcribed archeological accounts and old recipes involving red earth, hematite, ochres. One of the recipes that began to whisper and call me back to my senses was called "How to Heal Heartbreak," written in Egyptian hieroglyphs from the Ebers Papyrus (ca. 1536 BCE).

HOW TO HEAL HEARTBREAK

If you find someone sick

in the center of their being

and body shrunken

and tender with disease at its limit

if you touch them and you do not

find disease in their body

except for the surface flesh of their desire

make a spell against decay in their
inside-house[19]

Hematite (bloodstone) of Elephantine,
crushed

red flaxseed

wild gourd or carob (seed)

cook in oil and honey

let it be eaten by them over four mornings

to ground their thirst

and cure their sick-at-heart

This poetic recipe makes edible red oil paint. Healing paint for *inside* the body. The recipe calls for all the usual things in ordinary paint recipes: Red flaxseed is also known as linseed, the primary binder in contemporary oil paints. Honey is an important ingredient that helps paint disperse. Both wild gourd seed and carob contain additional binder or gum properties. And, of course, notice how the hematite is from a named and specific place—Elephantine is an island in the Nile River. More than anything, I love how the description of the illness, "sick in the center of their being," really feels like an illness of the heart or the weightlessness experienced in soul loss. As with this recipe, and elsewhere, I'm amazed how the remedy is clear: *Take red earth in.*

To take red earth in, as artistic material, as paint, as healing activity, is aesthetic: "to breathe in" or to "be full of the god" (or in this case, Earth).[20] When I tried the recipe, I imagined my intestinal villi (the thousands of little fingerlike things that line our guts) becoming like red standing stones. I felt so alert and awake, or "something between panic and arousal, not sexual but certainly an arousal of all senses," as a woman once described red ochre to me.

Close to where I live and work in rural Washington, I've heard about red ochre paint in use by healers in Skagit, Nooksack, and other Coast Salish nations. The paint is "taken in" but not as internal medicine—rather it is used in ritual dance and ceremony. In *Red Paint: An Autobiography of a Coast Salish Punk*, Sasha taqʷšəblu LaPointe tells her own story of stepping into her family's long lineage of female dancers and healers involving red earth paint. She describes seeing longhouse dancers who wore red paint, and asking her mother what it meant when she was a child. "The red paint is for healers"[21] was her answer. In Sasha's book, she is taught how to carry red ochre in her own cultural lineage. The Lushootseed word for red ochre is never shared, instead, her red paint discoveries are told within chapters named "hədiw," meaning "come in."

Come in, breathe in, take red in. Or another way to say this, perhaps, is *come back.* Instead of a heavenward transcendence, ochre calls us through descent, remembrance, a red love of what is at our feet and in our laps and hands.

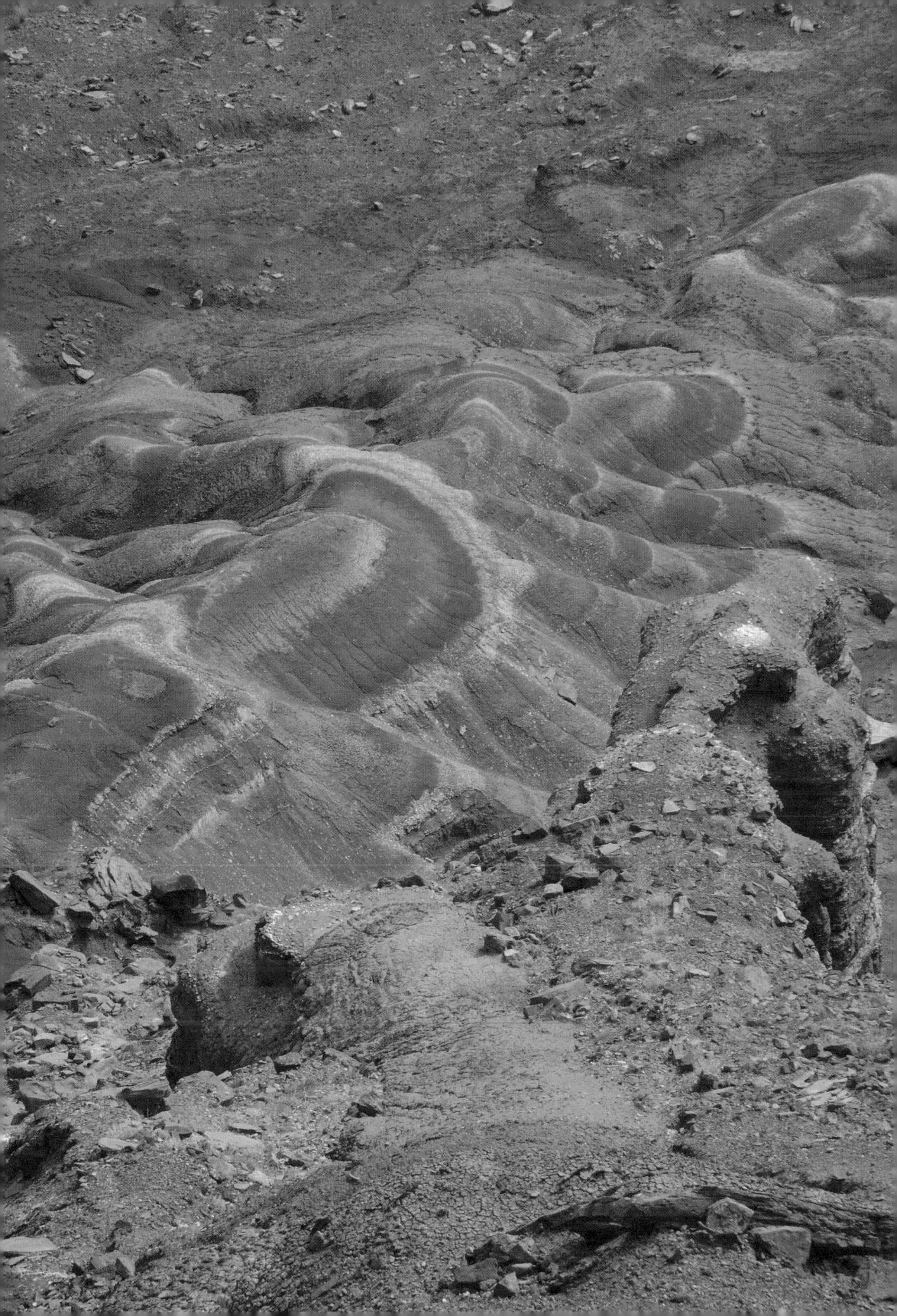

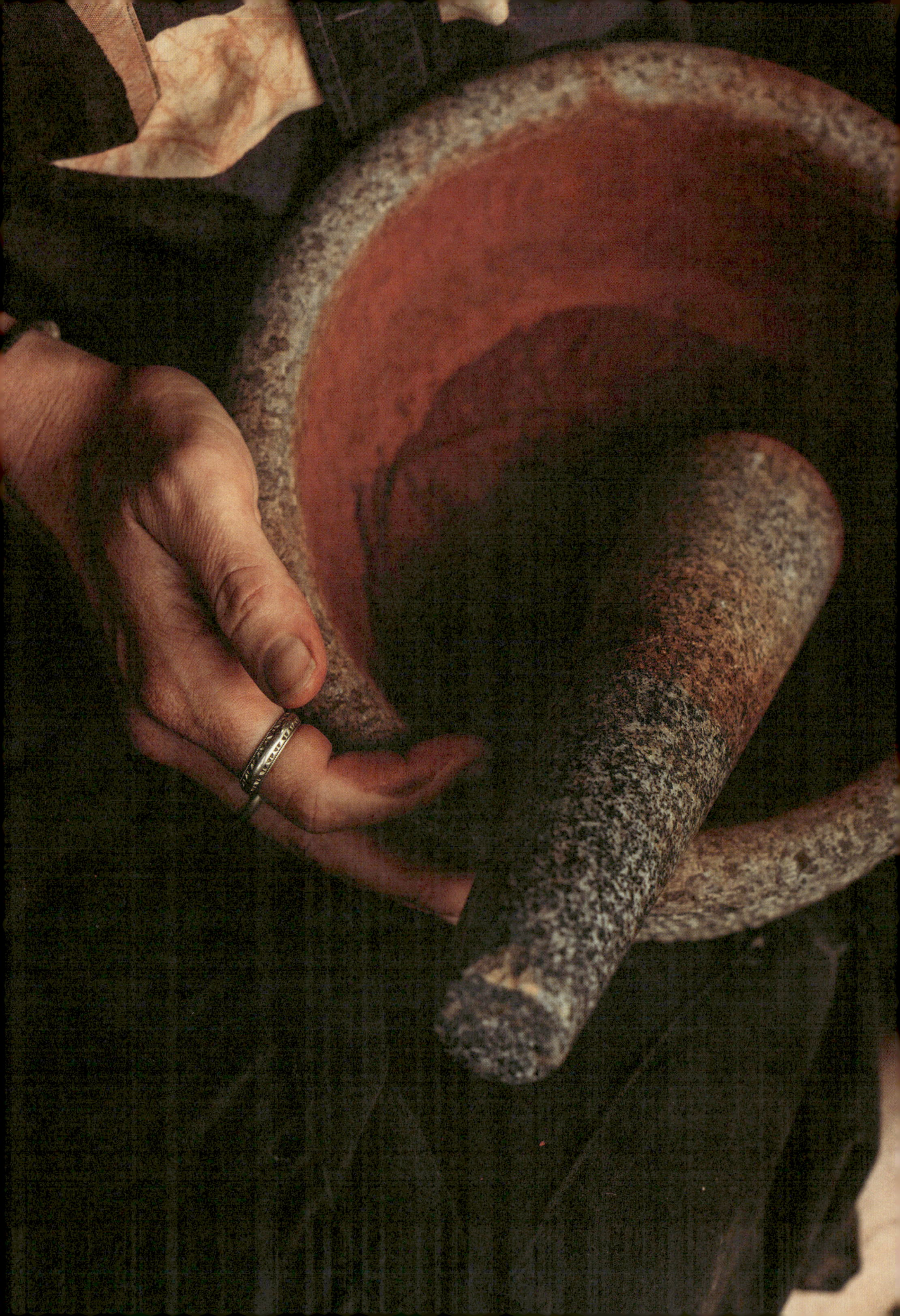

A NOTE ON PIGMENT

By the rock I rub against
I'm going to be tender again

—HARRYETTE MULLEN, *SHEDDING SKIN*

Wind-pestered, sea-sieved, pestled by the sun's long pulse . . .

—STEVEN CONNOR, *THE DUST THAT MEASURES ALL OUR TIME*

There's a stone mortar in my lap. Tucked into the space between my crisscrossed legs. I put a small earthen rock in the hollow. *Thank you.* In my body I listen. I listen for gravity. *Gravity is Earth's loving*, my baby said. Listen for sensation. Or whatever image enters the back of the mind. Sometimes earth and stones want to be honored before being broken open. I really don't know how to describe this. So maybe sing a little song. Or think about someone I gathered this with. Other times it feels as though a cold, brutal, invisible shield goes up around a rock and so I go no further. Like cooking, we develop intimate taste and respect for the creatures we will consume. Pounding pestle into the mortar activates the realms and releases the mineral. So here I am, with a stone bowl, a stone pounder, this stone and my body, clay? Rock on rock on rock by restless rock. Today I make a red pigment because my friend had a miscarriage and wants to offer earth back as a ritual to free the spirit and heal her womb. As I begin to pound and pound, I hear a line of a poem: *hurl yourself on the earth . . . hear the horse hoof in your heart . . .*[22] Up and down and around my hand goes, truly like a horse hoof galloping in slow motion over the earth. My heart and pelvis and bones and eardrums feel the striking vibration. Like a force of nature, stone opens stone. Sometimes a color needs hours of pounding to emerge, aching days, even. Others come so quick, in a mere minute or two. My lap is strong. Am I weather? What is sex?

& WAYS TO BREAK DOWN STONE

touch softly — gently rub with fingers — rub stone on something hard or harder — rub while wet or with spit grind — grind while wet — grind with balls or ball mills or machines — crush rock with anything — smash with another rock — shatter into dust — throw onto ground over and over — break using mortar and pestle — something becomes something else when it crumbles — scrape with antler — scrape with your fingernails — scrape with a blade — suction your body to a rock over and over — grate with grater — claw — score or engrave — pick — dig — dig with paws — dig with mouth — worm — pass through your digestion — burn — rust — pour over with acid — be lichen or fungi or insects or — wait for hundreds of thousands of years — wait for a volcano to breathe — wait for a meteorite to crash down — blow up a building — let water pass over — or wind — or shattering effects of frost — glacial movement — grow a body and then stop and let it harden and then soften and break down

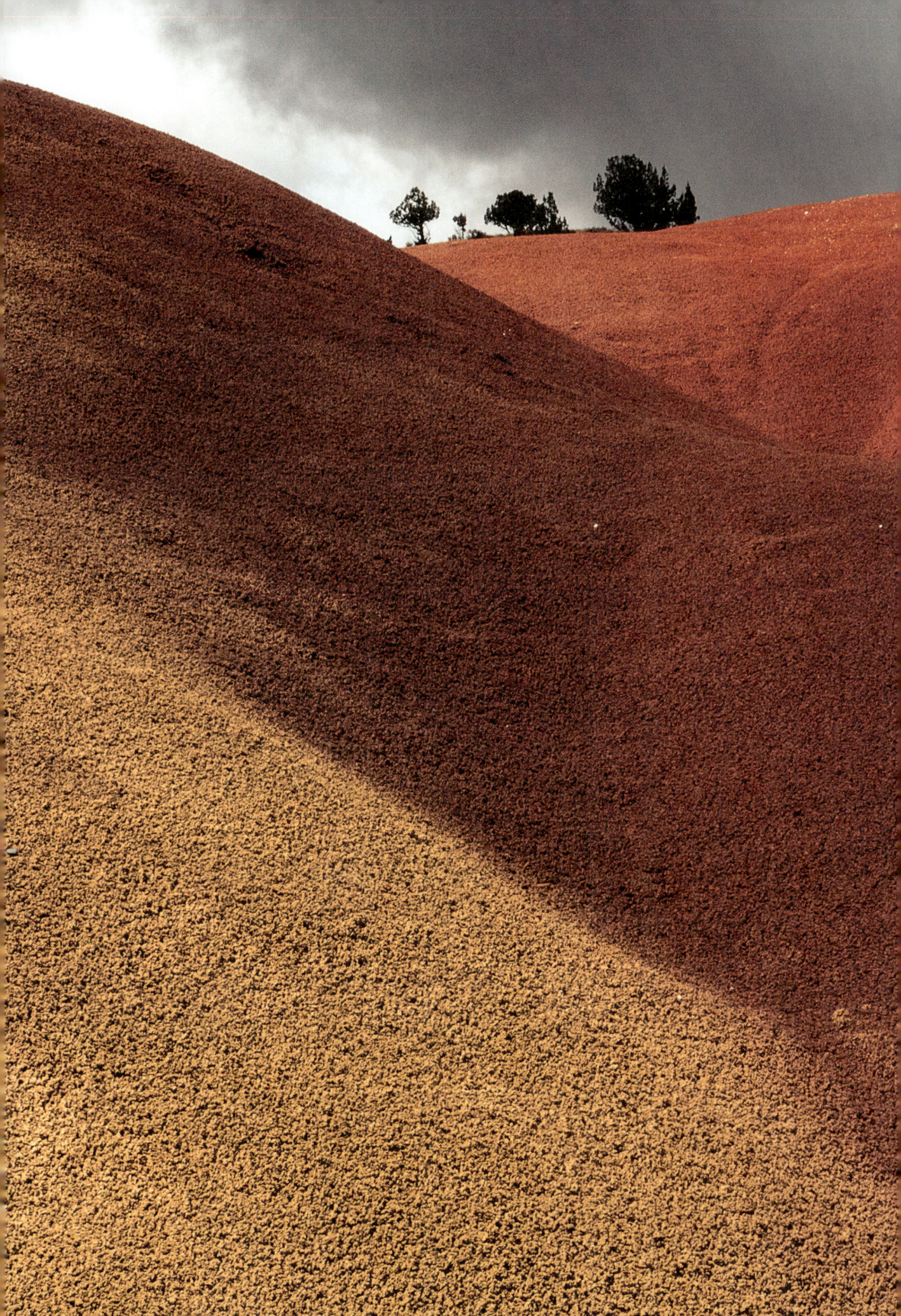

YELLOW OCHRE

GOETHITE, LIMONITE

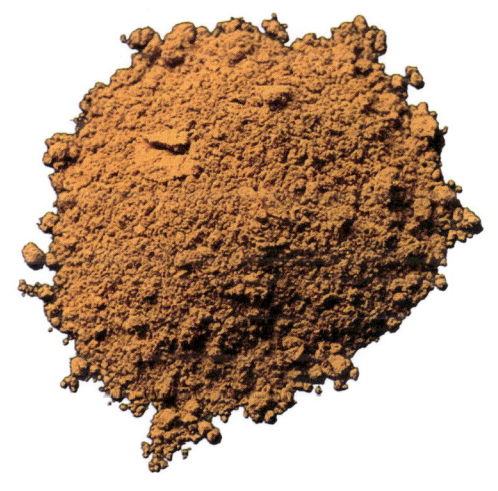

IRON OXYHYDROXIDES ♂

Etym. from *göte*, godfather, and from *leimōn*, meadow, "godfather meadow"

$FeOOH$, $FeO[OH] \cdot nH_2O$, $FeO \cdot nH_2O$ ($\pm MnO_2$)

PIGMENT SPECTRUM

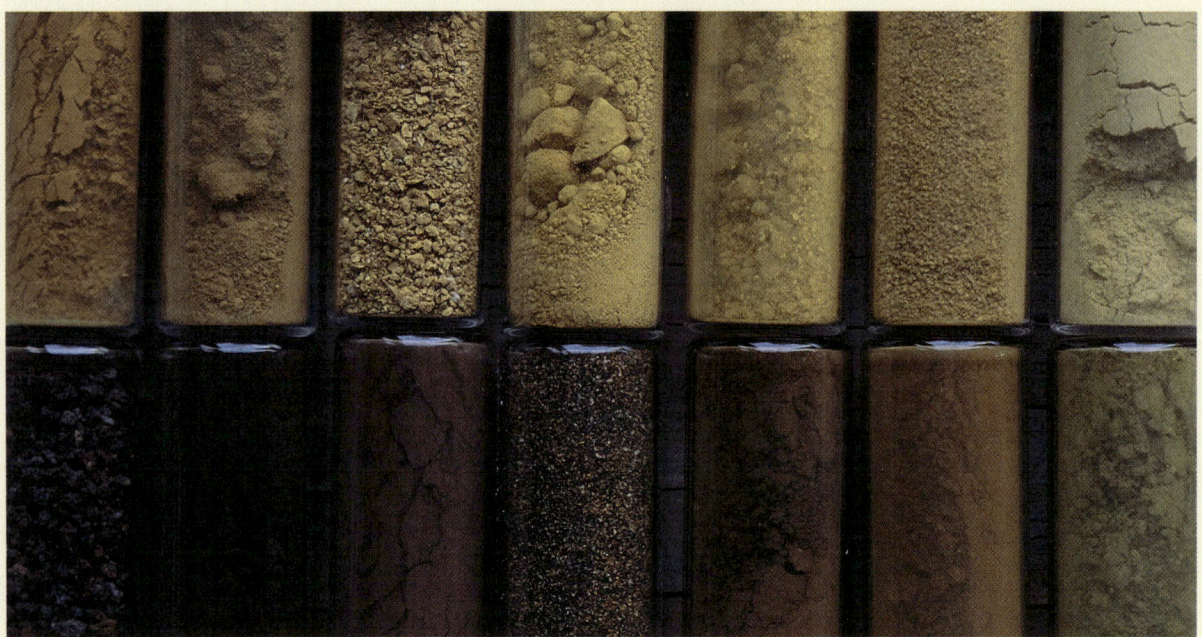

MATERIAL DESCRIPTION

A VERY STABLE HYDRATED IRON OXIDE, goethite often is the "first oxide to form or the end member of many transformations."[1] Rocks with goethite may appear everyday brown, but their streak and powder are usually sunny yellow, and if they contain other elements (carbon, manganese) can remain darker and more shadowlike. Hardly ever pure in crystalline structure, goethite, along with the older field term limonite, are used synonymously to mean imperfect ferric rocks found on the surface of Earth. Strong, persistent, and yet quite mutable, iron oxyhydroxides produce several well-known paint colors: umber, burnt umber, yellow ochre, raw sienna, and others. When roasted under a fire or in an oven, goethite dehydrates, or calcinates, to hematite (burnt sienna), or can turn brilliant red, even dark purple (at sustained, very high temperatures). Like the whole family of iron oxides, if intensely heated in a reducing environment with extra carbon, goethite will turn to pure heavy metal iron—the stuff of steel—and when molten, may be shaped into unbelievable forms.

NAMES

bog iron, pea iron ore, brown ochre, yellow ochre, ochre de rue, brown hematite, yellow hematite, ferric hydroxide, iron hydroxide, hydrated ferric oxide, hydrated iron oxide, limonite, lepidocrocite, xanthosiderite, lymnite, mineral yellow, sil, umber, *terra d'ombra* ("earth of shadows"), sienna, shadow ore

HABITAT

same as hematite (see page 61) — bogs and fields — wetness — underworld

CHROMA

METAL	PRACTICALLY BLACK
SHADOW	BROKEN OPEN HONEY
SHADOW SHADOW	YELLOW
DEEP DARK	YELLOW INCREASE
UNRED	ABUNDANCE
UMBER	NOT YET RUST
YAWN BROWN	GOLD
COLD BROWN	PEE
STEEL BROWN	SUN

CRYSTALLOGRAPHIC SYSTEM

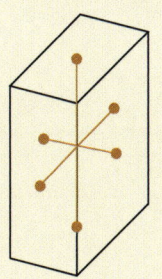

ORTHORHOMBIC

OTHER COMMON YELLOW & BROWN OCHRE MINERALS

Jarosite (iron sulfide),
$KFe^{3+}_3[SO_4]_2[OH]_6$

LOOK-ALIKES

Orpiment

Sulfur-based minerals
(they smell like rotten eggs!)

YELLOW OCHRE — Godfather Meadow

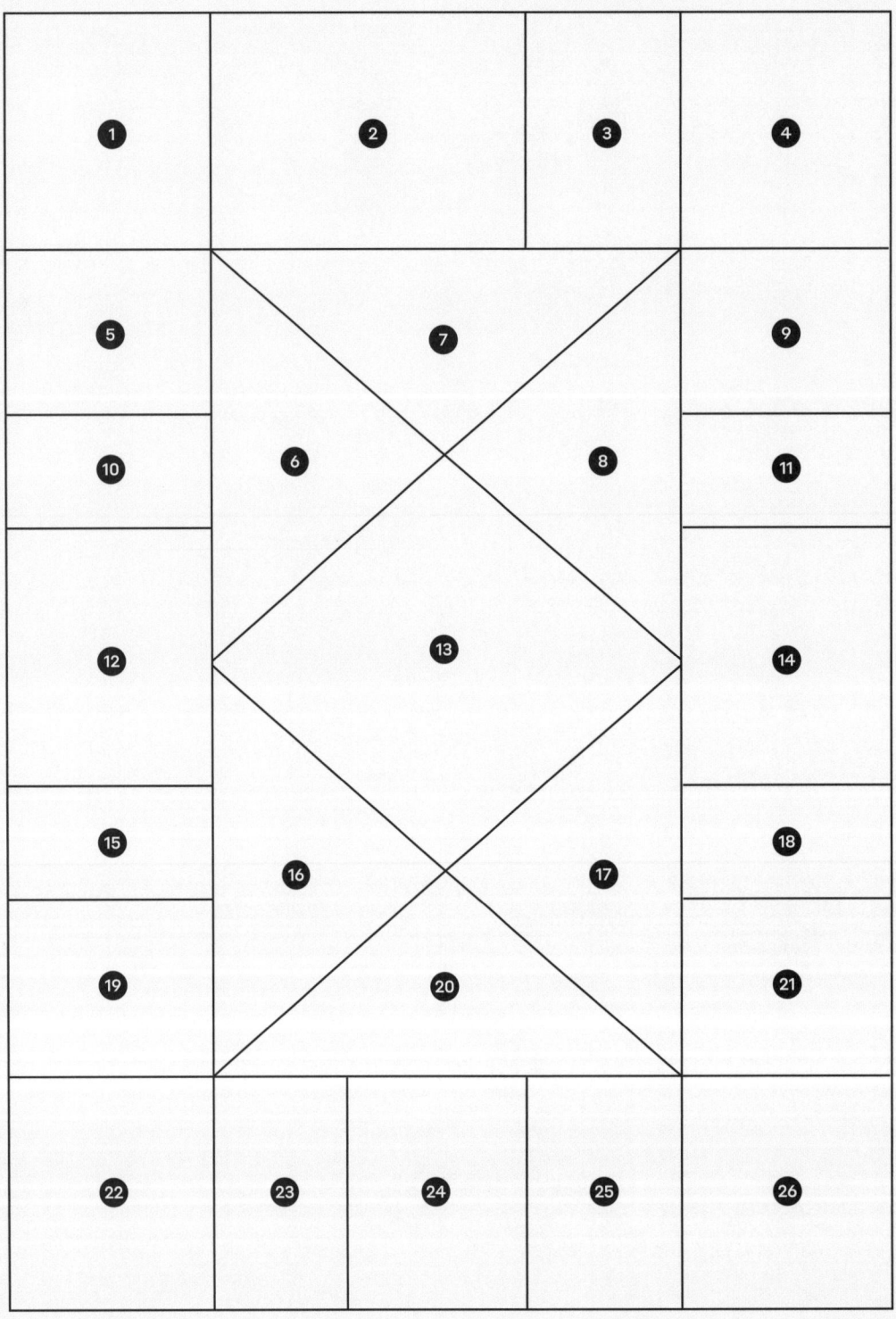

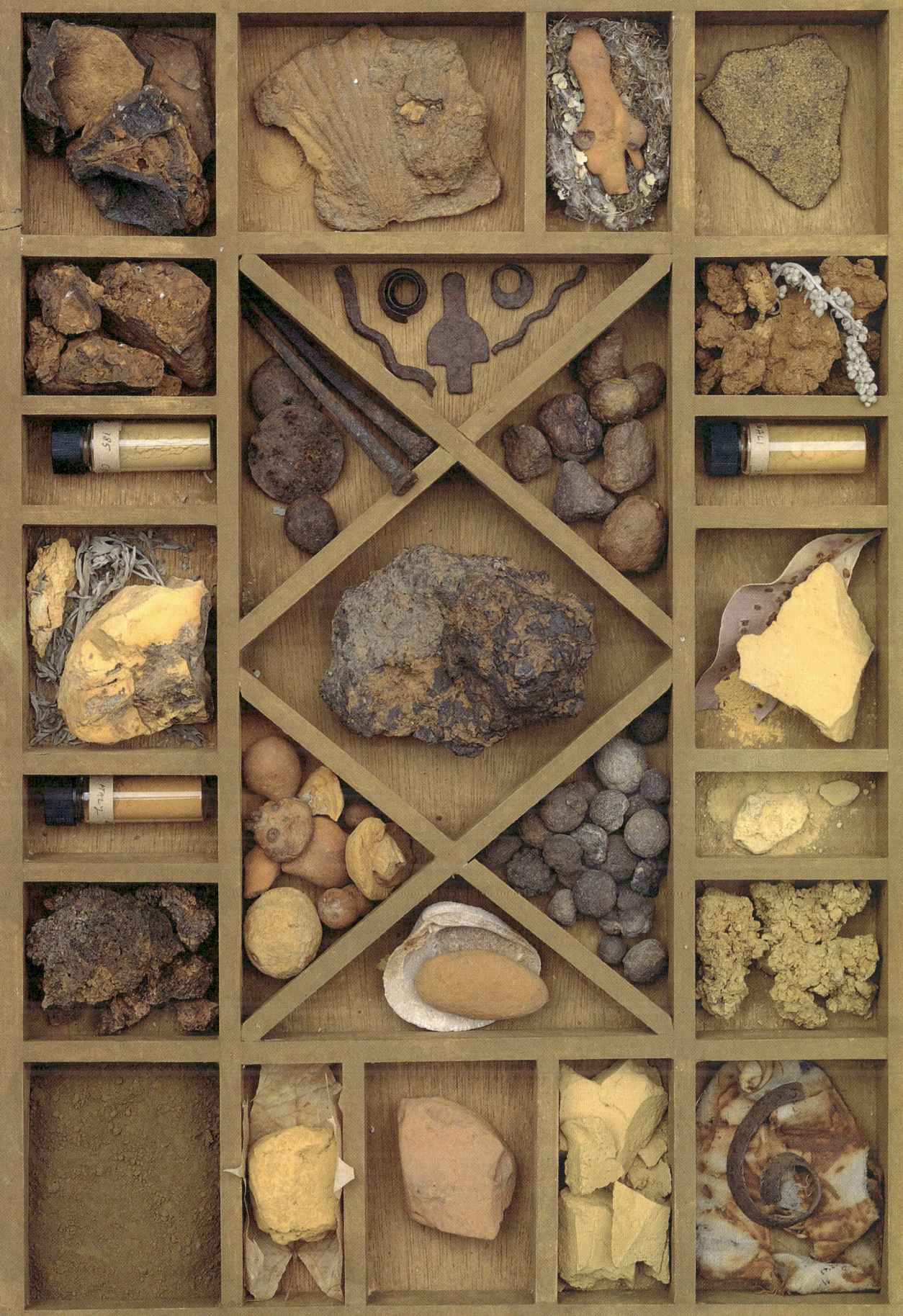

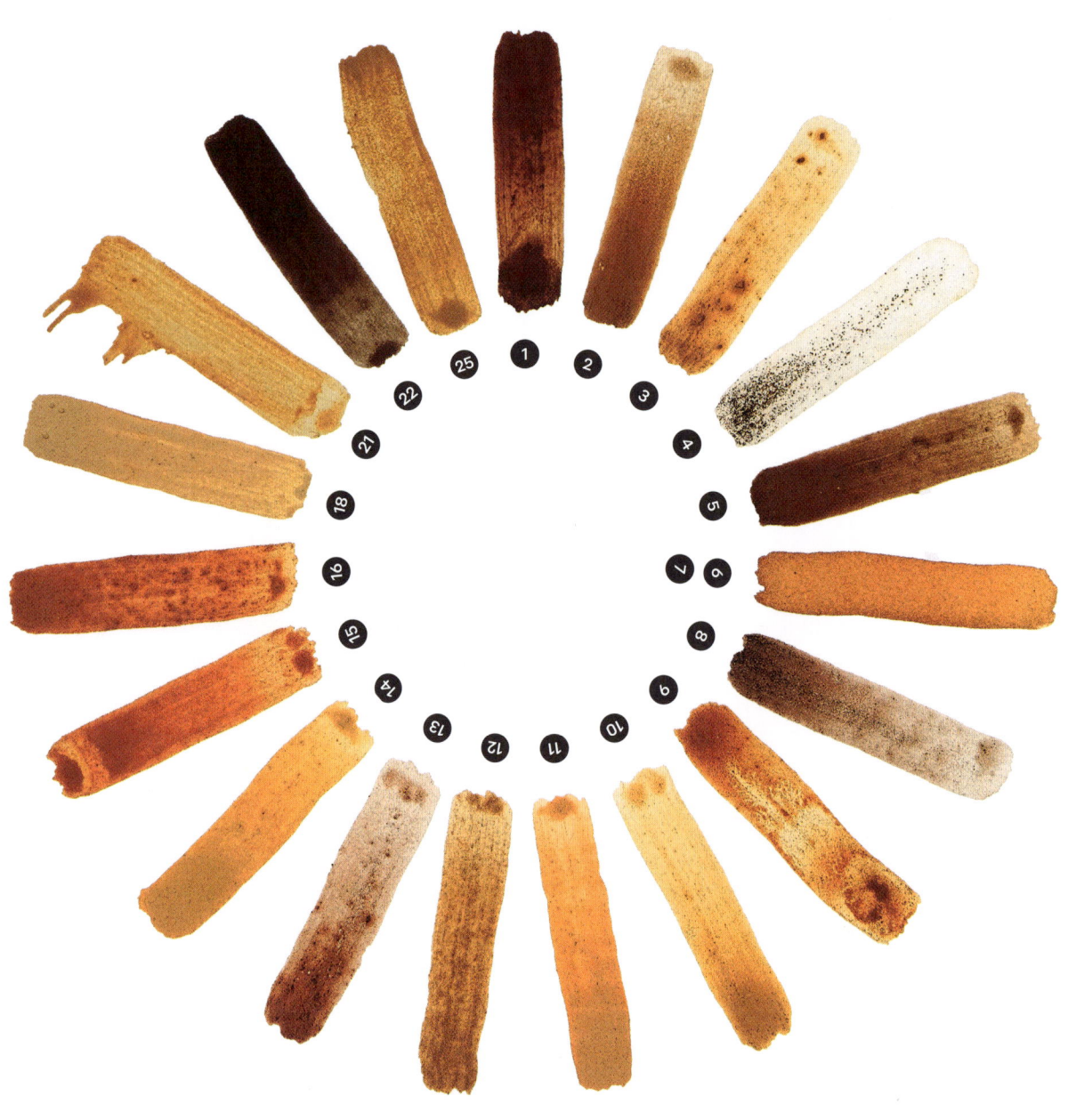

1. *Goethite*. Mined from atop the Earth's head, in the Kursk Magnetic Anomaly, Russia.

2. *Scallop Iron*. Four- to five-million-year-old "pig iron" fossil of a scallop, from ancient sea beds near Virginia shores. Gratitude to Rebecca Coffin Anderson.

3. *Body in a Nest*. Mineralized iron casts of underground plant roots.

4. *Iron Lens*.

5. *Tidal Bog*. Bog iron found washed up by the sea, near Irondale, the earliest steel processing plant in northern Washington, traditional Chimacum, S'Klallam, and Coast Salish land. This stone escaped their fate as steel and has been swimming in the saltwater bay tides for 150 years.

6,7. *Rusted Iron from a Shipyard Railroad*. Bay Area, California, traditional Ohlone land.

8. *Iron Hearth Nodules*. From pre-Revolutionary iron furnaces, long abandoned, near Thicketty Mountain, South Carolina.

9. *Popcorn Clay*. That's a geologic technical term. But yes, it is probably edible. And melts (or paints) like butter. Central Oregon.

10. *Desert Ochre*. Colombia.

11. *Yellow Ochre*. Pigment gift from a friend, bought from a local market in Mali.

12. *Limonite*.

13. *Bog Ore*. Mineralized biotic goethite, from someone's front yard in rural cow milk county.

14. *Yellow Ochre*. From a beloved place in Australia, gift from Lorraine Brigdale (Yorta Yorta artist), along with eucalyptus leaf.

15. *Dark Sienna*. Goethite transforming to hematite, from traditional pigment soils near Mount Amiata, Italy.

16. *Beads Made by Time*. Iron concretions found buried in ancient muck in coastal Oregon near the clam flats, with Scott Sutton.

*17. *Spheres of Iron*. Gift from Hephaistos.

18. *England*. Ochre gathered by wild walker Caro Ross.

*19. *Volcanic Iron*. From a strip mall roadside, guarded by mountain lions, vultures, yappy dogs, surveillance cameras, doves, children, and ancestors. Ohlone land.

*20. *Paint Stone*. Found in the riverbed of Chaguayanga, a Tataviam ancestral village site, gathered and gifted by Mona Lewis and Alan Salazar (Tataviam/Chumash elder and storyteller).

21. *Morraine*. An iron hydroxide-rich soil from Germany.

22. *Umber*. Traditional artist pigment from ocre mines in Luberon, France.

*23. *Yellow Ochre from the North*.

*24. *Shale*. Not technically a real yellow ochre, but for some reason I still think it belongs here. From South Africa by way of Tammy Hodgskiss.

25. *Raygor Yellow*. Bandlands clay for use as a slip. This ancient clay originally formed probably around 55.8 million years ago. Gift of rock art scientists Evelyn and Robert, from a private homestead near the Hoodoo Paint Mines, Colorado, traditional land of Plains Apache, Comanche, Ute, Arapaho, and Cheyenne peoples.

*26. *Rust Offerings*. Strip of ribbon dyed with rust from old iron, from Cara Maria Piazza.

Swatch not pictured, see raw material/color on page 85

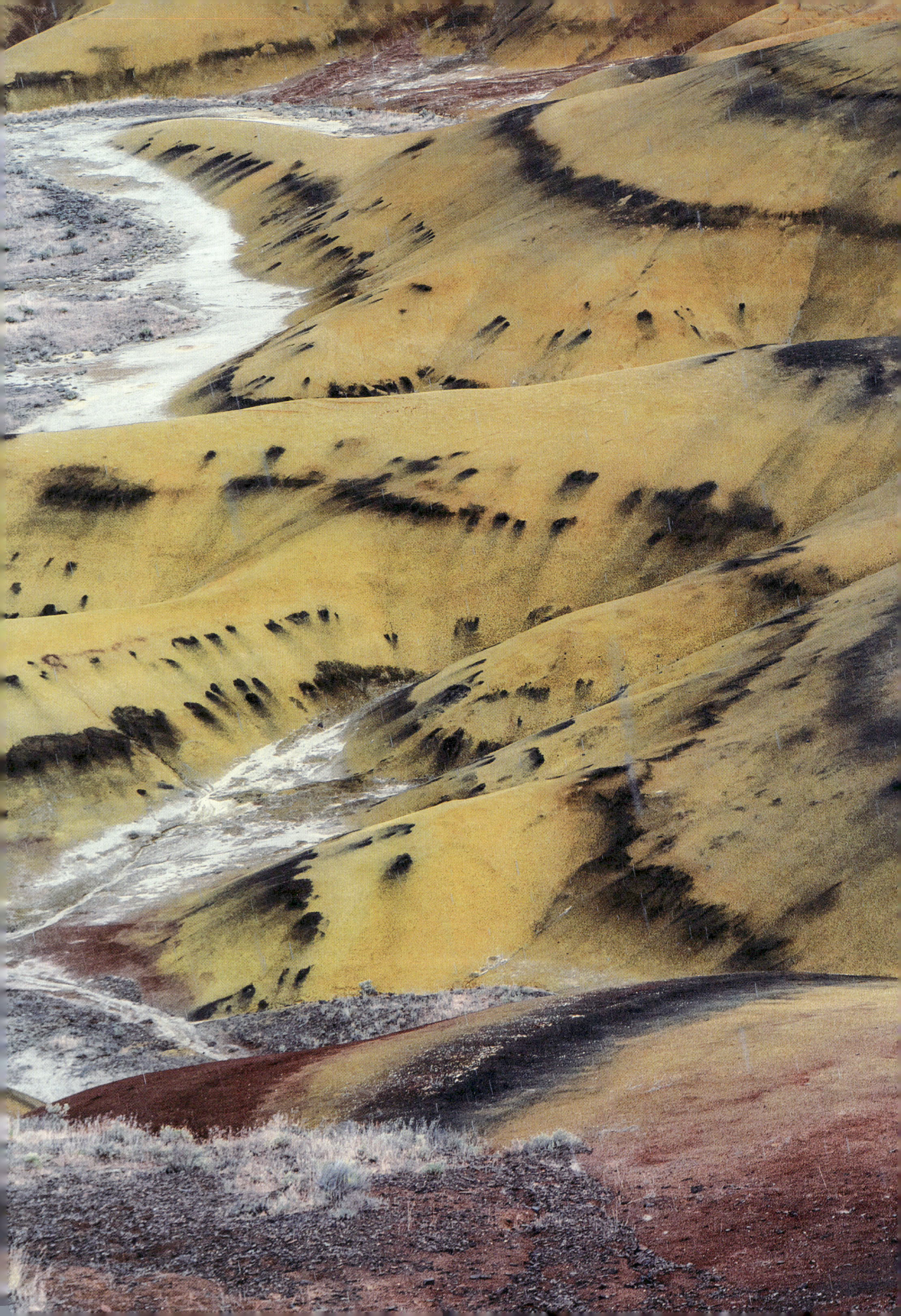

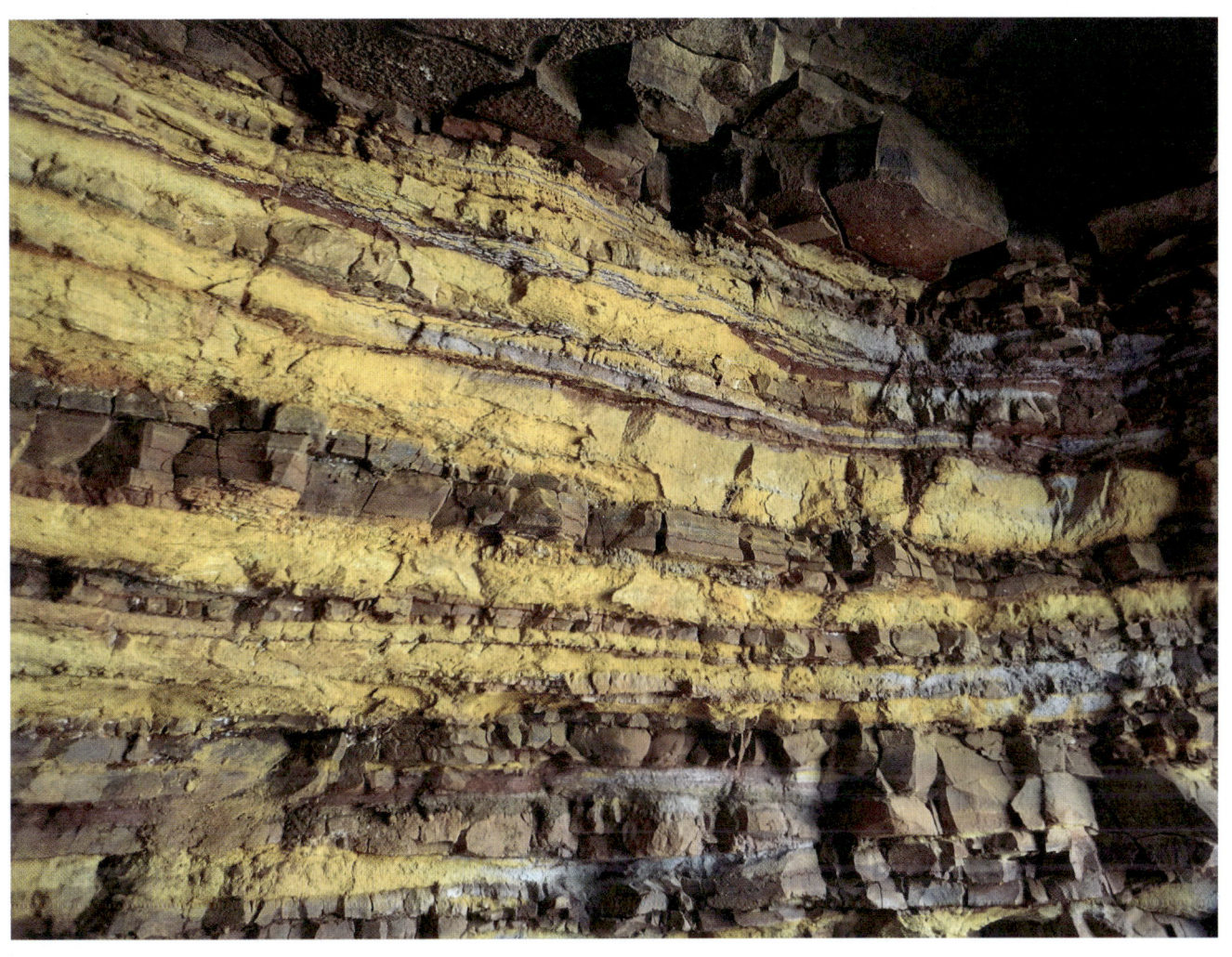

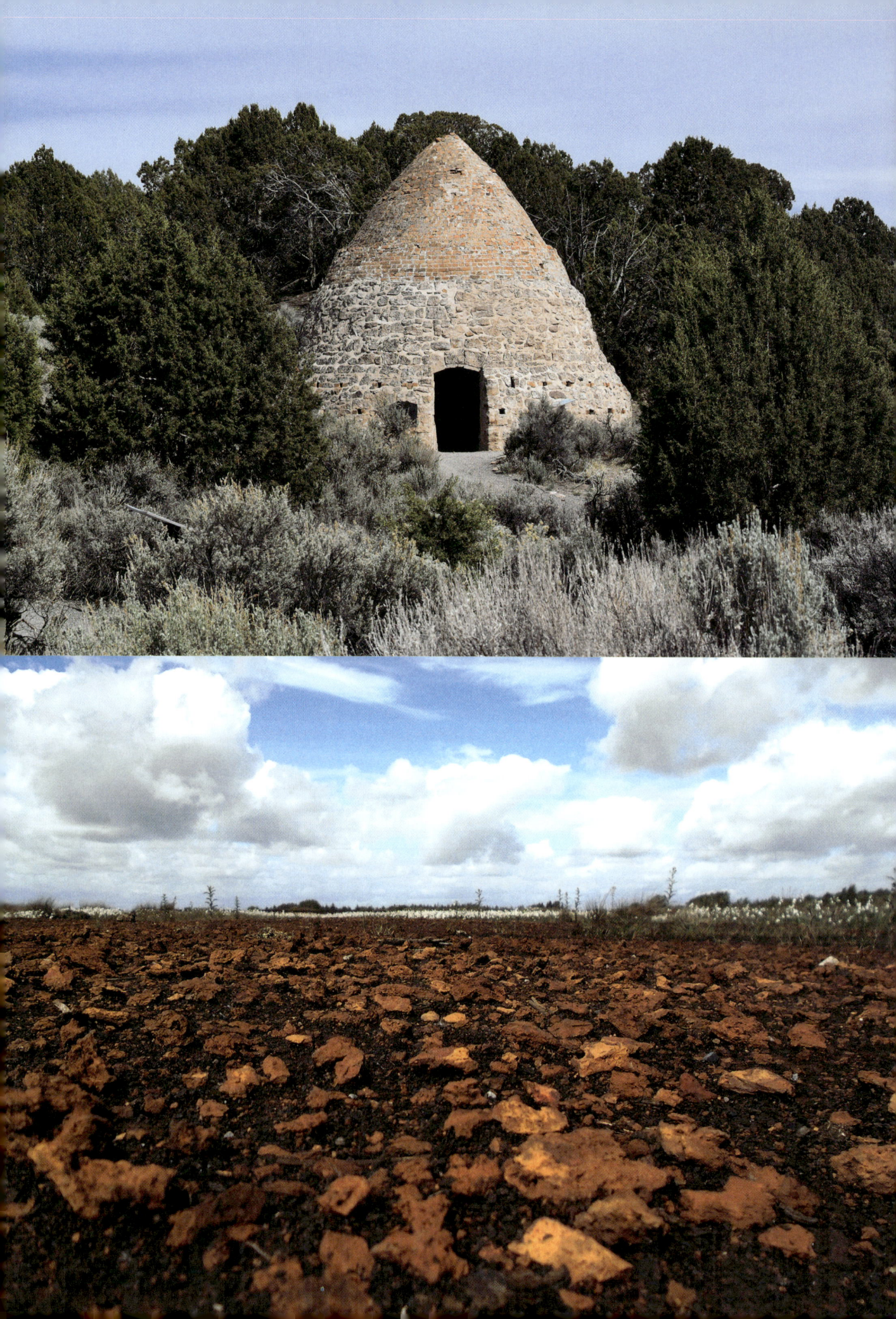

Shadow Ore

Goethite colors both light and dark. From tender yellow ochre to shady brown, iron hydroxide's spectrum lends warmth and shadow. In painting, we use the term *umber*, from the Latin *umbra* ("shadow")—a "cheap, somber" necessity in even the most limited color palettes, according to Kassia St. Clair. Umber, a smoldering shade, is made from goethite mixed with a bit of manganese oxide. Goethite, a "profoundly unglamorous" mineral, says St. Clair, gets the name from naturalist Johann Wolfgang von Goethe, who keenly observed: "Color itself is a degree of darkness," and thus "allied to shadow."

And that pretty much sums up art histories on one of the most archaic, pervasive ochre materials in the world. A *blah blah* mineral that gives us common golden, glazed-over earthy and shadowy tones for ever and ever. I read all that and shrug. Goethite feigns to be the *last* color we're supposed to care about.

Shadow, in a psychological sense, means stuff kept hidden from consciousness—most easily felt when in agony (or dreams or bored drowsiness) where we're caught between the known world and an unaware, unconscious gravity. Depth psychologists—those in traditions of psychological science that honor the soul, following from Carl Gustav Jung, James Hillman, and others—consider the shadow a hefty paradoxical energy inside of us—competing convictions, repression, the dark side of the moon, a big little transgression, buried ancestral memory, a lost truth, an old lie, the most basic habitual physiological process we do without thought, like blinking. Plus, shame and all that.

The whole concept of shadow conjures contrast: a powerful impulse held back by an upright ego. Personally, I imagine a discerning boss, who takes on forged qualities of being unbreakable and untouchable: an iron-fisted, nerves of steel, heart of iron, steely eyed, stone-faced man, maybe.

To track these elusive buried energies, we try to begin with what *is* accessible to the psyche. Materially speaking, the most readily available form of goethite ore, known as bog iron, thrives throughout cooler, wetter, northern latitudes. You can dig it out of the ground with your hands, a stick, a simple rake, or a shovel. Remember that beloved biogenic "mother of ochres" who hardens into rocky worlds? Bog iron dominates as her eldest child, usually the first to form solid mineral chunks out of sludge (see bottom image on page 90). Bog iron's body even retains, up close, fossilized evidence of microbe sheath and tube architecture.

Bog iron, if ground into the finest dust, offers a lovely golden sunshine hue of abundance. If left rough, it remains a chestnut brown, and if heated, bog iron quickly turns dark, sometimes blood warm red (burnt sienna), steely purple, or brownish black (burnt umber), and eventually just melts into pure metal iron, a source of cast iron and steel.

There are several paths I want to follow from here. On the one hand, does anyone really get super excited or care about steel these days? Like the mineral goethite, steel appears pretty humdrum, the boring chatter of metal tycoons. I don't want to put us to sleep here, but I do want to claw more deeply into what steel does and what it's made possible for human life. Why? Because we're still very much of the Iron Age, "on whose far edge we still teeter."[2]

In general, steel is made from iron ore (iron oxide and hydroxides), from the *same exact* geologic places and material as ochres. Steel's many capabilities liberated humans into modern life. A modern life where "none of its great accomplishments—its surfeit of energy, its abundance of food, its high quality of life, its unprecedented longevity and mobility and, indeed, its electronic infatuations—would be possible without massive smelting of iron and production (and increasingly also recycling) of steel," notes renowned economist Vaclav Smil.

In a sense, we are bound by "a world using iron for every part."[3]

There's, of course, a known extractive cost. To make steel, humans mine iron ore. Over three billion tons of it annually or *94 percent of all metals* mined on Earth. Iron ores are then either: crushed finely into various ochre pigments and mixed with water to become colors and vessels of our imaginations, or crushed roughly, heated to very, *very* hot temperatures with old-growth forest hardwood fire or fossil fuel—charcoal, peat, coals, coke—and a little carbon-rich limestone or oyster shells (flux), and become metallic infrastructures for our industrial lives.

When ochre touches fire in a crucible—no water, no air—pure heavy metal comes forth. That inflamed smelting process removes the oxygen (and hydrogen and water) from iron hydroxide. Isolated iron coagulates into what's known as a bloom. That bloom iron is considered reduced and "freed" from oxygen (echoing the acid mine waterway process, where iron becomes unbound and separated from mother geology, as I mention in "Vulture Mud," page 45). In the meantime, formerly rock-bound oxygen affixes to formerly tree-bound carbon, and rises up as atmospheric CO_2—freshly spent fossil fuel.

Before steel arises, we first *suffocate* ochre.

And we exhaust the spirits of wood and water and earth.

Ochre must be deprived of oxygen to release steel iron and carbon dioxide, and that is how Iron Age civilizations have spread across the globe and then gradually expanded into Industrial Revolutions, where, to put it succinctly, "people started burning coal and other fossil fuels to power factories, smelters and steam engines, which added more greenhouse gases to the atmosphere. Ever since, human activities have been heating the planet,"[4] science journalist Julia Rosen writes. Today, steel production is directly responsible for a big percentage of the atmospheric CO_2.

But what of the *indirect* responsibility?

For creating a world using iron for every part?

I find that a bit more difficult to imagine.

When ochre reduces—transitions into hard metal iron—cool steel erects endless forms and shapes, which unlocks key functions of colonial expansion, setting the stage for globalized capitalism and future offsprings, accelerated climate change, where we liberate so-called new worlds—new worlds, conjured from lowly pieces of bog iron, *burnt umber. Goethe*, after all, means godfather (*göte*). So to echo the paternal, economic stance: There are undeniable innovations from anthropic (human-born) iron.

To me, the taxonomy of steel's epoch of things could be perhaps categorized as *overwhelming* and *indispensable*: sword, knife, helmet, armor, nail, screw, staple, chain, cage, lock, key, fence, anchor, cannon, gun, arrow point, sewing needle, sewing machine, zipper, wood stove, cauldron, an individual knife and fork, plough, spade, hammer, hoe, horse-shoe, railroad, train car, engine, car, plane, battleship, ocean liner, cable, drill bit, metal cutter, loader, digger, earth mover, rig, pipeline, storage tank, shipping container, beams, columns, rebar, wire, reinforced parts of reinforced concrete, elevators, massive bridges, transformers, transmission towers, electrical wires, key components for automation, paperclips, watch springs, printing presses and metal type, nuts and bolts, infernal iron shades.

"Like the candle, wrought iron and steel swords have long been surpassed by more technologically advanced means . . . yet they have not been superseded as fundamental materials of the imagination," architect Daniel Willis recognizes in his essay "The Valor of Iron."[5] Military, agriculture, economy, industry, city, society, domesticity—steel's fundamental temperament tends to globalize, expand, ascend. People want power. In fact, one of the earliest, largest megacorporations in the United States, fueled by the forced labor of enslaved and imprisoned people, is the United States Steel Company, whose Nasdaq identifier is just "X," and iron-possessed founders include some of the wealthiest families in America.[6]

Maybe now's a good moment to circle back to the origins of steel production in early colonial America. Where did it come from? Throughout Europe and early northern Iron Age settlements, village sites were synonymous with bog iron marshes near big old woodlands. Ferric swamplands. Settlements preferred their own boggy iron deposits, forest-derived wood fuels, and stone ovens, making it easy to extract, mine, smelt, and smith iron. (Example of beehive kiln used in colonial ironworks at Iron City, Utah, page 90.) Bog iron bloomeries supplied all of Europe's first iron, the essential hinge and gate and bell of churches, palaces, and monasteries. Ocean-faring colonial ships were outfitted with steel anchors, nails, shovels, awkward, heavy body armor, chains, and guns. No matter which version of the Bible colonists read en route to far shores, Deuteronomy's vision of a thriving promised land stayed the same: "a land whose stones are iron."

If we could take a close look around early colonial settlements, at the birth of America, you'll find an eerie repetition: bog iron creeks near big old woodlands. Why? England was starving for iron. The "motherland" had already deforested its entire country, used up for hardwood charcoal to smelt iron. People ran out of wood for fuel, forced to convert entirely to coal by the 1700s. Thus, huge investments in ironworks with access to forests were made by early colonists in Virginia. Jamestown set up the very first facilities near bog iron deposits in Falling Creek. In a letter to investors, a 1621 colonist wrote, "The company is assured there can bee no fitter places of Mines, Wood and Water for Iron than there."[7]

The first use of *landscape*, the Dutch loanword, enters the English language around this time. A few months later, in 1622, Indigenous Powhatan inhabitants of the land, strong spiritual people for millennia with a population over fifty thousand, swiftly destroyed the budding iron facilities, slashing bellows and animals and killing all but two ironworkers and their families. There's clear intention and purpose. Historically it is seen as retaliation for settler land occupation, raided food supplies, broken agreements, and the murder of their previous tribal chief and his family, including his infamous daughter Amonute (Pocahontas), and several others within their community. As far as I imagine, the Powhatan also hit the colonists where it could hurt their agenda most: *iron*. And that effort was totally successful for over a hundred years. Colonists stayed up north, where instead Boston developed an influential bog iron ironworks on the Saugus River.

Back in woodless, plague-filled, iron-hungry England, Isaac Newton declared color a property of light—a *property*! that could surely be owned—and the old ways, where iron clings to ochre and color is substance, dissolves into wavelengths, optics, ownership. Color flees from matter. Ochre doesn't return to the mother bog. Land becomes chained to land*scape* and picture and map. A separation from the living ecological place occurs. Reverence with nature becomes a painted image, an escape: a romanticized media for show-and-tell-and-sell, and that, in practice, helps us leave the actual Earth behind. This shift can perhaps be summed up in two contrasting opening lines of poetry, one from a dead American poet, and one from an Ojibwe woman (enrolled at Turtle Mountain):

The land was ours before we were the land's.

—Robert Frost (American), *The Gift Outright*

We were the land's before we were.

—Heid E. Erdrich (Turtle Mountain, Ojibwe),
The Theft Outright, After Frost[8]

I was told that in the Pacific Northwest (where I was born), Haida bands called settlers "Iron People" (Ya'aats' Xaatgaay), and the bitter conflicts during early European contact are called "the time of the Iron People." Iron, in this case, speaks to both a specific material output (guns, armor, nails, chains) and a culture as intractable as iron: unyielding, formidable, and often rather devoid of humanity.

I feel shame and tension. I'm from northern Iron people. Iron chains clink and drag in my soul, even during moments of total creative joy, and especially if painting with goethite's "cheap, somber" umber, raw sienna, or yellow ochre. Perhaps this is why my inner ochre guide once said to me, "my ochre body, my land, *is* your shadow." Another time, while holding an ochre spirit stone from urban colonized Ohlone land, I saw it transform into a two-faced wrathful being who spoke shyly, saying, "I could kill you from inside out."[9]

And rocks speak truth. Iron, as a universal principle, gathers until everything implodes. Thus, Iron People's neurosis, by nature, is a self-annihilating fantasy. "Metal's hostility is thus its primary imaginary aspect," says Gaston Bachelard, elemental magician. I think of Óðinn, the mythic Norse god, when hung upside down and wounded by his own spear of enchanted iron, who said: "I offer myself to myself (sjálfr sjálfum mér)."

And I think innovative writer Maggie Nelson also got it right:

Goethe wrote *Theory of Colours* in a period of his life described by one critic as "a long interval, marked by nothing of distinguished note." Goethe himself describes the period as one in which "a quiet, collected state of mind was out of the question." Goethe is not alone in turning to color at a particularly

fraught moment. Think of filmmaker Derek Jarman, who wrote his book *Chroma* as he was going blind and dying of AIDS, a death he also forecast on film as disappearing into a "blue screen." Or of Wittgenstein, who wrote his *Remarks on Colour* during the last eighteen months of his life, while dying of stomach cancer. He knew he was dying; he could have chosen to work on any philosophical problem under the sun. He chose to write about color. About color and pain.

Color and pain and dying. Goethe's fraught namesake iron initiated an enhanced quality of life for billions of people and gave us petroculture, worlds that could rely on oil—and yet iron always carries its own destruction at heart. In our own cells, iron storage is fanatically regulated; iron ions are never allowed to be "free." Free iron ions have the highest potential for internal biological toxicity (that's why antioxidants aim to keep iron absorption in balance).

I encounter many ochre places—iron quarries and veins and old bogs alike—that feel long overharvested, way out of balance. Their bodies extracted centuries ago, dismembered into steely bits and spread across the planet. To me, I experience these places as profound evidence. Evidence of Earth undergoing iron toxicity—the problem of iron set free. So much of ochre's rainbow-like spectral wonders now hide, embalmed behind metal fences, buried under buildings, covered in the invasive garb of superstitious humans who know better than to ever utter the word *ochre*—ochre, earth, thus becomes our shadow of shadows. In *The Nature of Substance*, alchemists of the twentieth century warn of exactly this: "Iron becomes a mummifying agent when it overshoots the mark" and "overwhelms the vital processes."[10] Umber, burnt out.

Iron overwhelms, overshoots, multiplies, and "seeds the imagination," as Nor Hall says. Francis Bacon, the "godfather" of empirical science, also knew a powerful seed within iron, and observed "a Kind of Iron, that being cut into Little Peeces, and put into the Ground, if it be well Watred, will increase into Greater Peeces."[11] This lifelike germinal impulse of iron echoes in Ovid's Metamorphosis, where "rocks naturally grow from their roots."[12]

For me, this hints at where metaphor and spirit and science merge together into Earth's inherent wisdom. In Western science today, bog goethite is still known, technically and literally, as "seed ore." Like plants, iron bog ore will grow from a small seed of itself. Over the years, they get larger and larger and harder and regenerate and have children and rocks prove themselves to be lifelike, if not, simply, alive. But they do not grow or regenerate if they are over-harvested. A swampland will replenish ochre stones that people take to transform into steel, but only if harvested about once each generation.[13] So, maybe every twenty to thirty years? If you follow the bog's harvest protocol (similar to those of plant medicine): you are allowed to take enough iron to make steel for a nail or two, maybe a sword or little shield, by the time you come of age. That's practically the same amount of iron in a human body when cremated—which contains enough iron to smelt into one small nail.

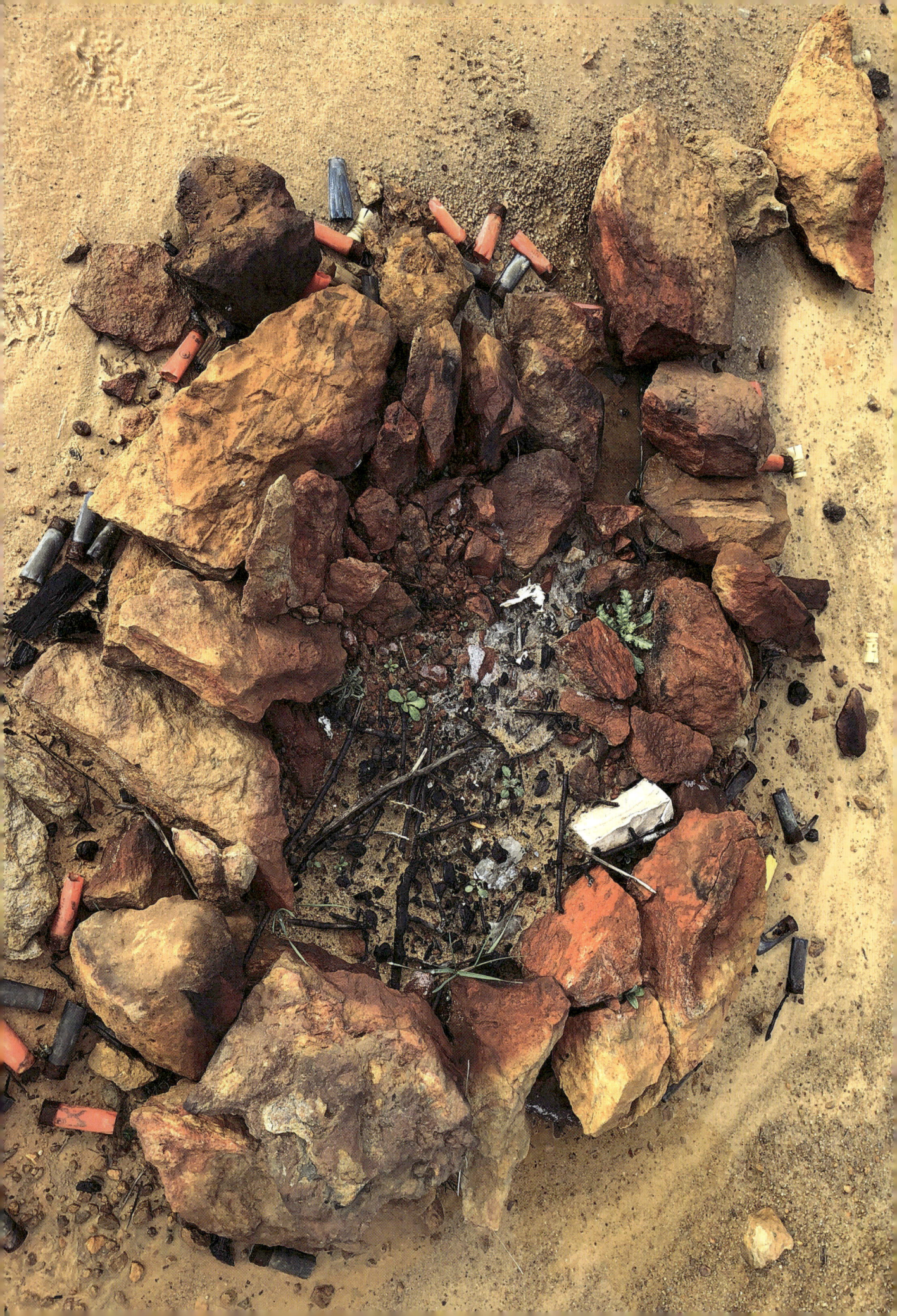

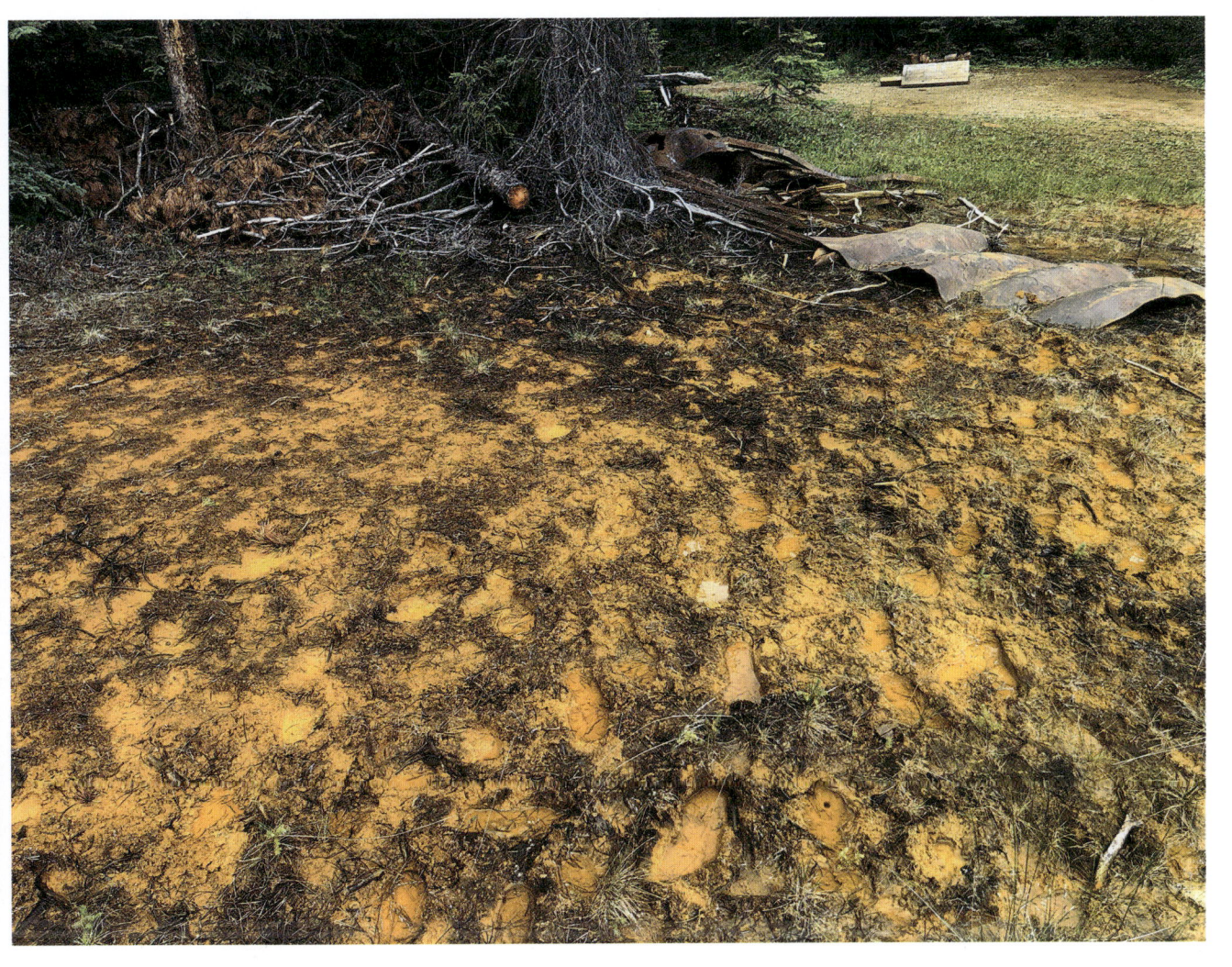

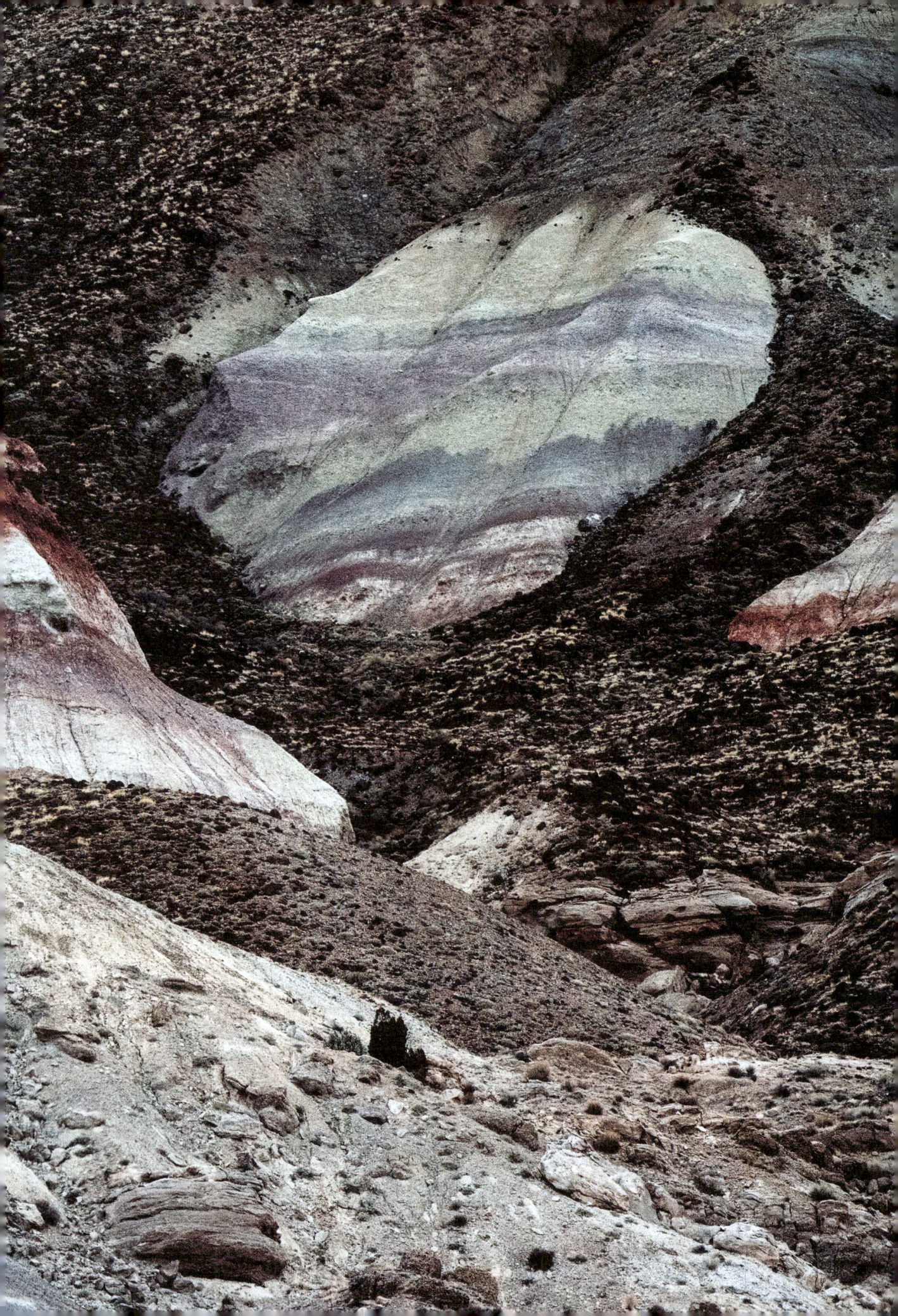

GREEN EARTH

CELADONITE, GLAUCONITE, GREEN CLAY

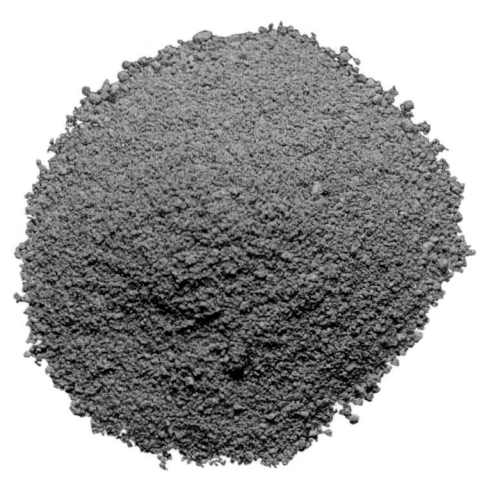

**IRON SILICATES, FERROMAGNESIUM SILICATES,
SHEET SILICATES (AKA PHYLLOSILICATES)**

Etym. from *céladon*, booming sound or sea-green, from *glakous*,
glimmering, "glimmering booming sound"

$$K(Mg,Fe_2)(Fe_3,Al)(Si,Al)_4O_{10}(OH)$$

$$(K,Na)(Fe,Al,Mg)_2(Si,Al)_4O_{10}(OH)_2$$

$$(Ca,Na,H)(Al,Mg,Fe)_2(Si,Al)_4O_{10}(OH)_2 \cdot xH_2O$$

PIGMENT SPECTRUM

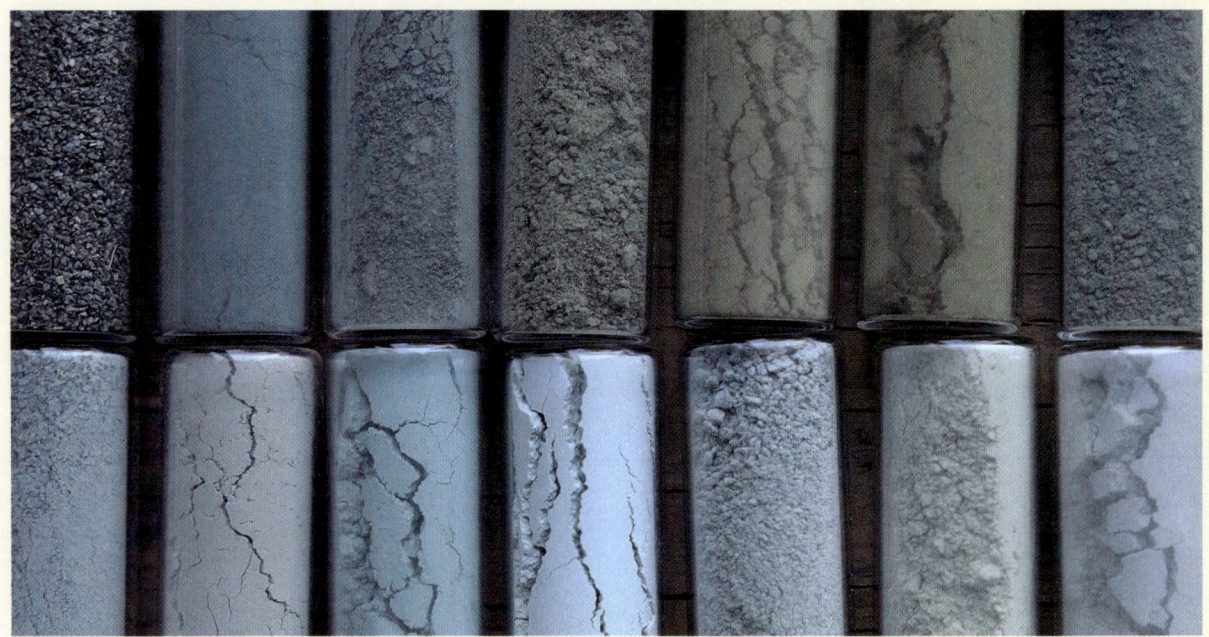

MATERIAL DESCRIPTION

G REEN EARTHS COME FROM distinct minerals and may be difficult to identify without microscopic or geologic knowledge, so it is common practice to consider them a unified family. The principal source minerals, glauconite and celadonite, both have similar chemical compositions containing iron with doses of magnesium, potassium, and aluminum silicate. Their terms distinguish their different geologic origins: glauconite (or greensand) for marine sedimentary environments and celadonite for altered basaltic volcanic places. Green smectite clays, or bentonite clays, also called "French green clay," are formed by ash raining down from volcanoes and sinking into lakes, lagoons, and shallow seas—a middle ground of celadonite and glauconite. Green earths are often found in exposures along long-lost shores, on organic farms utilizing glauconite's slow release of nutrients, or revealed on hillsides whose visibility is seasonal depending on obscuring vegetation and undergrowth. Earth needs only a very small percentage (as little as 2 percent) of celadonite to color an entire mountain in green. You could also encounter greenish clay in bulk as cat litter, as hunter's camo face paint, or healing face mask powder found in bulk in co-op grocery stores.

NAMES

terre verte, terra verde, verdeterra, greensands, French green clay, celadon, prason or prasina "leek green," seladonite, sankir, sea green, verdetto, theodote, creta cirina, creta viridis, stone green, celadonite, glauconite, green clay, green bentonite, holy green

HABITAT

extinct volcanos — weathered basalt — marls — volcanic ash in seawater — accidental trickles — near the deadliest — neighboring — radioactive burials — graves

CHROMA

GREEN	ZOMBIE
FOREST	PALE MOONLIGHT
FIELD	PALE MOLD
DRAB	SICKLY
PLANTED YELLOW	NO SEA
CHLOROSIS PALE	BODILESS
CAMOUFLAGE	LIFELIKE
DEAD LAYER	GRAVEGRASP

CRYSTALLOGRAPHIC SYSTEM

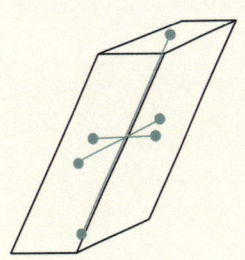

MONOCLINIC

SECONDARY GREEN EARTH MINERALS

Chlorite group
$(Mg,Fe)_3(Si,Al)_4O_{10}(OH)_2 \cdot (Mg,Fe)_3(OH)_6$

Serpentine, Olivine
$(Mg^{2+}, Fe^{2+})_2SiO_4$

LOOK-ALIKES

Malachite

Verdigris

Copper-based green minerals

GREEN EARTH — Glimmering Booming Sound

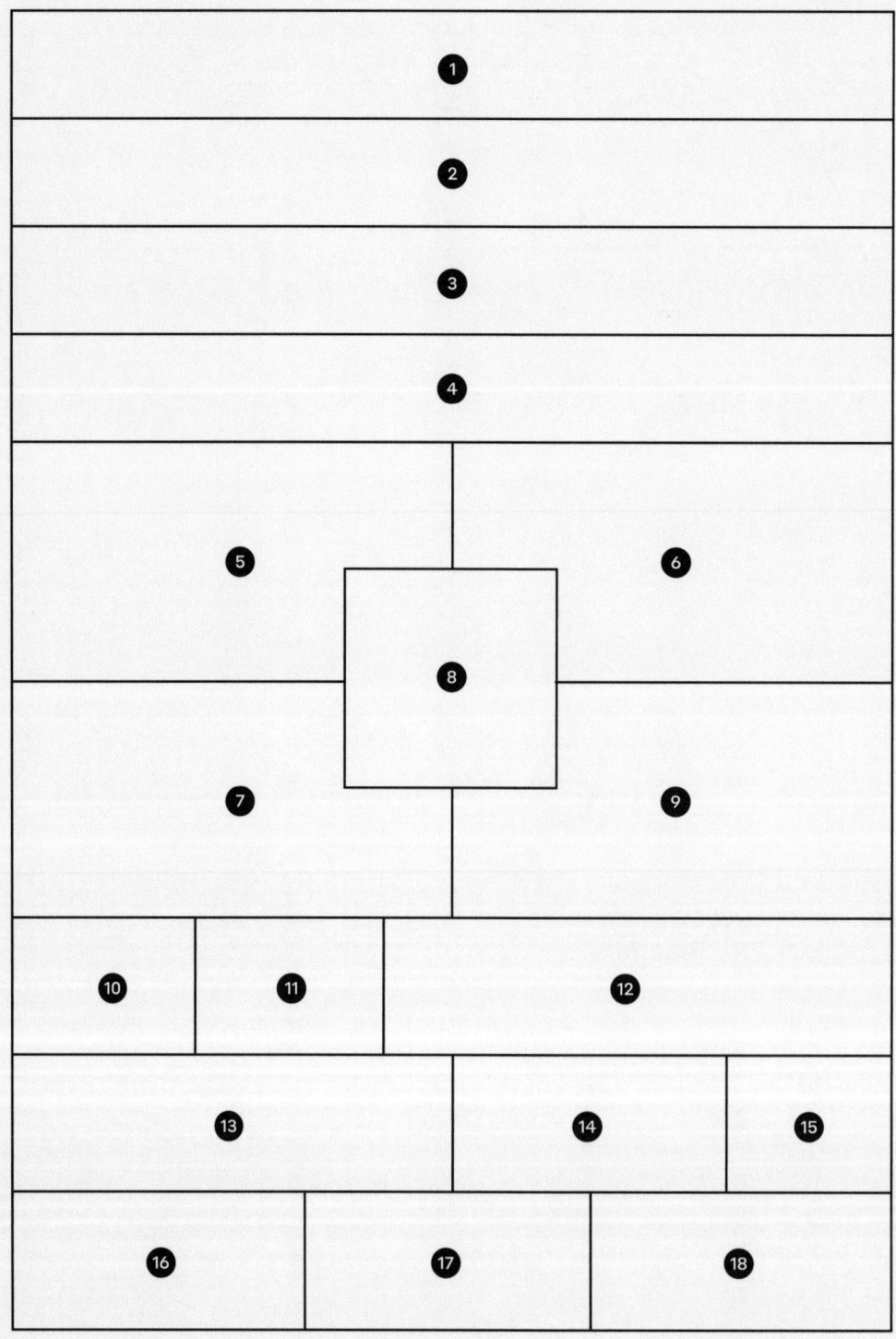

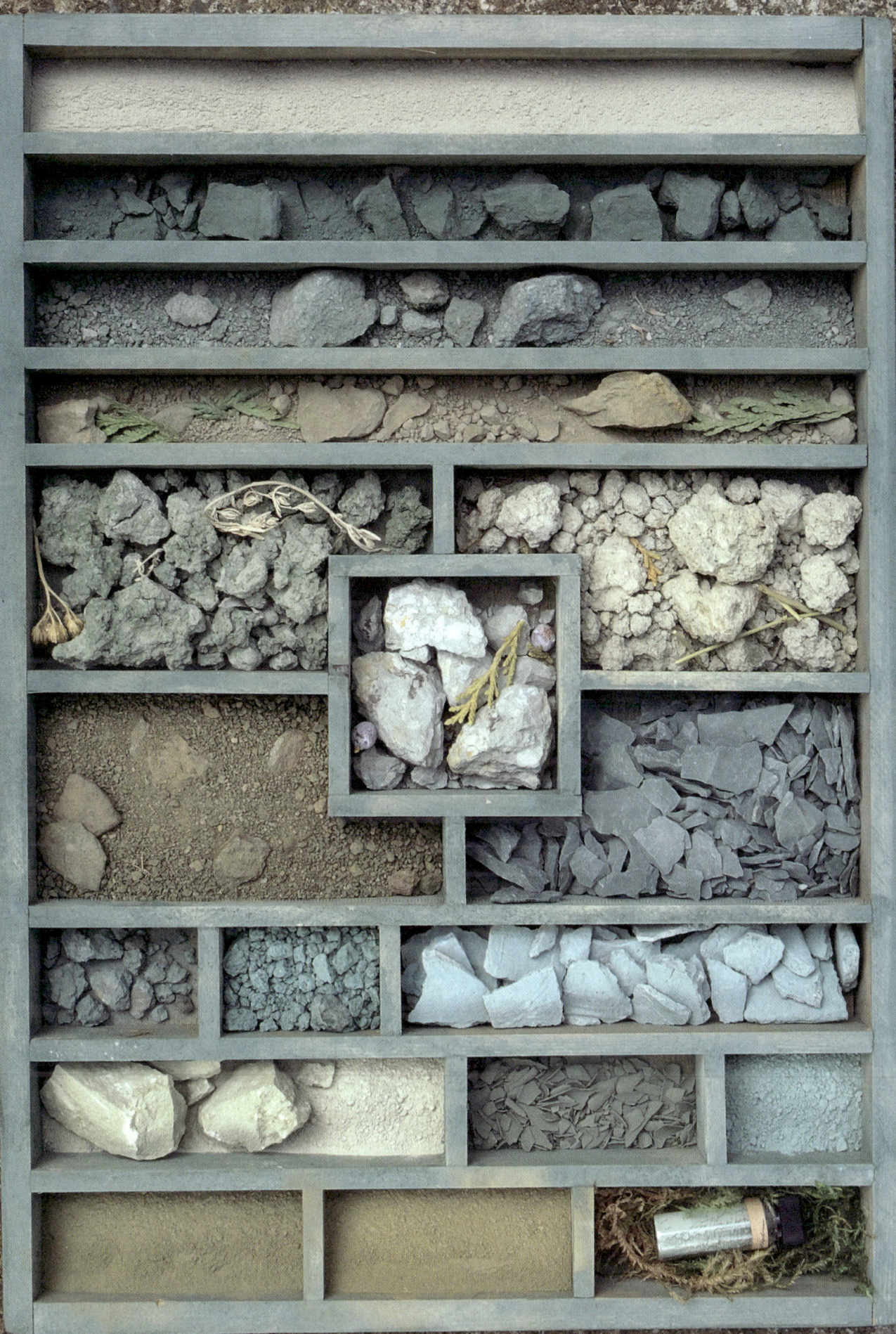

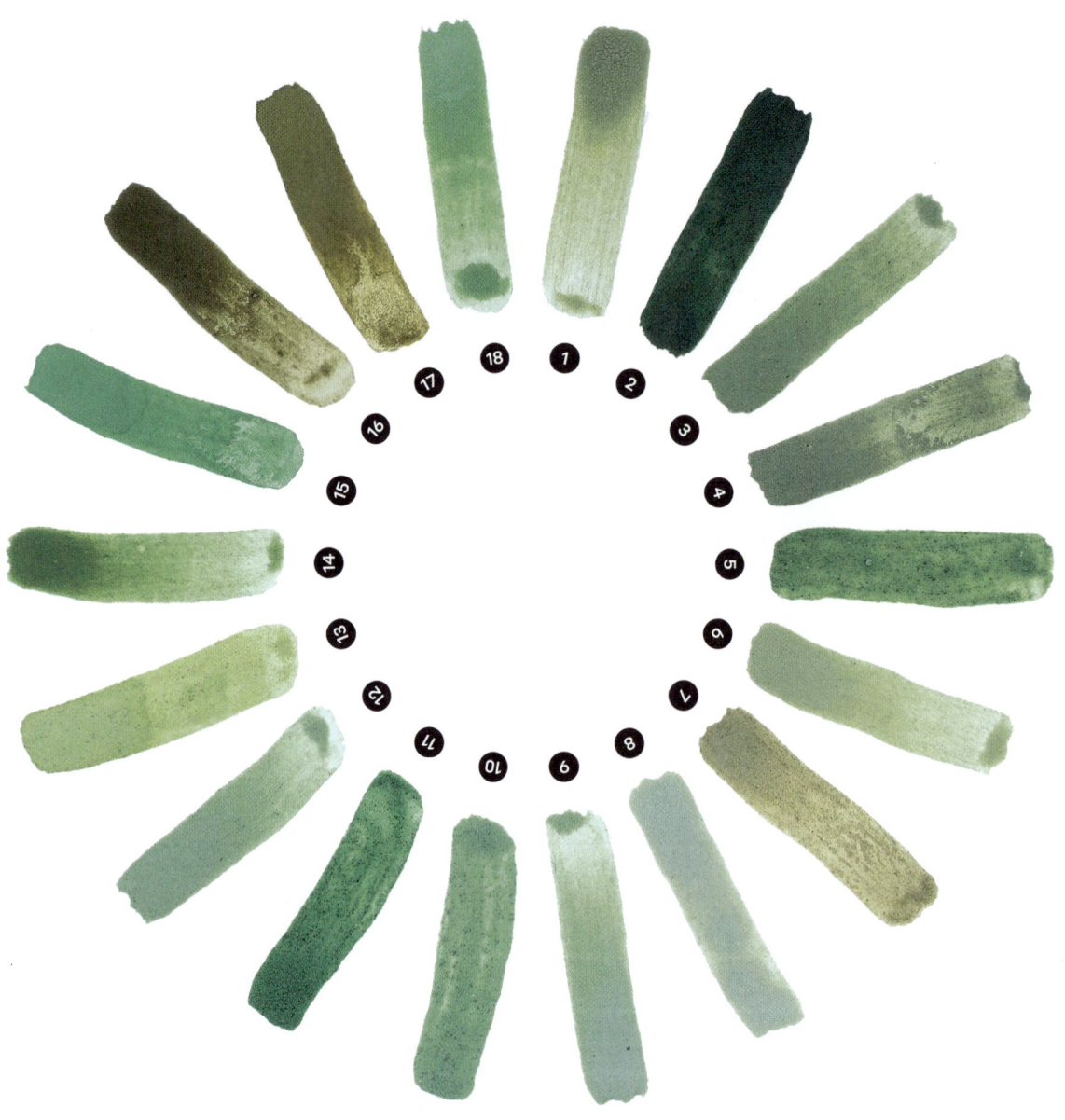

1. *Facial Clay.* No clue where this green bentonite is from, but likely somewhere in Europe, maybe Greece. Foraged from my local co-op grocery store's bulk section, where it is sold as a beauty and medicinal product premixed with rose essential oil.

2. *Glauconite.* Gathered from northwest Russia, milled, processed, and distributed for artists pigments by Colibri. Soft as a mullein leaf.

3. *Greensands.* From a truck stop ditch up north from where I live. Weathers down into more of a clay after about a year, if left outside under the seasonal stars.

4. *Cemetery Ancients.* Green earth from Chuckanut mudstone formations from the Eocene (33.9 to 56 million years ago), exposed along a creek that flows through the Bellingham, Washington, cemetery. Looks deceptively blue when seeing it from a distance atop the dead.

5. *Sage.* Clay from an eroding hillside off the side of a farm road down south from where I live. Smells of wild sage, paints the color of fresh spring sage, and that, too, is the actual plant that holds the remaining soil together on this hillside.

6. *Green Earth.* Utah. Traditional lands of Diné (Navajo), Ute, Hopi, and several other nations with cultural ties to the Colorado Plateau.

7. *Glauconite.* Utah. Traditional lands of Diné (Navajo), Ute, Hopi, and several other nations with cultural ties to the Colorado Plateau.

8. *Earthgrip.* From where the Uranium is held across time and space.

9. *Shedding Celadonite.* From the Morrison formations (near the Four Corners).

10. *Whilamut Green.* Gathered and gifted by Tilke Elkins in conversation with Kalapuya elders from traditional Kalapuya land.

11. *Green Earth from the Hearth.*

12. *Snowlight.* A gift sent to me in the mail. The green was discovered on a long winter run in the snow, where the green mound was so bright in the sun against the snow.

13. *Celadonite.* From a special place.

14. *Watered Greensands.* From **3**, except levigated to make a refined pigment.

15. *Green Earth.* From traditional mines in Cyprus, Greece.

16. *Ancient Green Earth.* So very very very "ancient" that I had to buy it from a commercial distributor, from an unnamed geologic place (probably, from the looks of it, Eastern Europe?).

17. *Sankir.* God might have made this green earth, in Russia, less than 10,000 years ago.

18. *Sankir.* From the Tavush region of Armenia, as processed by Agulis Pigments, which works to protect historical pigment mines, especially known elsewhere for their Armenian Bole (red clay).

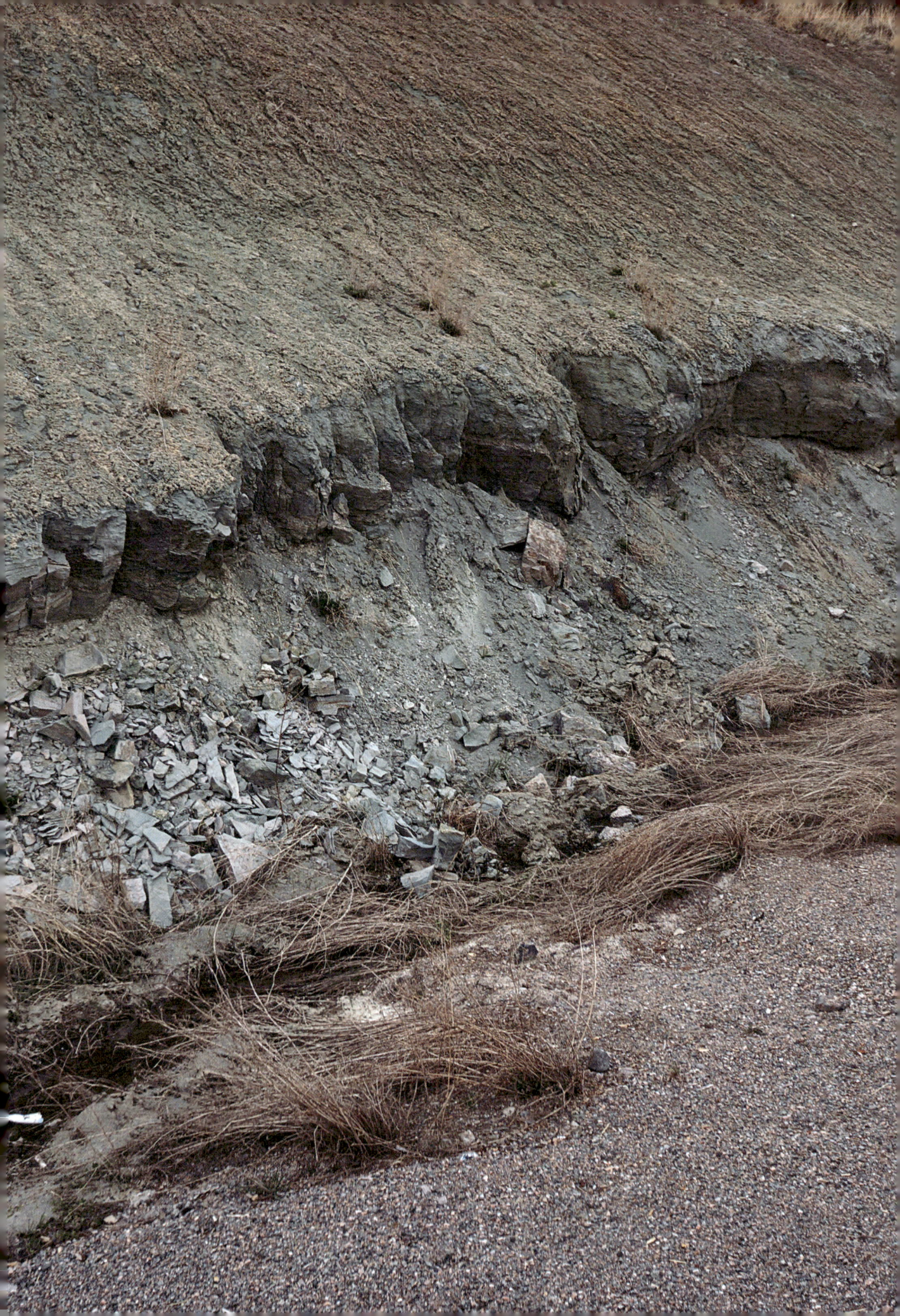

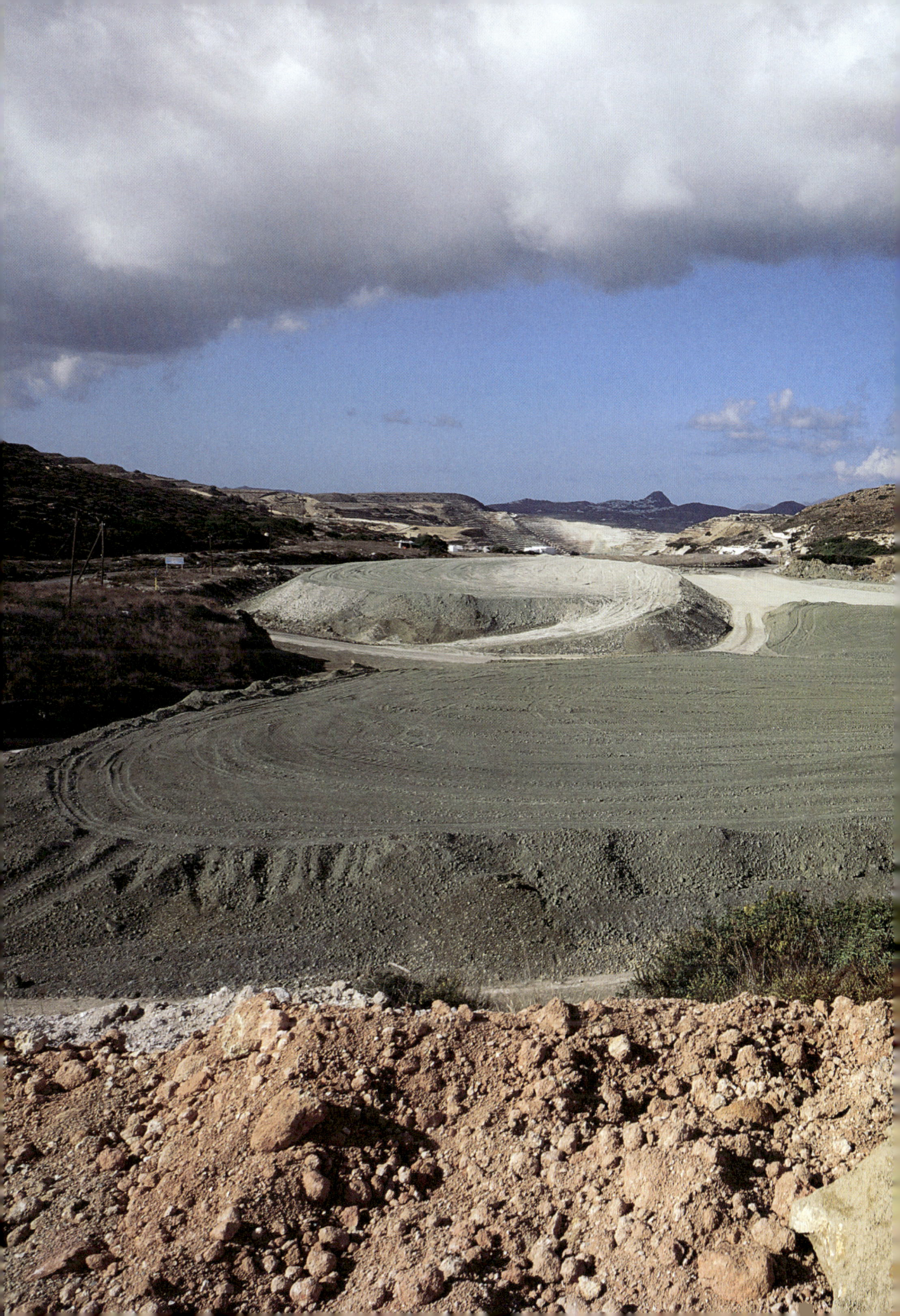

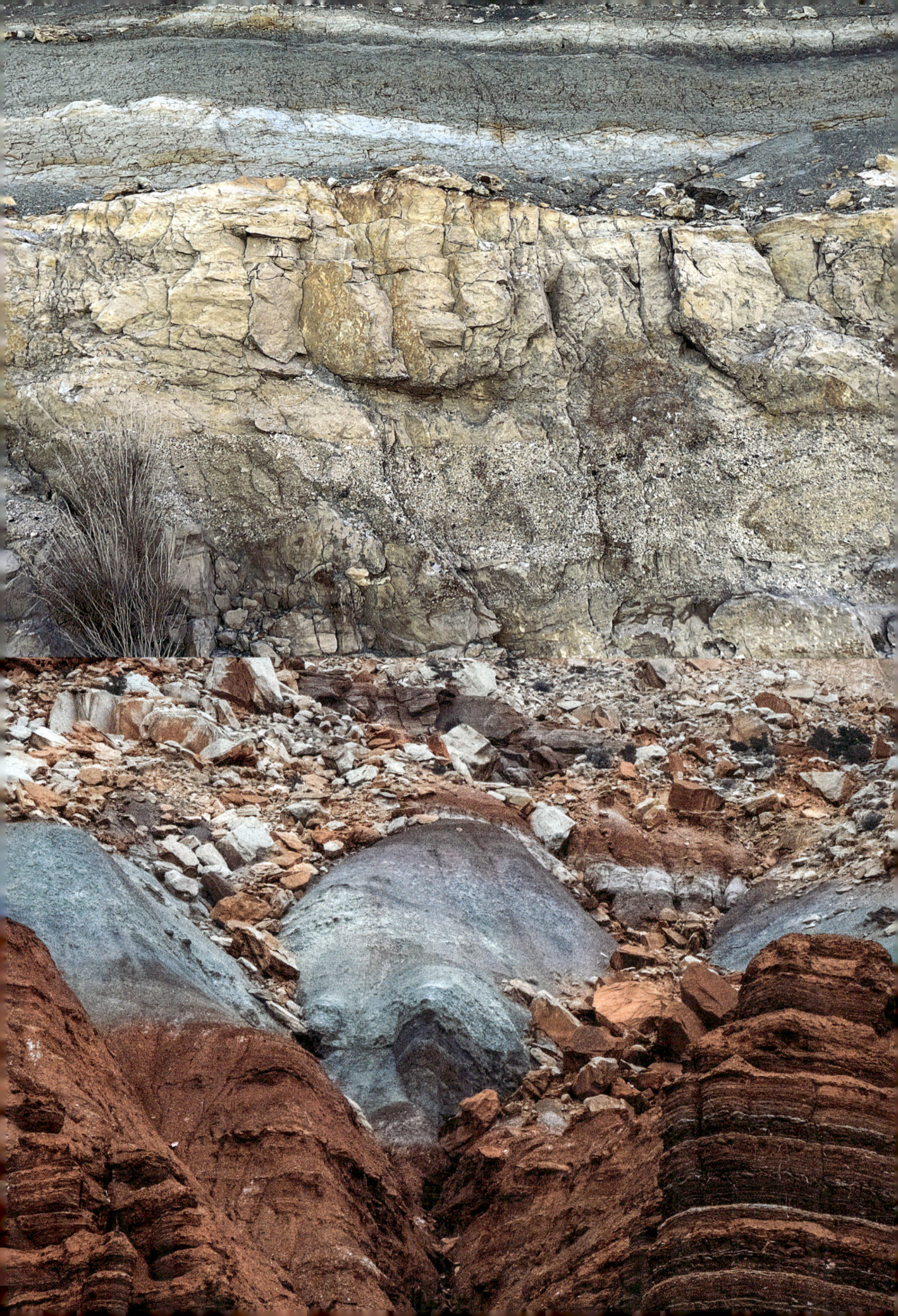

Earthgrip

GREEN EARTH INVOKES ECOLOGICAL HABITAT, a shrinking green world many of us live in. Green earth is a classical name for natural green pigments made from several minerals and rocks (but not trees or plants). They are a family of minerals known as sheet silicates because they are structured in layers, piled up as foliage, folia, folios. As deforestation, drought, and smog uncolor Earth, what was once a delicate pigment used to envision lush forests and plentiful grasslands, earthy "green" colorants today tend to describe disease: chlorosis, gangrene, green sickness, green jaundice, sickly shades of nausea, and puke.

Yet in Eastern Orthodox iconography traditions of Russia—where painting is not painting but writing spirit into a visual image through prayer and mineral pigments mixed with egg yolk and dry wine or vinegar, becoming endowed with miraculous capacity, even autonomous healing powers—green earth is central to a sacred logic. Special green pigment mixtures, called variously sankir, terra verte, verdaccio, seladonite, or green earth, are painted underneath every flesh tone and lay the incarnate ground

of existence. Green earth is the basis or matrix onto which spiritual beings come alive or are able to exist. I find it very beautiful that the idea of "existence," across several languages, is defined by and appeals "to being in a place, by invoking ground, dwelling, sitting, standing."[1] As if existence and verdant earth are inseparable.

In iconography instruction, green earth gets applied as "the initial underpaint tone, which covers the faces and other parts of the body; leaving the sankir (green earth) exposed."[2] There's similar classical painting instruction in a workshop manual by Italian master painter Cennino Cennini—the thirteenth-century European instructor of alchemists and painters—for "How to color a dead man," where you simply lay down "several shades of *verdaccio*"[3] over pale ochre, and, he says, "you must use no rosy tints, because dead persons have no color." Even further back, encaustic Egyptian mummy portraits—faces of a dead person painted on linen wrapping surrounding an embalmed body within—generously use greenly earths as the integral first layer. In realist painting schools today, that corporeal *terre verte* retains a more somber name, "dead layer," because of the sickly greenish skin cast in the underpainting stage. I've read some cynical artists even call it the "zombie layer" or, as my pigment chemist friends put it, "best for the lurid pallor of corpses."

Personally, I prefer the term earthgrip (eorðgrap), the oldest Anglo-Saxon term for "grave."[4]

Generally, in portraits of people, dead people, saints, or angels, an expertly applied nethergreen achieves a kind of special glow from below, "an effect of pale moonlight."[5] We are said to contain this fragile lifelight, reflected in how delicate our skin layer is. It's notoriously difficult to convey lifelikeness in paintings, especially without celadonite, glauconite, or others with green earth's "weak concentration of colour," as art historian Michel Pastoureau points out.[6] Contemporary Russian iconography lineage leader Vladislav Andrejev says the green glow "symbolizes the human, mystically 'the inner human' upon which the ascending layers of light must be built. And it is also what cannot evolve, it is ours, what is given. Underneath, inside it all."[7]

Outside of classical painting traditions, green earths go by largely unnoticed. So inconspicuous that even in material color histories or pigment compendiums, natural green earths *maybe* get a couple paragraphs,[8] while bolder synthetic poisonous greens—like Scheele's Green, a beloved copper-arsenide developed by Germany's chemical corporation IG Farben, which also supplied Zyklon B (used in gas chambers) to the Nazis[9]—get far more fascination and attention.

That said, during World War II, the United States Army did seek "valueless," humble green earth pigments. The reasons are clarified in 1942 letters sent between the Bureau of Mines (on behalf of the U.S. Army) and their geologic experts, requesting information on earthy pigments, including specifics such as "In the event of air raids, large quantities of certain classes of natural mineral pigments may be required for camouflage to hide cantonments and other areas subject to attack . . . green rocks are of special interest because their colors blend most readily with colors of vegetation," and "pigments should be granular, and not have a fine body or texture and in fact materials that would be valueless in commerce may be used." Geologists responded with locations of green earth minerals (glauconite, celadonite, and clays along with olivine) suitable to match shades of deciduous trees and dense conifer woods.[10] Official color swatch names include: very dark drab, dark drab, olive drab, field drab, medium drab, medium light drab, yellow drab, light drab, sand.[11] *Drab*'s a synonym for "dull" or "dirt" or stuff drug through mud.

Perhaps their humble value proves a little more worthwhile than their flippant color swatch names imply. After green mineral pigment camo was employed, no cantonments were destroyed by air raids. Instead, in 1945 the United States detonated nuclear bombs in Japan, killing more than 200,000 citizens. Atomic weapons initiated a new era, an unfathomable technosphere that makes green earth camouflage or paint, again, feel instantly quaint and nostalgic. I questioned it myself. And then a few years ago, I encountered a startling power of very dark drab *terre verte*, whose close affinity with nuclear war and atomic power is, as you might expect, a well-hidden story. Before I leap into that, I want to try to remember and carry with us what the "atomic era" unleashed, as if gathering clues to help discover the secret life of green earth.

Human Shadow Etched in Stone is the name of a permanent installation I saw as a teenager at the Hiroshima Peace Memorial Museum. It displays chunks of cement with shadows of incinerated persons, suspended in time, captured on city walls, steps, and sidewalks. Actually, they are a kind of inverse shadow. When the atomic bomb exploded, the flash burnt everything, and these human beings' bodies blocked the searing light, leaving behind an unexposed spot on the cement that was not scorched, a bit like a cyanotype print, where a flower is placed on a pigmented surface and exposed to sunlight, leaving behind a lighter negative image where the physical flower was placed.

After the nuclear explosion in August 1945, *Time* magazine printed a very short article on the nuclear bombs, which ecological activist Joanna Macy alerted me to. The article is poignant and raw and says that forever after that day, "all thoughts and things are split," opening "a bottomless wound in the living conscience of the human race."[12]

Americans released a toxic consequence: a split within people was born when the atom split open. We "broke the strongest binding force known in the universe," a force known chiefly by uranium (U) itself—the central keeper of nuclear power. The half-life of uranium is 4.5 billion years: almost the same age as Earth. That's how long it will take to settle or "decay" back into Earth's body. Between now and billions of years in the future (will the Earth even spin that long?), it's clear that radioactive material, mine tailings, and manufacturing waste will need to be dealt with (i.e., buried somewhere or else ejected into space). Otherwise radioactive decay will continue to bring rare cancers to humans, and who knows what to other life-forms that emerge after humanity probably goes extinct in the deep future.

Today, engineers develop systems of land burial for uranium and radioactive waste. Systems designed to last one million years (!?) and keep nuclear radioactive waste from leaking into groundwater and topsoil. So, what's a *key technology* of these crucial landgraves?

Dull green earth. Drab dirt.

And why is a "valueless" deciduous color so integral in high-level nuclear waste disposal? Because green earths create a very responsive and self-sealing barrier. Green earth barriers have a special capacity to seal off waste and guard water and soil from contamination. I've read and heard technical and mathematical discussions on how they work from nuclear engineer briefings. My dear friend and soil scientist Dr. Morgan Williams helps manage and heal a few of these government sites. He confirmed that the important properties of green clay or mudstone (specifically green bentonite) are: self-sealing or self-healing ability; durability over the long timescales (because they are already old-ass cosmic rocks); and a low permeability or breakability when saturated, meaning radioactive waste can't creep out or escape.[13] An integral camouflage of sort, a soil that contains and hides. (The image on page 112 is engineered soil with green earth used in active nuclear waste containment.)

I find that capacity to "self-seal" or even maybe "self-hide" fascinating. Green earth does this in other ways, too. As a healing clay, it's well documented that "bentonites have a great capacity for absorbing many times its own weight in toxins." And like a swollen sponge, "bentonites can absorb pathogenic viruses, toxin, and pesticides and herbicides."[14] I also see that ability more commonly in fancy green clay masks—simply add water, spread on your face, let it draw out toxins. I could even eat green earths (lots of people do!) and would experience a similar cleansing, absorptive power and effect. And in the context of war (and mining), field doctors use green earth mixed with petroleum jelly (instead of water) to act like a swelling poultice for the most common cause of death: severe bleeding from vital organs through laceration and wounds. Green earthy clays are so archetypally intertwined with flesh, interfaces, and layers of life that they function as a kind of second skin, able to clot blood at the source and facilitate significantly faster wound-healing.[15]

All these qualities tell me something about how green earth likes to act. Yet it tells me very little about where they actually form and exist. What are they like in their own environment? How could I find them? How do I participate with and honor them and their capacities beyond "using" them? How might I need to behave in their presence, and in relationship to their guardians, ancestors? When I drove to remote terraced canyons of Utah, I went to feel my way further into these questions, and other curiosities about how green earth, ochres, and even uranium ore live in their natural habitats.

In some ways, seeking to be with minerals in their own ecological systems overlaps with prospecting. Throughout the twentieth century, uranium prospectors in Utah would also "read" the land using gut feeling and observation, including plant knowledge and evidence of deep-rooted juniper, who uniquely and efficiently uptake uranium from soil. Using botanical prospecting methods, along with unethically extracted knowledge from local Indigenous Navajo guides, prospectors found, mined, and extracted uranium for weapons development. As was discovered later, uranium mining (and subsequent waste flushed

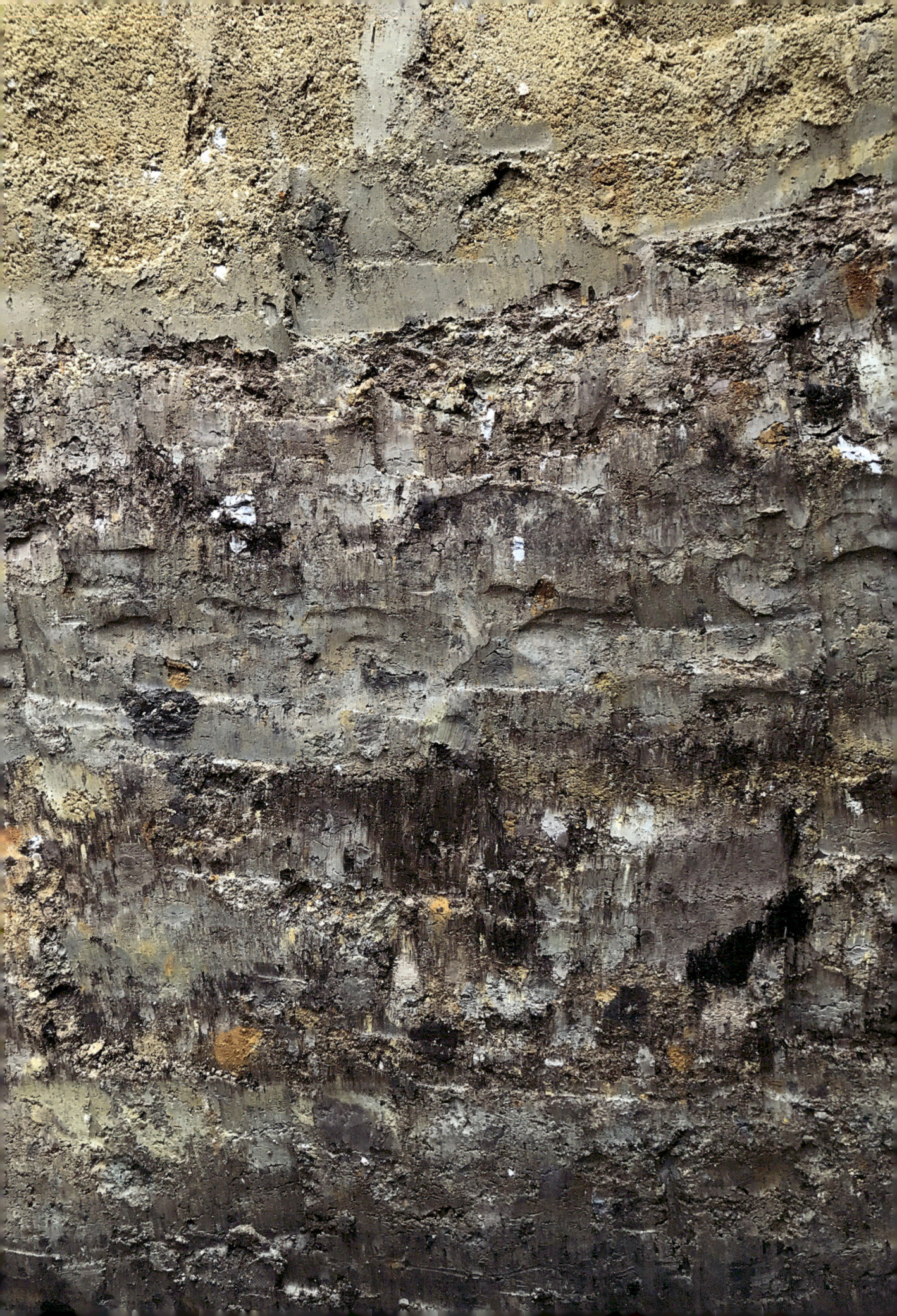

into riverways) devastated water, air, and soil in the Four Corners area (also known as the Colorado Plateau of Utah, New Mexico, Arizona, and Colorado). These vast places are homelands of Diné (Navajo), Hopi, Pueblo, Ute, and Apache Nations "with continuous pasts and futures that refuse the apocalyptic timeline of nuclear logic," as historian Alicia Puglionesi reflects.[16] Mine tailings and nuclear waste dumpsites were, and still are, forcibly put on their ancestral land (not far from where the minerals were extracted in the first place), a further extension and act of far greater, longer ongoing war and genocide on our "own" soil. "Nuclear colonialism" is what Anishinaabe activist Winona LaDuke calls it.

When I arrived in Utah, I wondered about green earth as a wound healer. At first it felt bad to do so. Everything in Utah was so unfamiliar. I grew up in wet swamps, with cedars, ferns, and moss, and this place is full of scraggly desert trees, snakes, and dry cow patties I didn't recognize. My friend Elpitha Tsoutsounakis, an artist and educator who lives and works in this area for her own research on sites of mining extraction, met me there. We walked up an old, abandoned mine road in the side of a tall, undulating cliff, a road miners probably built to access uranium in the higher layers. I noticed something subtle, almost invisible. I noticed it again on another walk elsewhere. And again, when driving among the enormous cliffs and giant ship-like rocky masts. In strata where uranium-bearing ore would likely occur (often inside dinosaur bones), there's a layer of green earth *surrounding*, above or below, sticking with the uranium-rich layers. I observe this pattern in other places more and more, as I look for it. As if suddenly the world shifts valence, like noticing a duck blind or a camouflaged soldier only after they've been pointed out, or if you stumble upon them. Although "companion plants" is a common term in organic gardening, herbal medicine, and naturalist observations, "companion *minerals*" is kind of a non-term. But they're here. Green earth and uranium, natural lovers, skin to skin, together where they formed over 155 million years ago (as shown on pages 108, bottom, and 114).

I find it eerie that green earths, just as they naturally occur, hold super atomic energy suspended and undisturbed in liminal silence. As if green earth already knew what to do for all time. It feels rather strange and self-indulgent of those who claim to "discover" and "engineer" novel sealants of green earth for depleted uranium burial. Green earth and uranium are two minerals who for time immemorial were *already* held close, stuck together, and who—in an extractive instant—we tore apart. In *Homeland Elegies*, a book reflecting on acts of terrorism, Ayad Akhtar mentions a metaphor that resonates with me here: "The process was counterintuitive, akin to restoring the incidental sedimentary layers on a piece of extracted ore. The mind recalled the essence and discarded the dross, but the dross was what swarmed with generative life."

What does it mean to restore sedimentary layers on extracted ore? For me, I start by staying with sediment, by holding and listening to the rocks, the portals and pieces of traumatized ancestral place themselves. I read in an essay on war by cultural psychologist James Hillman that "the nuclear imagination is without ancestry. Nothing to look back on and draw upon. History provides no precedents."[17] This human hubris of "discovery" that lacks acknowledgment of ancestry feels like a major part of the wound. Maybe for me, green earth's ability to draw "nuclear imaginations" and literal nuclear energy back down to earth is a way of re-tethering us to ancestry, a social responsibility. Perhaps this response is similar to how nuclear guardians—activists who fight to stop nuclear terror—advocate that "by storing the waste where we can keep an eye on it, we also keep the danger, and the guilt it generates, from being suppressed. It is much better to stay aware and deal with it as best we can; the real peril lies in ignoring these dangers."[18] They further suggest that "places with the greatest potential for destruction the world has ever known acquire, in this way, a certain spiritual significance. All the great religions remind us that besides recognizing and accepting our mistakes, the path of freedom lies in owning our own failings, rather than projecting them onto others."

I imagine looking down onto those remote, deep geologic burials from a drone, or vulture or angel's eye view—from a spiritual point of view. I

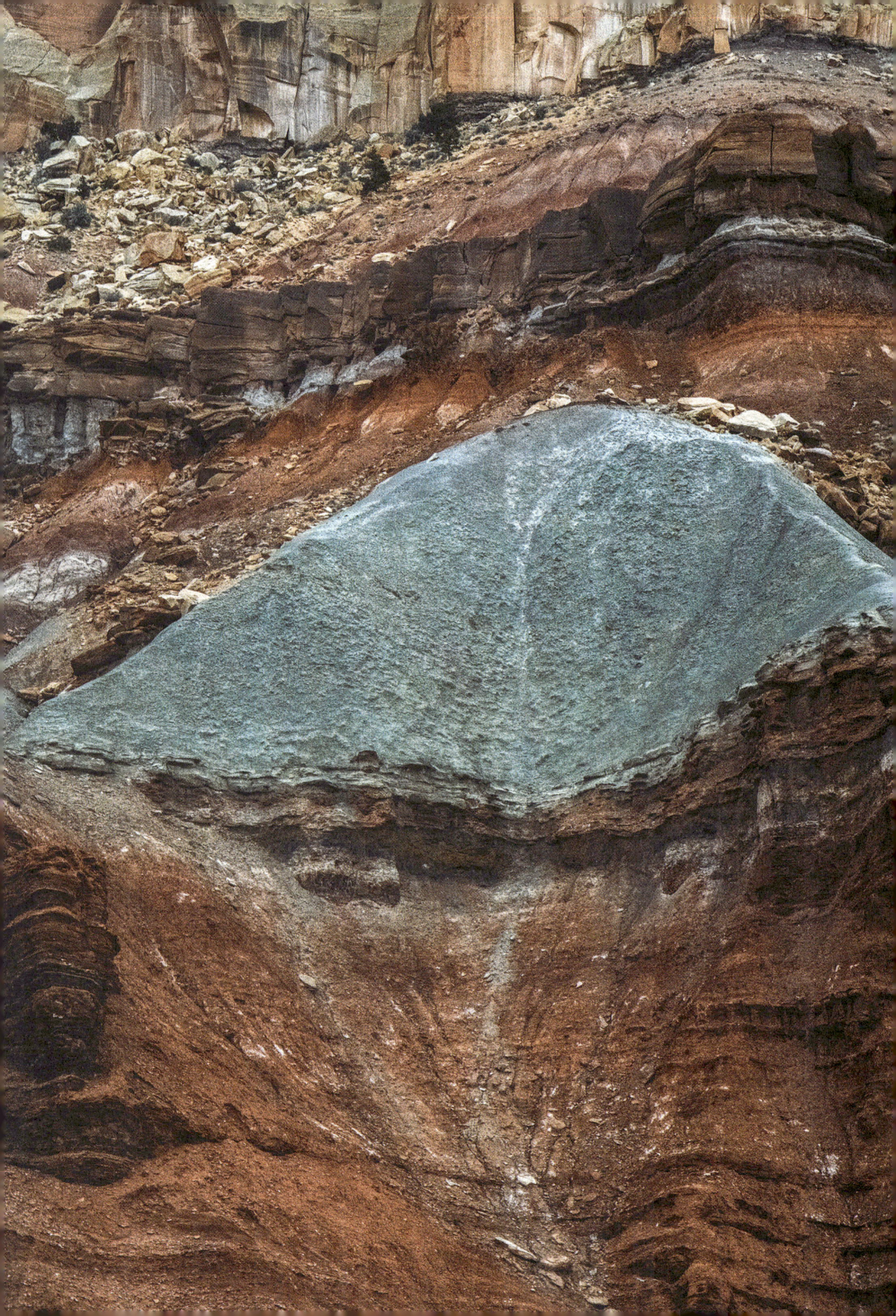

begin to see nuclear waste sites as spiritual pictures. As layered paintings, writ by engineers, that define a twisted icon of our culture, a deep strata of our cultural soul, "what cannot evolve," to repeat Andrejev's words about what green earth (sankir) layers indicate.

Is there a creative impulse shared by the art of making nuclear graves and religious icons? On impression and aesthetic gut, I'd say yes. But in material principle, I'm not so sure. In icons, green earth is an *under* layer; in atomic waste burial, a *sealing* or *masking* layer. While both practices metaphorically signal a "dead" layer beneath, do they actually have anything else in common? I sent some questions to my iconographer friend, and longtime student of Andrejev, Nick Maione, asking for his opinion. His response was coy. I could feel him smiling. He told me a little professional secret. There's an iconography practice far less known in secular art histories: "We float, as one of the very last things, celadonite in the center of the eyes of the icon, reaching back (through all the layers) to sankir, floating the naked truth of the human being back into the irises at the end," and, according to his teacher, that means, "we return to self." My impression was that this final act was an essential part of what makes icons feel alive and sacred. Green earth seals the soul so it can be seen—we are able to keep an eye on it.

I think it is fair to observe that green earth invokes a similar meaning in nuclear graves, where green earth seals an atomic demon, a cultural spirit. I can't help but hear novelist Clarice Lispector, who wrote that ". . . coagulated color, violence, martyrdom, are the beams that sustain the silence of a religious symmetry." Her visionary sense of a "coagulated color" feels precise to me. There's something sticky and yet untouchable, completely profound. As there is when I feel clayey green earth in my hands or on my skin or in my spirit. Clay's root, *klai*, means "to stick," or *klei*, "stick together"—a deep root shared with *cleave*. *Cleave* is weird because it carries two opposite meanings: *to split, part or divide by force*, and *to adhere, cling*. And here I am again, trying to find the right word, a tone, to describe how I feel, how I hear these rocks speaking and humming. It feels a little unfair to treat green earth as only the "valueless camouflage" or a "zombie layer" or "best for the lurid pallor of corpses."

I still think the most accurate term is "earth-grip." Maybe I could whisper it over and over and over until my split spirit glows back into juniper bushes and trees and forests and a billion resting stones.

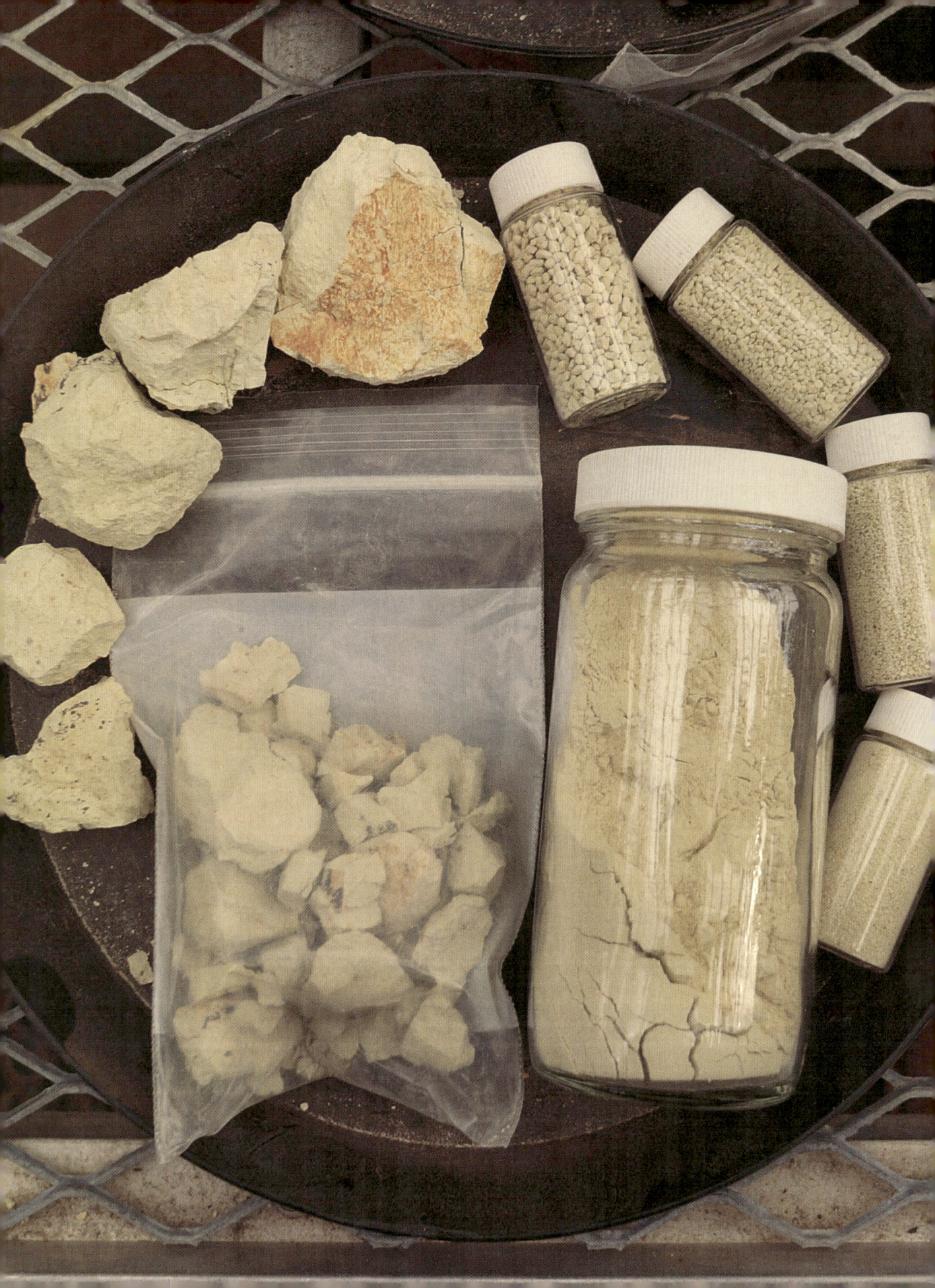

A NOTE ON SCALE

I break down stone. What remains is a green cloud of dust, nearly blowing away in my own breath, my wind. I guess I am weather. In between my fingers I feel an atmosphere, personality. Each particle feels alive, suspended, and heavy still, alive with memories, qualities, tendencies. Do my dactyls sense deep time sounds, a spinning hum? There's a foggy notion of wide waters passing centuries ago. I wonder if my fingers will ever be able to register the single-cell amoeba's ornate house whose diameter is a mere 150 µm?[19] Or maybe could I feel inside the pigment's smoothness that memory of a seasonal creek flowing over their rock body at home? Or sense heat from a long-gone wildfire? Will my human touch remember how to feel the immense pressure from forces of sudden uplift? How will this dust move in liquid? Will it run to the edge or cling to the middle? Who breaks down who?

Atom	0.0001 µm*
Nanoparticle	0.001–1 µm
Bacteriophage	0.225 µm
Coronavirus	0.1–0.5 µm
Wildfire smoke	0.4–0.7 µm
Fine pigment	0–62 µm (mesh #400 to #300)**
Silt	2–62 µm
Bacterium	1–3 µm
Clay	<2 µm
Dust particle	2.5 µm
Red blood cell	7–8 µm
Dust particle	<10 µm
Bee pollen	15–200 µm
White blood cell	25 µm
Grain of salt	60 µm
Coarse pigment	62–200 µm (mesh #200 to #100)
Sand	63–2,000 µm (mesh #10)
Mica flake	1,000+ µm
Gravel	3,200–5,000 µm
Crushed stone	5,000–101,600 µm

*
microns (µm), *smikros* ("little, petty, small") is the unit of measurement that artists, colormakers, and scientists use to define and arrange grains of pigment by size.

**
Mesh size indicates the fineness of sieve (which is associated with a number when you look to purchase a filtering sieve) that one must use to achieve that pigment particle size.

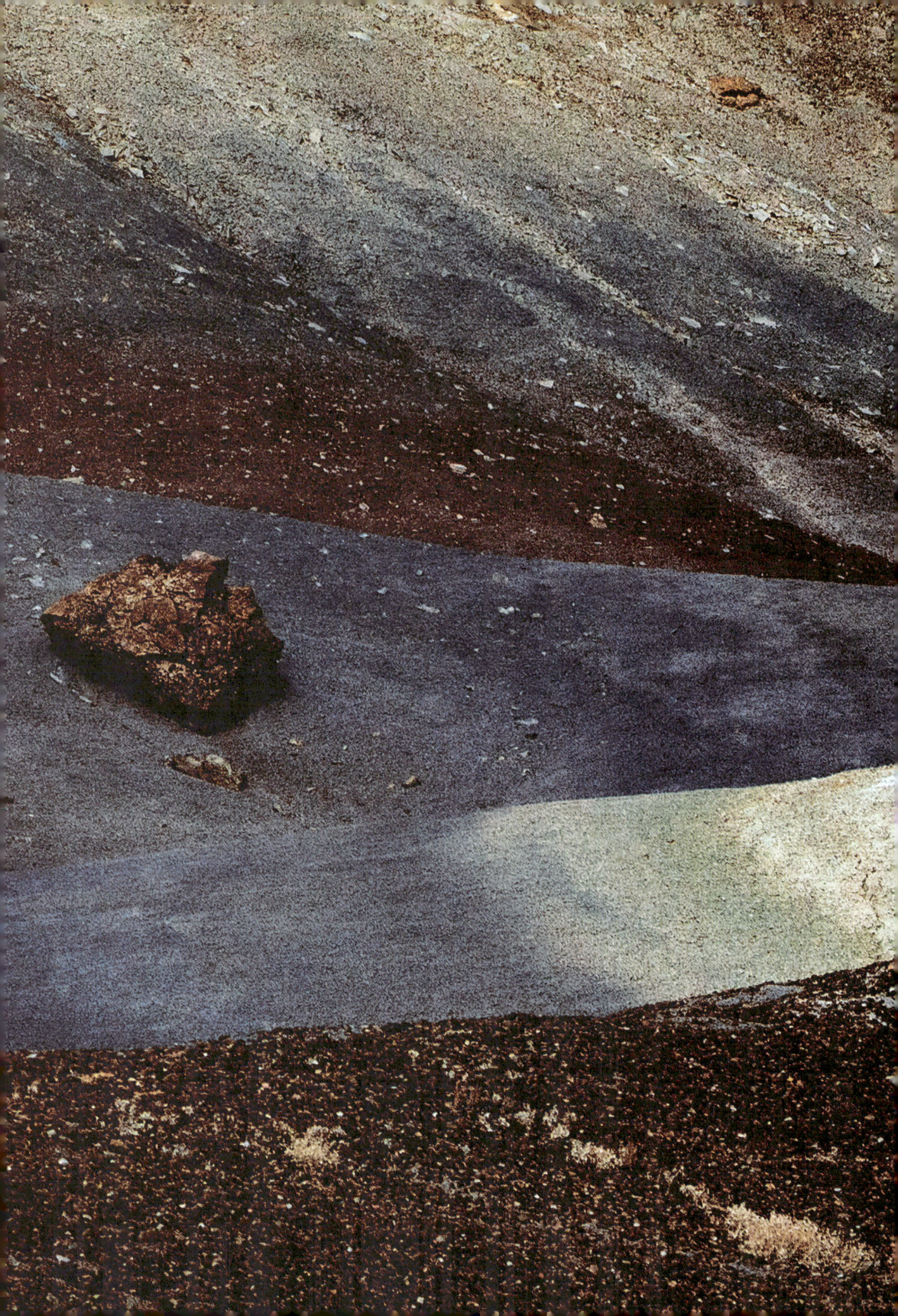

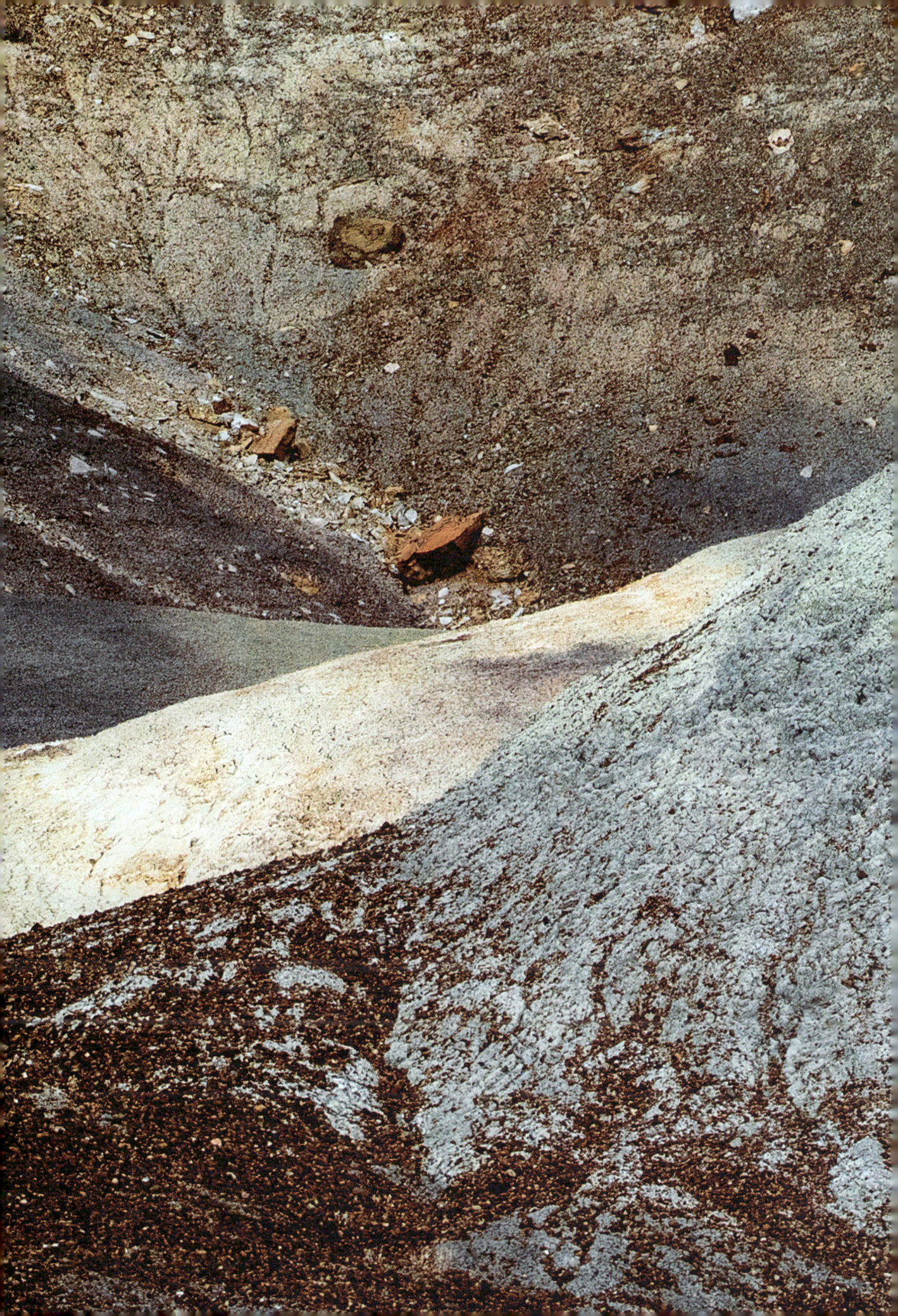

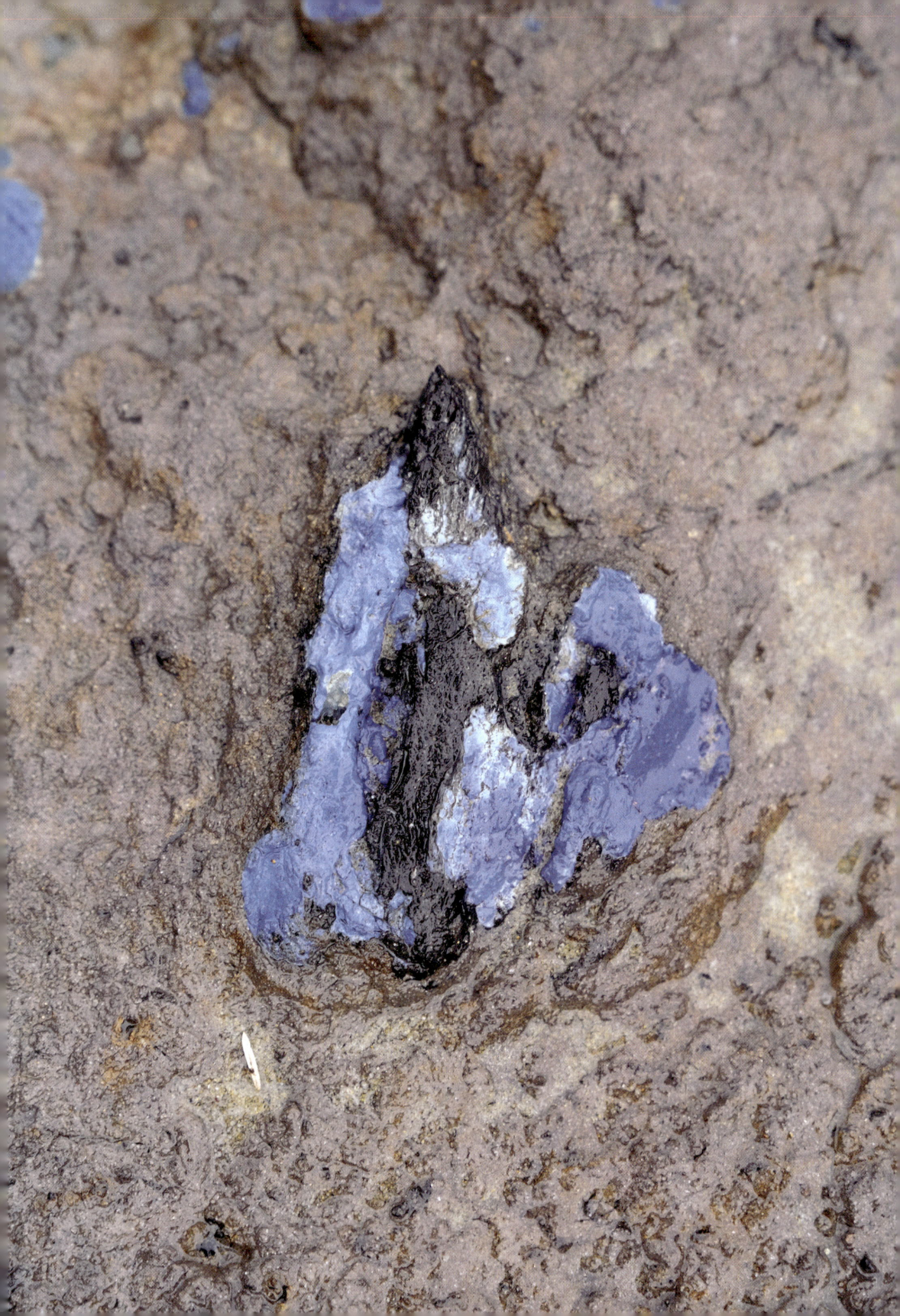

BLUE OCHRE

VIVIANITE

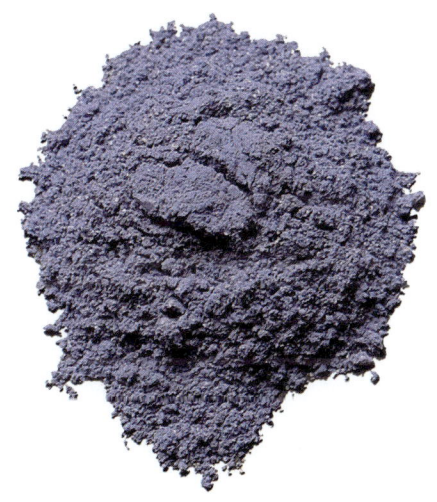

HYDRATED IRON PHOSPHATE ⚷

Etym. from *vivus*, living, and *niht*, darkness, "living darkness"

$$Fe^{2+}_3(PO_4)_2 \cdot 8H_2O$$

PIGMENT SPECTRUM

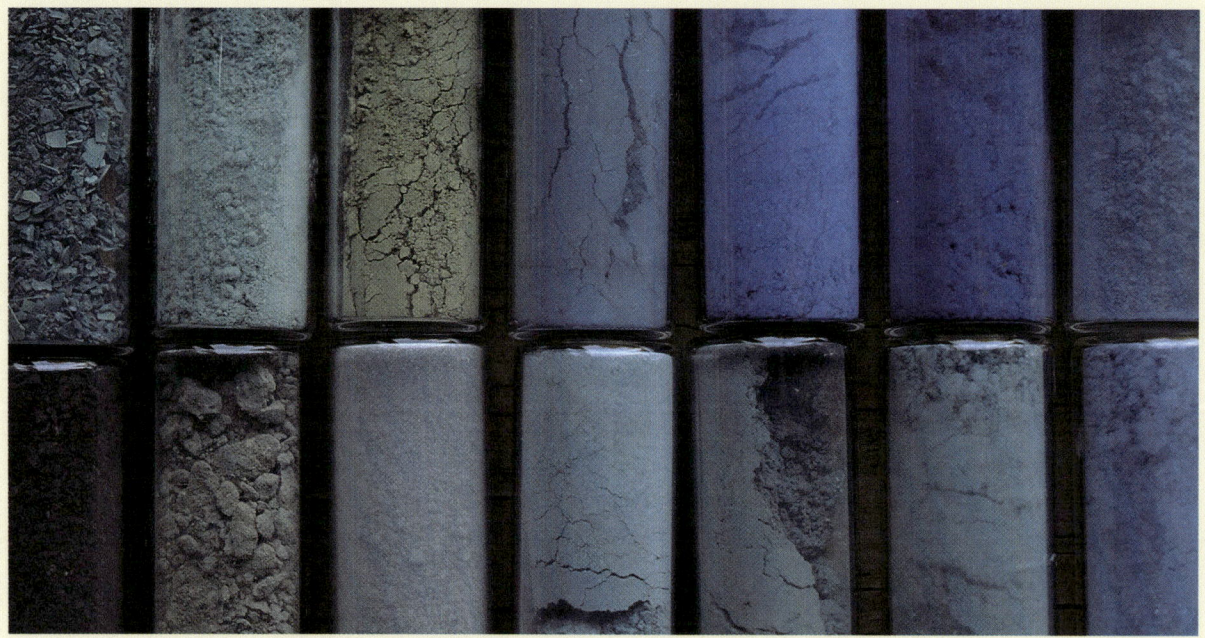

MATERIAL DESCRIPTION

As an enigmatic hydrated iron phosphate, vivianite knows tidal places, estuaries, rhythmic coastal sediment infused with salts of the sea and an "unplaceable magical discomfort."[1] Prone to invisibility, blue ochre clings to places where excess phosphorus in water runoff seeks to bind with iron, and in anaerobic environments such as butchery pits, wastewater pipes, and peat bogs rich in organic matter, soluble iron, and phosphorus. When formed below ground, vivianite is usually white or colorless unless exposed to sunlight or air, where the mineral begins to rapidly turn, as if gasping, from pale to deep blue in a matter of hours or days. Referred to lovingly as a diva, vivianite transforms quickly upon contact with fingers, oils, lipids, heat, light, gases, spirits, ghosts, lost souls, and local gaps in the etheric fabric. Prone to deep transformations, vivianite will easily change to olive or dark green, as well as darker bruise-like colors, and may become lush matte black over time. Because of that mysterious process, blue ochre darkens into what is technically called meta-vivianite, which can be seen as misplaced black on Northwest Coast bentwood chests and shaman masks, traditionally stored in dark places, and if exposed to UV light, the deep blue becomes an unrecognizable shadow tone. A similar transformation can be found with Dutch master painters such as Vermeer and Rembrandt, whose pale "blew clay"[2] paint today look moody and soil dim.

NAMES

qesuuraq (Yupik), o'tal (Haida), neixinté (Tlingit), pukepoto (Maōri), blue ochre, blauer ocker, blue clay, blew clay earth, blue ashes, blaú as, blue iron earth, cœruleum patavinum, ochre friabilis, cœrulea, cendre blue, Bergblau ("mountain blue"), terra de Harlem, Harlems blaauw

HABITAT

exhausted peat moss — lacustrine sediments — peatlands & bogs — anaerobic or reducing soils — interglacial clay — iron-impregnated water — mammoth throats, brains, tusks — animal hide — human skin — the suddenly dead — burial grounds — with calciferous materials — marine sediment — wastewater sludge — rotting organic matter — horsetail homes — butchery pits — liminal realms — little people poop — poop — doo

CHROMA

TRANSPARENT	HUMIC
WHITE OR COLORLESS	STORM
SKY BLUE WITH EXPOSURE	WINEDARK
	FADED MOSS
BLUE EYES	STICKY YELLOW
CADAVER-COLORED TWILIGHT	OLIVE GREEN
	VELVET BLACK
CARBUNCULAR BLUE	INVISIBLE
REGAL BLUE	GORE
BLUE-GREEN	LOST
BRUISE	

CRYSTALLOGRAPHIC SYSTEM

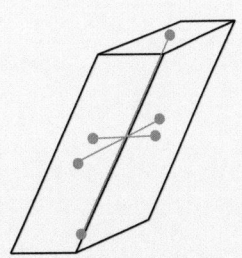

MONOCLINIC

PRISMATIC

LOOK-ALIKES

Iron sulfides or reduced soil
(may appear blue at first but fade to gray)

Azurite
(copper-based minerals)

Indigo, woad
(plant-based compounds)

Mayan blue
(indigo + heated clay)

BLUE OCHRE — Living Darkness

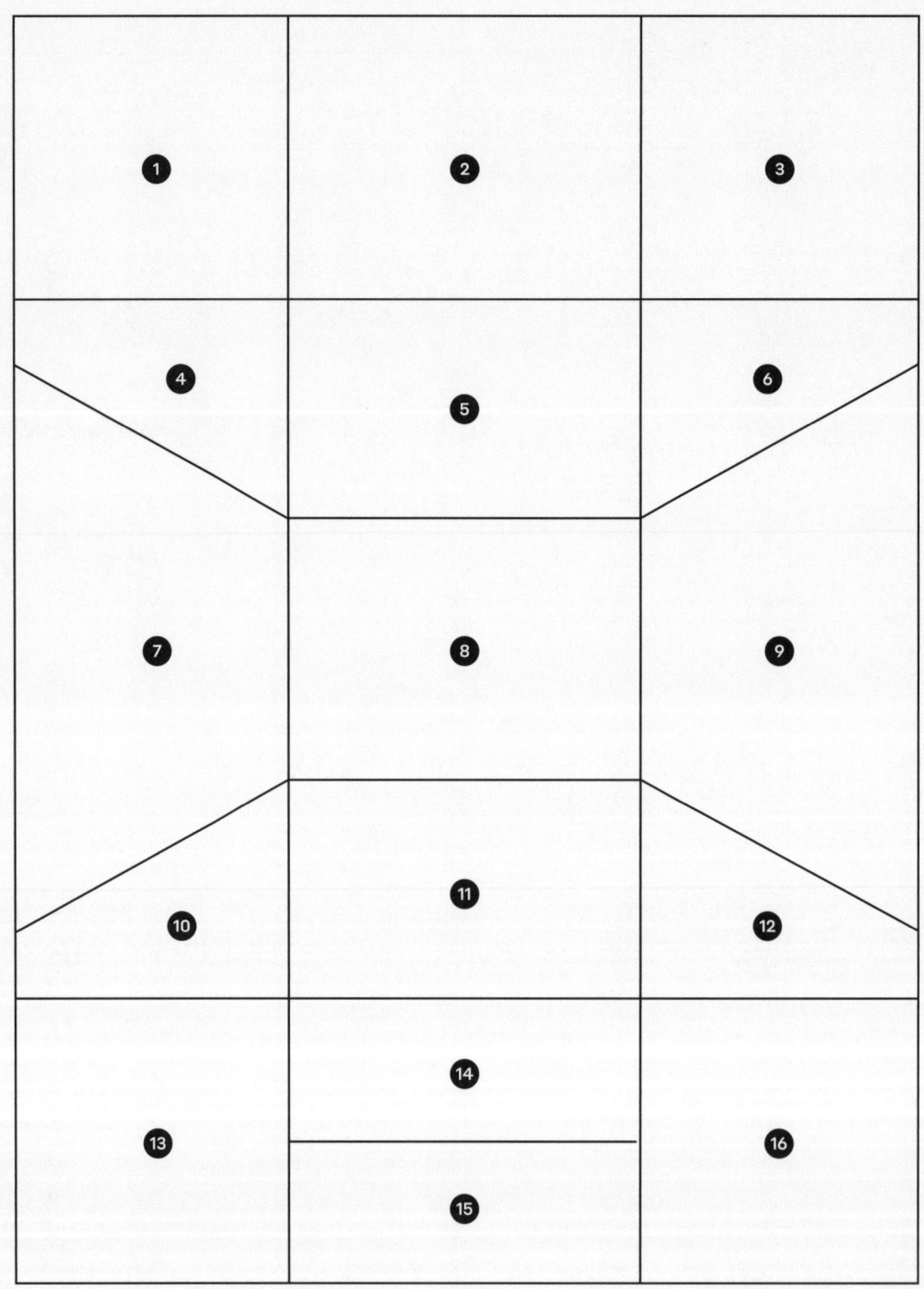

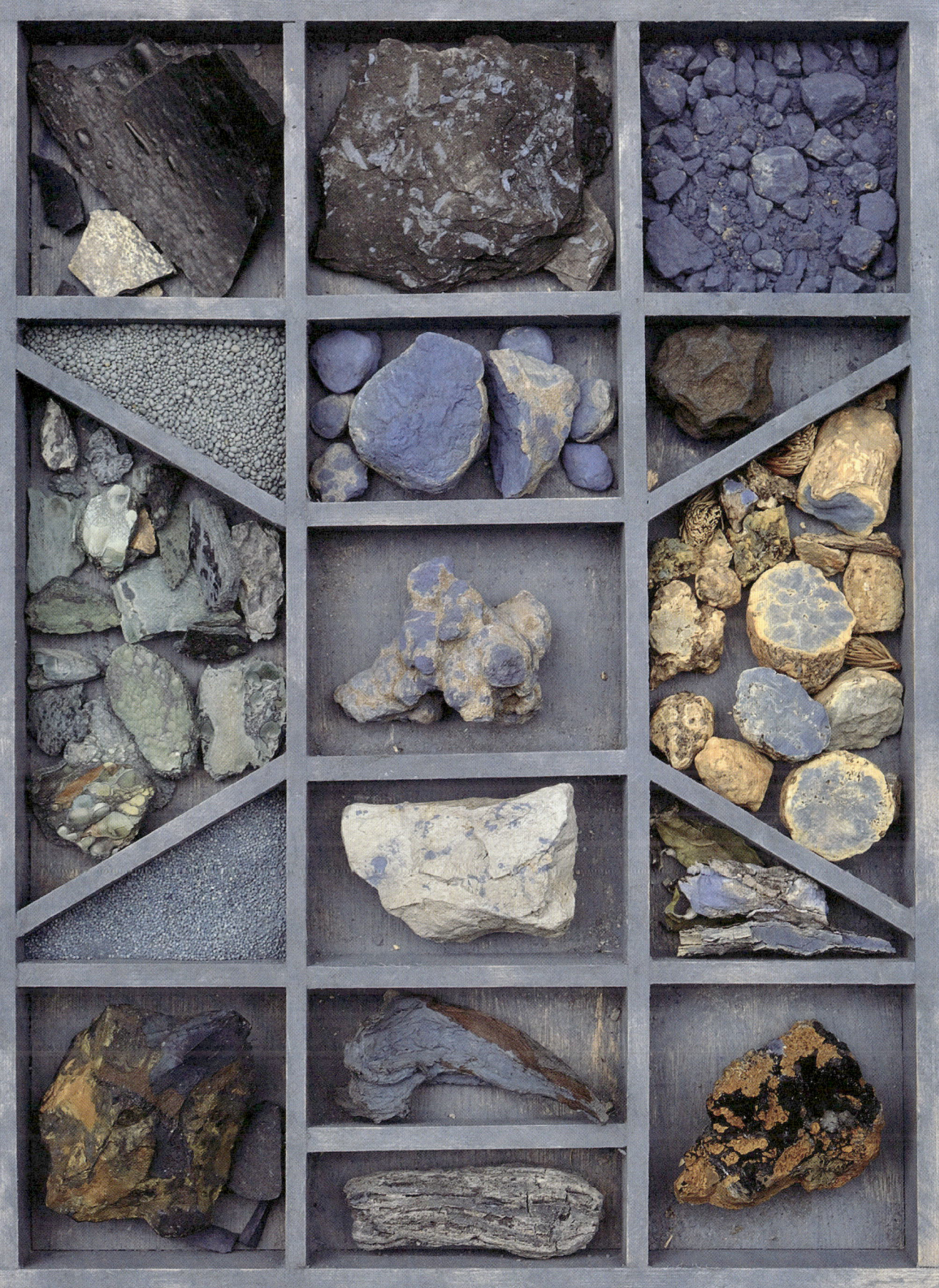

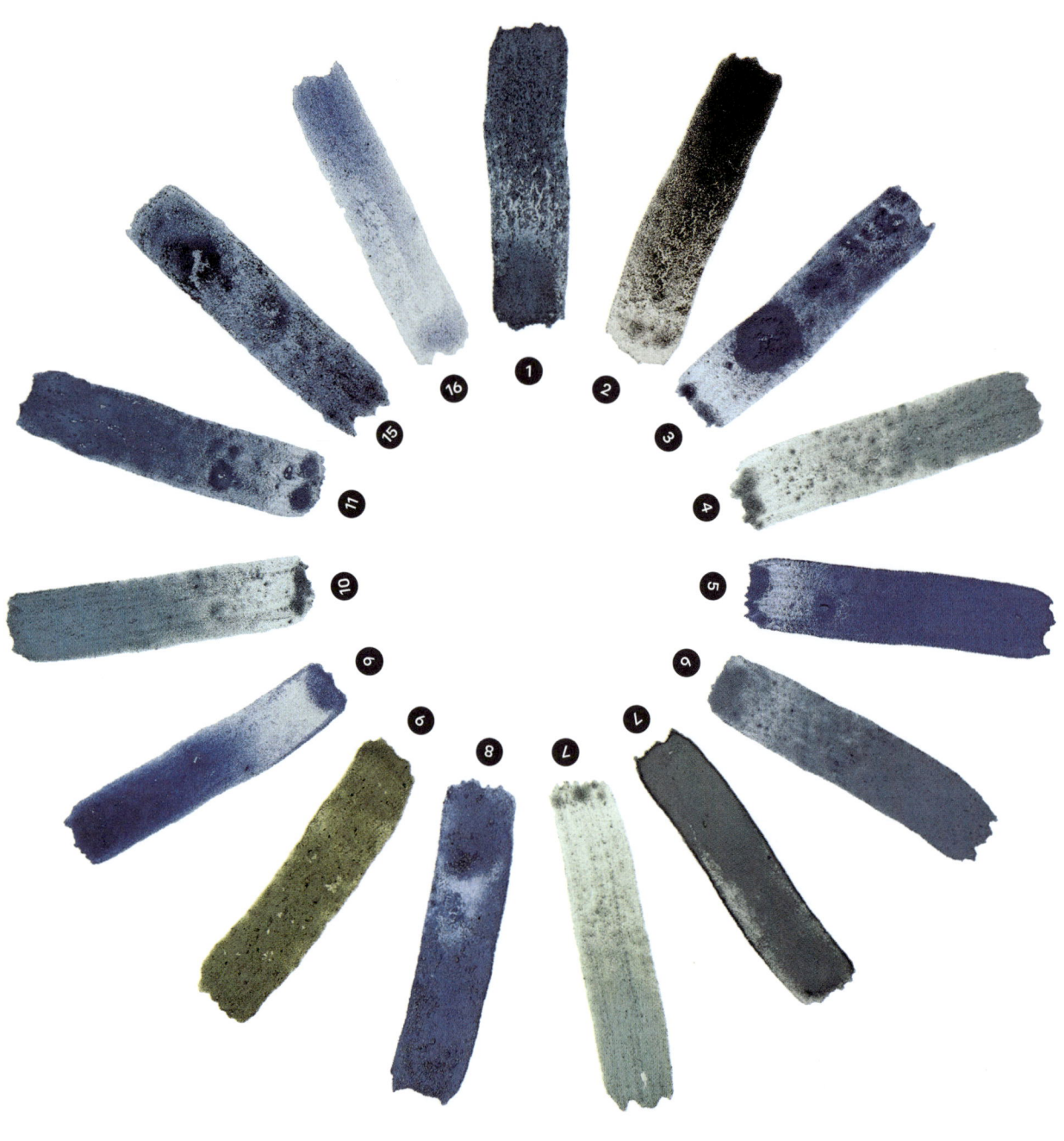

1. *Night Soil.* Wastewater sludge from pipes in the Netherlands. Gathered by Wetsus scientists for research.

2. *Ice Age Vivianite.* Horsetail peat with blue vivianite, 80 to 100,000 years ago. From near my grandparents' beach on the coast.

3. *Bog Undersoils.* Vivianite soft clay gathered from somewhere in the peatbogs of northwest Russia, commercially for artist-grade pigments.

4. *Night Soil Seeds.* Vivianite silica seeds cores extracted from wastewater in industrial automotive industry.

5. *Blue Clay Between Worlds.*

6. *"Fool's Vivianite" Nodule.* From Demon's Bluff, Australia.

7. *Dragon Scales.* Wastewater chunks aged 4 to 5 years. Extracted from clogged pipes that were replaced in the treatment plant, Iowa. Two color swatches show different pigments that can be made from the outside and inside of mineral.

8. *Root Viva.*

9. *Spruce Cone Turned to Stone.* Ripe seed cones buried by tsunami and transformed into vivianite over ~2,000 years, Oregon Coast. Gathered with S near clam diggers at full moon tidal times. Two color swatches show different pigments that can be made from the outside and inside of mineral.

10. *Night Soil Seeds, Smaller and Bluer.*

11. *Glacial Clay with Little White Polka Dots.* Unearthed vivianite (white turned blue) in a foundation dig for the new auto parts store in my rural town of Everson, Washington.

*12. *Wood with Blue Insides.* Twenty thousand to 80,000 years ago, from peat bluffs on the coast where I played as a child.

*13. *Seastone.* Vivianite rock used to make jewels, from the dangerous beach in southeast Australia.

*14. *Blue Ochre on Cedar.* Via Melonie Ancheta's research collection, from the Olympic Peninsula, Washington, ca. the Pleistocene, traditional Coast Salish lands.

15. *Dark bone.* Mineral meta-vivianite found embedded in Lawton clay and peat. Probably a being—but plant? Animal? Who?—suspended in time.

16. *Underflower.* Crystalline vivianite with weathered powdery parts, mined from Kerch, Crimea, 1970 to 1980s. I bought on Instagram.

Swatch not pictured, see raw material/ color on page 125

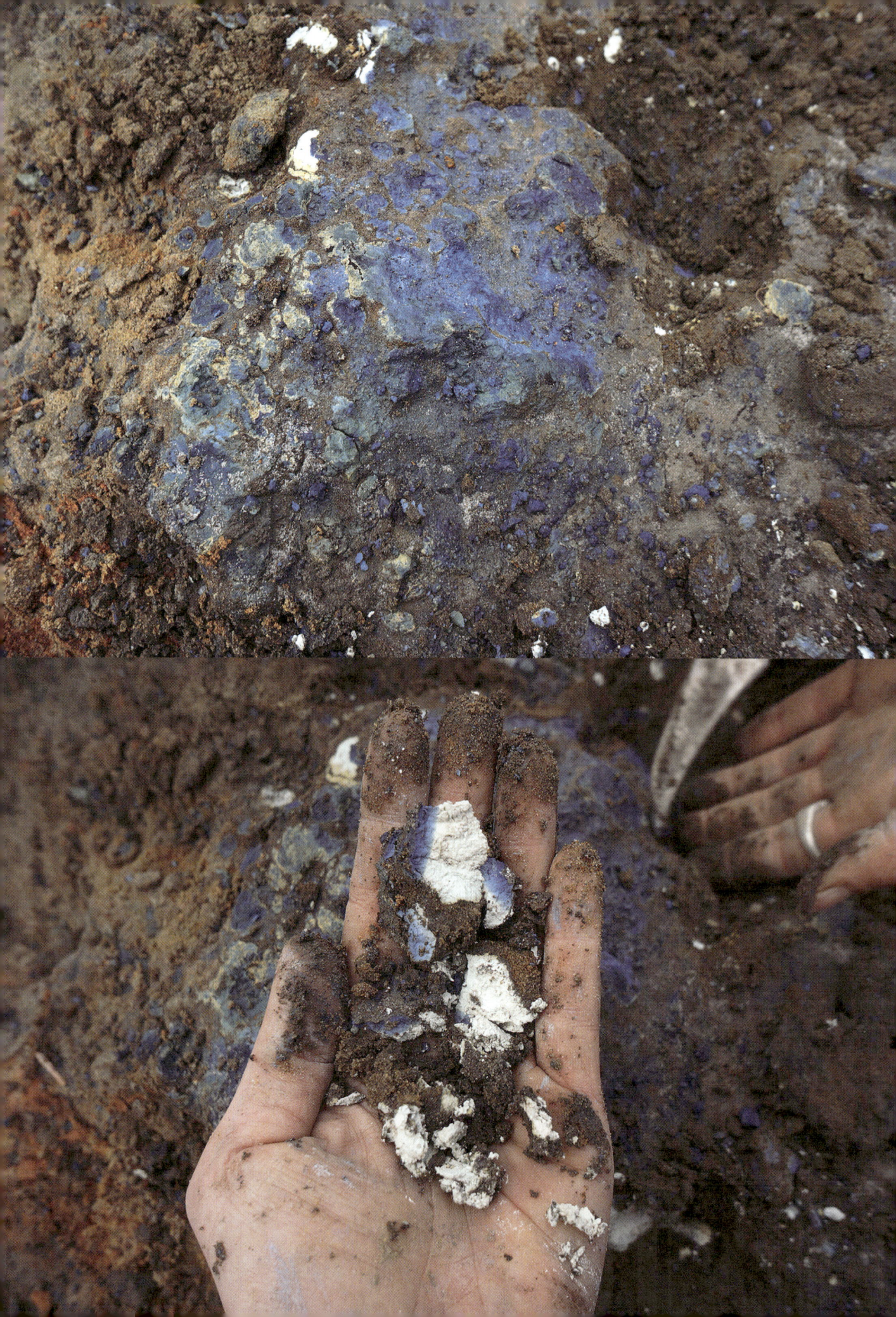

Divinity escorts us kindly, at first with umblemished blue,
later with clouds . . .

—FRIEDRICH HÖLDERLIN, "THE WALK"

Blue Clay

A HEALER AND MOTHER, MEG, sent me a message and an image. She found an eroding piece of coastal bluff with blue clay. She tells me a story, which I then read in the news: A father took his two young children onto the bluff in a winter storm. A sneaker wave came up over the land, taking them down into the sea. The two children drowned.

They both died, but their father lived. The blue clay is near where the children were standing when they were swept away. I read the daughter's name is Lola. Names are partial oracles, often as clear as a trumpet or a bell: *Lola* means "sorrows."

I phone Melonie Ancheta, my close collaborator and local Native woman, a Northwest Coast traditional pigment expert and arguably one of the world's hidden vivianite specialists, and describe the area and get her feelings on going to gather. "Sounds like viv, what do you think?" I ask. Her lifetime of research gives her unique knowledge on the material, and one of her aims is to protect and revitalize traditional use of vivianite pigment along the coast. Asking her is a barometer and a step toward approaching the

material and the place appropriately. Could it be of benefit to gather? Should I gather this? "Yep," she says. "If vivianite calls, go."

At four the next morning, I go south a half-day's drive. I feel I hear a big blue wave: blue blue blue, she's gone dark blue. I think of *Conflict Resolution for Holy Beings* by Joy Harjo:[3]

I will see you again, is one of the names for blue—

A color beyond the human sky of mind—

I know from medical studies that in the days and weeks before people die, powerful dreams of crossing an ocean or going "beyond the human sky" become premonitory, more lucid, relaxing even.[4] Then people do die. Their bodies go bruise pale; color leaves the face when we are sick, afraid, haunted, and hunted too.

What's weird is vivianite *begins* this way—colorless, pale, and corpselike—and then *whooooooshhhhhh* becomes vibrant, endless deep blue. Blue vivianite ochre undergoes color transformation. I sometimes find vivianite pure neutral white, at first. Slowly she

129

reveals her sky and then wide blue self. After days or years breathing in the world, or mingled with the oily sweat of my hand, vivianite may turn dark olive brown or matte black. I've also witnessed this spooky change happen in minutes. White to blue to black. In alchemical traditions, they say that true blue "always carries *mortifactio* (death) with it,"[5] a strange place somewhere "between black and white." Poetic master of blue Maggie Nelson says:

> It calms me to think of blue as the color of death. I have long imagined death's approach as the swell of a wave—a towering wall of blue. *You will drown*, the world tells me, has always told me. *You will descend into a blue underworld, blue with hungry ghosts, Krishna blue, the blue faces of the ones you loved. They all drowned, too.*[6]

Melonie has told me of local Indigenous stories that connect vivianite and underworld waves. There's a Northwest Coast story, *The Finding of the Blue Paint*, recorded in 1909 from Dekina'k!u of the ḳóok Hít (Box-house) people,[7] that tells of a supernatural blue clay found in treacherous sea places. It involves hunters who take refuge from a huge storm while hunting by canoe. "They discovered a rocky cave or overhanging cliff where soft blue stuff continually dropped. They knocked off a big piece, rolled it up in their clothes and hid it away . . . the sea began to get rough . . . they were not drowned, however, and reached the shore safely. . . . To this day, they say that, if you take anything from there the weather will be stormy and people are still afraid to do it, but take the risk because the thing obtained is so valuable." Tlingit blue-green mineral, neixinté,[8] is also associated with places (according to at least one story) where "breaking the stone incurs the anger of the sea spirits thus making it rare and difficult to obtain," and "if mixed with grease will turn green or lose its color altogether."

In far north Yup'ik lands, their word for blue vivianite found near cliffs, *qesuuraq,* comes from qesuir, "to lose color." "Qesuuraq was also hard to obtain," according to local Paul Jones, "in the cliffs . . . and since it's difficult to climb, it's hard to reach, but I've heard people say they have gotten qesuuraq that

broke off." He attributes the "origin of qesuuraq to the ircenrraat [extraordinary persons taking both animal and human form] that reside there . . . perhaps from their dump sites."[9] I heard from a contemporary Yup'ik artist that vivianite also translates as "little people."[10]

Little people. Shapeshifting beings. Lost color. Drowning. Rough waves. Sea spirits. Blue underworld. These things are on my mind as I drive.

I remember Melonie's statements about how Haida "believe blue to be a portal, a portal between sea and forest, earth and sky, and animals and gods. Vivianite is always in a state in which it is in between," as noted in her published academic work, "providing physical and spiritual protection from its genesis in death and decomposition."[11] I've heard that among nearly all Northwest Coast Native cultures, including Coast Salish, Tlingit, Kwakwaka'wakw, Tsimshian, Nuu-Chah-Nulth, and others, vivianite was used to color the background spaces (also called "tertiary fields") of special objects used in ritual and ceremony such as shaman's masks and rattles, and protective gear for warriors.[12] According to Melonie, the use of vivianite on these objects indicated their special status as supernatural protection for safe travel and transformation in liminal realms and places beyond the human world, as further hinted at in a Haida word for vivianite, o'tal ("sky").

Hours later, I arrive to meet Meg on the coast. Even though she is a total stranger, we greet one another as if old friends. We walk together down to the beachside cliff, after paying homage to the cliffside makeshift memorial. We descend into a chilly, trickling estuary. I glimpse a hint of blue on an eroded piece of bluff that's broken off. Bruised clouds above signal a coming storm. As we approach, I begin to see several shades of blue. Our backs are toward the ocean, a rising tide looms, and I heed the signs and realize we need to be sort of careful, a bit fast. She and I get on our knees. As we begin to touch the clay, a black dog comes out of misty nowhere and begins to dig into the blue. Underneath the blue, I now see raw white vivianite. The dog licks the clay over and over. My heart moves into my fingertips. My fingers are up to my nose. White vivianite clay smells unlike anything else. I can't describe it except as a faint hint of something unrusting, undead—do spirits leave scent in their wake? I can't imagine how this blue underworld tastes to that black dog.

As we gather bit and bit, tear after tear, of Lola and her brother's footprints, I start to feel these two beautiful children are suddenly now somehow dancing with us. She's waving, *hi*. Is he playing? Splashing her and laughing?

In many traditions, laughing is a ritual technique for banishing—to release or free.

In this moment, both Meg and I felt as though we were in an altered state. One that mirrored vivianite's powerful ability to shift tone, quickly. As we sat in this altered realm, gathering up a few last nodules of sweet white dancing blue, in a quiet voice, Meg told me a memory:

> I met another blue being once. In my twenties when I was struggling immensely, falling into a darkness that felt endless. My father had passed away a few years before. Somewhere around midnight, I was alone and walked with my dog, Canoe—I also call her my shadow—down from a parking lot that overlooks the beach. I needed to be in complete darkness: no lights, no humans, nothing but myself. After some time, several rocks came down from behind me and startled me. I figured it was a deer or something. Immediately my dog started to growl and got very close to me, I could feel her hair raised up. I was trying to calm her and tell her it was just falling rocks. I noticed she wasn't looking up at the cliff, but out at the ocean.

> I looked out and saw this luminous, vibrant blue being coming from under the water at the shoreline. The blue being stood up—they were very tall, maybe seven feet— and in a seamless motion, emerged out from under the edge of the water and stood up, in a clearing human form, then walked south until they disappeared.

> I felt electrified. I vividly remember the sense that even though this form did not turn to look at me, the being was presenting himself, making their presence known.

Somehow the rest after this is a blur. I travel back home, sleep, wake up, and deliver la lola viv viva ("the sorrow of life alive") to Melonie's hermitage, her heritage. She's got bins of vivianite—*a color "beyond the human sky of mind"*—tucked around all kinds of hidden places: on her porch; underneath cedar cordage, seed starters, bird feeders, mortars and lichen and vines and things strewn about; by a new bucket of willow branch and grape vine cuttings for a drawing-charcoal fire someday. She lays out a big roll of butcher paper to cover her tiny kitchen table.

I bring out a bag of sandy *lola* clay, still soaking in saltwater, and we sit in her kitchen for several hours in a kind of psychedelic reverie, washing, brushing, cleaning the beach sand off of each little childlike blue tear and root and wonder, one by one. Did we even eat? What did we say? Did we drink tea or water? Did we piss? Did time pass? All I really remember is singular note, a high and low together, a true blue preserved only for when "the sky is so blue, things sing themselves."[13]

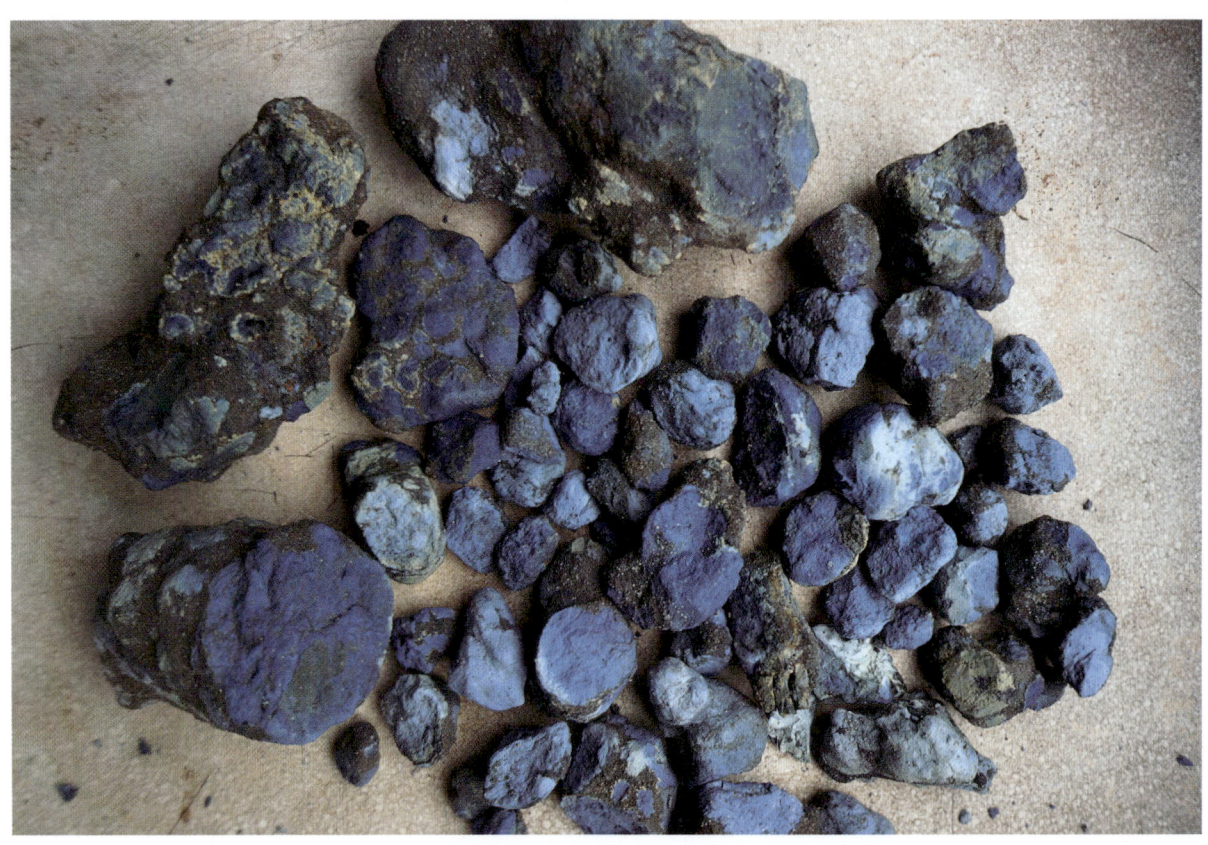

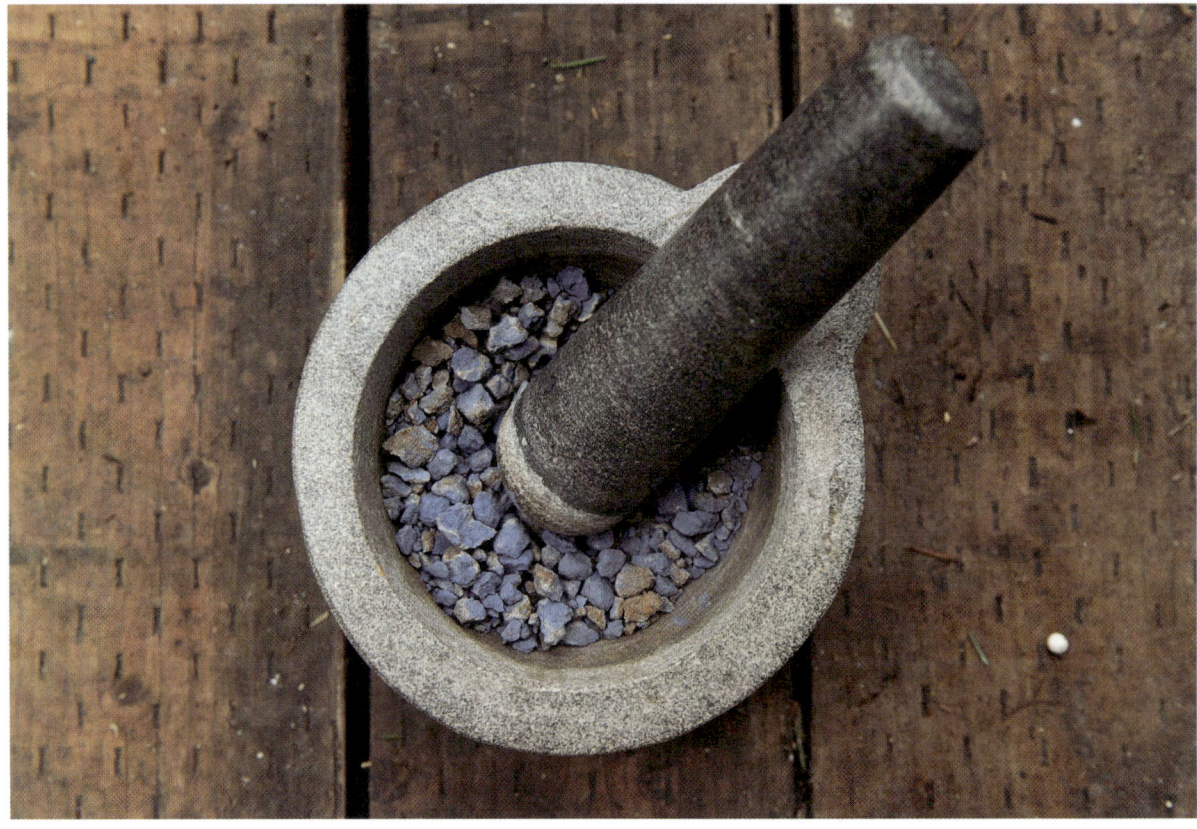

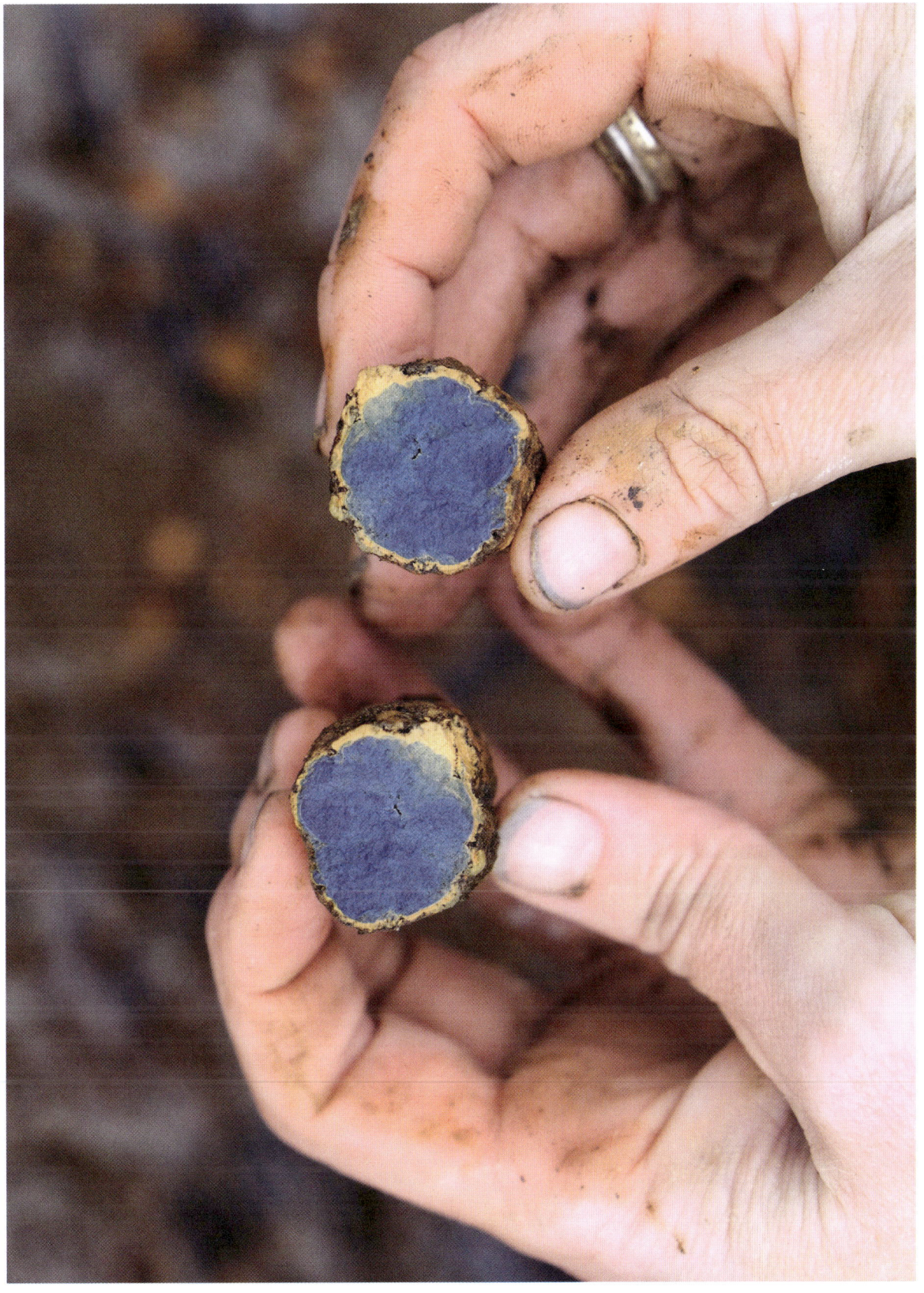

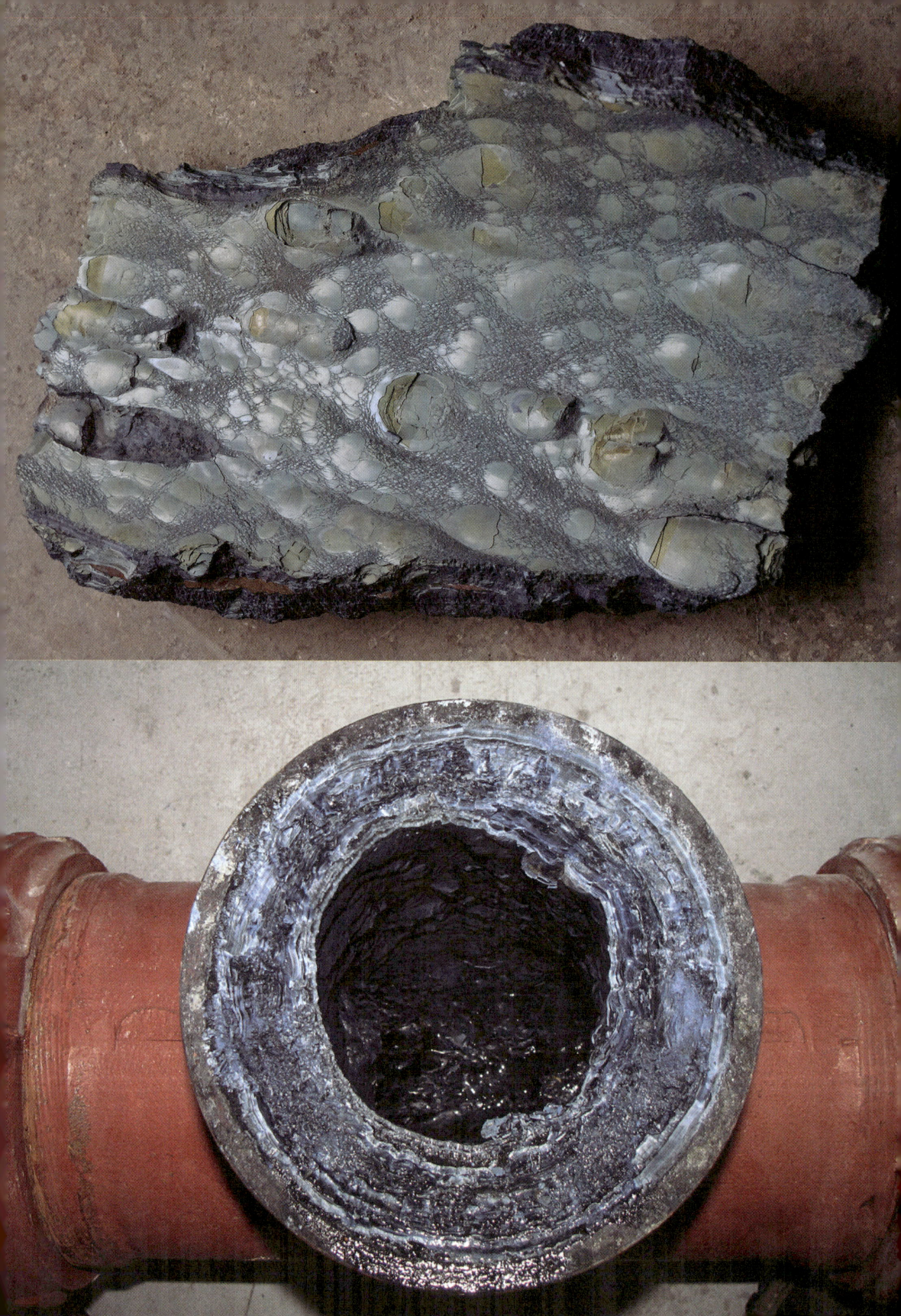

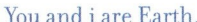

Night Soil

WASTEWATER, HOLY WATER, and their patron saint, vivianite, love incarnate matter: blood, bone, salt of the earth. Some say she is reborn from the dead, in seaside cemeteries at full moon, or kept buried under ancient sea ice, kept alive in mammoths and other extinct megafauna.

I found her spirit thriving not that long ago, in Iowa, near where forty million pigs are slaughtered in a single year. These pigs exist and grow inside "farms" that are really large ventilated steel boxes. They are raised only for their bodies—for organs and meat: ham, shanks, pork chops, spare rib, rolled roast, trotters, hock, tenderloin, baby back ribs, whole head, ears, and of course, kidney and heart (which can now surgically function to replace our own).

North Americans tend to call hogs by their beloved afterlife names—bacon, or bakkon ("back meat" in Proto-Germanic). In 2020, during the beginning of the COVID-19 pandemic, several disturbing reports came out about big Iowa farms mass suffocating their herds overnight due to lack of workers. Hundreds of thousands of pigs, just dead.

No future. Not even as meat. Imagine the dump trucks piled to the brim with dead soft fat. "Did you know that mother pigs sing to their children?" asks Derrick Jensen in *The Myth of Human Supremacy*. "And pigs dream. And pigs have a good sense of direction, and can find their way home from great distances. They learn from watching each other."[14]

Their lifeless renderings get sent to industrial wastewater treatment plants to be blended, salted, and processed through heated sludge lines. An operator at one of the largest wastewater plants in North America, who helps monitor those pipe flows, told me the details: "The pipes mainly contain primary sludge and thickened waste activated sludge . . . waste loads are high in biological and chemical oxygen demand . . . animal rendering, vegetable canning and processing, dairies, biodiesel plants, pharmaceutical productions, and area restaurant grease trucks."[15] They're basically all called volatiles.

It is very wild. Sewersheds overflow with corrosive and mummifying chemicals, oil, gas, flammables; not just pigs but several other dead animals in pieces like cows, birds, bunnies, baby goats, fish, whatever else you put in fancy cat food; grain, corn and soy plant remains and their fertilizer-laden leaves, roots, stems; a gazillion kinds of opioids and drugs; vats of grease that fried all those frozen curly fries and popcorn chicken for days and days and days; not to mention the city's ordinary garbage, every morning post-coffee plop with intestinal microbiota intact, along with the excrement and prayers of a million neighbors and their teeming everything.

Night soil is an industry term for this surreal excreta.

And night soil grows things. Operators don't like that. In particular, night soil grows blueish crust-like scalings and crystals that build up in the treatment pipes, like a clogged artery. This strange glittery thickness disrupts everything. Water doesn't flow, sewage backs up. Operators consider the fatty clogs a nuisance at best and at worst, a very expensive problem.

I think about all the dreaming pigs: how their fat is a special coagulation too, how mammal fat stores medicine of plants and energy and flavor from whatever they eat, "in season."[16]

When I received a chunk of the crystallized night soil in the mail, it came loosely wrapped in a pile of wastewater industry magazines with a note. Instead of some gnarly gross crap, I pulled out what I could only describe as the profound bakkon of a mythical dragon, scalings of an esteemed gigantic snake goddess, a wild sacred stone message. (See top image, page 134).

Wastewater vivianite contains three basic elements: iron, oxygen, and, most critically, *phosphorus.* In ancient burials and butchery pits, vivianite taints all kinds of unearthed bones and skin (of 5,000-year-old mummies frozen in ice,[17] or soldiers buried in jungle mud). Why? Bones and teeth are where bodies store phosphorus. The other big phosphorus supply (15 percent in humans) is in soft tissue and blood (i.e., skin and arteries). I remember being particularly moved when discovering there are nodules of vivianite found inside brains of baby mammoths who drowned in mud. They entered what's known as "diving mode" (where you hold your breath for a long time) right before dying, which causes

surges of oxygen and blood to the brain, creating perfect conditions for mineralization of blue ochre. Imagine little blue pearls of a last mental glimmer of light at death—fossilized crystals of drowning trauma. Even an ostrich who suddenly died was found with two swallowed, undigested, iron nails encrusted in blue vivianite.[18]

To me these end-of-life processes indicate that vivianite knows something about not just death, but *capturing rapture:* that light at the end (or beginning) of the tunnel. A solid blue place between beginning and end.

Phosphorus means light-bringer. So what is important and unique about phosphorus (element symbol P)? P forms the primary architecture of life as we know it, essential to cell membranes, DNA, and our skeletons. We also rely heavily on phosphorus for agricultural fertilizer because it encourages and stabilizes growth. So no phosphorus means no cells, no animals, no crop stalks, no agriculture, no food—you get the idea.

While vivianite is a critical carrier of phosphorus, most P that humans use is extracted from phosphate rock, geologic places controlled by a select few countries, currently overmined or near depletion. Consider it akin to peak oil (a critical point at which the rate of human extraction outpaces the available material reserve, indicating decline and/or a limited, threatened Earth resource). Phosphate rock gets crushed, separated into smaller particles, and used in farming, predominately on cereal crops that sustain populations of people and their domestic animals (like pigs, cows, chickens). When we eat plants (or animals who eat plants), we absorb the plant or animal body's phosphorus, digest what we need, and excrete the rest as "waste." In reality, a large amount of excess phosphorus passes straight through us, probably by natural design. Thus human waste is another important deposit of phosphorus. Ideally, wherever we poop, land could and will receive P back as nourishment. That's why manure remains so valuable as a fertilizer. Yet what do people in developed countries do? We flush it down the toilet. In our "modern" civilization, what's essential to earth and field and body mostly gets flushed down the drain, through the wastewater systems, and into the larger ocean dumping ground, "reshaping a loop into a one-way pipe," as Julia Rosen succinctly says.[19]

So that's a clue. Phosphorus is used as fertilizer and, often instead of recirculating via excrement, gets flushed into the ocean. There is so much excess phosphorus in our agriculture systems it now overflows into water supplies like groundwater, oceans, lakes, and ponds, where it acts as an *aquatic* fertilizer, or bad steroid. Algae gobbles up the unexpected (and unnecessary!) phosphorus feast, blooms and blossoms into large anoxic dead zones. So P's ability to grow life is very active, but at the same time, these algae begin to devour water's oxygen (and from a fish's perspective, darkens the sky, slowly, like an endless storm), and, in effect, *suffocates ecosystems into collapse.* The risk for humans is enormous, as anoxic dead zones not only deplete fish, they majorly mess up drinking water sources, like reservoirs. Phosphorus cycles are places where life makes or breaks itself.

Antonin Artaud, a transgressive avant-garde artist, seems to have foreseen this whole problem quite clearly, associating sewage systems with initiation temples in his controversial book, *Heliogabalus,* published in 1934:

> The blood from the sacrifices up above cannot be lost down the usual sewers; *it must not* [emphasis added]—mingled with the usual human evacuations: urine, sweat, sperm, spittle, or excrement—*find its way back to the primitive waters of the sea* [emphasis added]. And so beneath the temple . . . there is a system of special sewers, wherein human blood rejoins the plasma of certain animals. Through these sewers, coiling into broiling corkscrews whose circles diminish the further they descend to the depths of the earth, the blood of those sacrificed according to the needful rites will find its way back to the sacred recesses of the earth, reaching toward the primitive geological seams, the congealed cracks of chaos.

Maybe he was high on vivianite when he wrote this. Or channeling her soul? When I read this, it feels like a description of vivianite's initiatory and descending

power. With vivianite, life-force can "find its way back to the sacred recesses of the earth." Vivianite is one of very few minerals that is formed by "holding onto" phosphorus. This act allows phosphorus to rest, and to stop overflowing into oceans and lakes.

In fact, scientists—falling just short of using the words *miraculous* or *mystical*—now recognize that vivianite is perhaps the *most important burial sink* for phosphorus in wastewater and coastal systems worldwide, that has been largely, and detrimentally, ignored.[20] Burial sinks are places where elements circulate, become sediment, settle down, rest. In the case of vivianite, the iron helps "bury" phosphorus.

Europe's leading wastewater research team, Wetsus, based in the Netherlands, currently develops crucial technology to honor and utilize vivianite's burial magic. They've developed regenerative, circular technology to rescue vivianite from wastewater sewage sludge[21] (before it clogs up all the pipes). In sum, my friends gather and swoop vivianite up from the wastewater.[22] They then separate and grind her blue gore into ochre powder, aka fertilizer. Vivianite gets painted on fields, crops grow again, and are then fed to animals, and in some way, the long-lost spirits and "blood from the sacrifices" is not "lost down the usual sewers" and instead finds a way home, to perhaps still dream, sing, and life-bring.

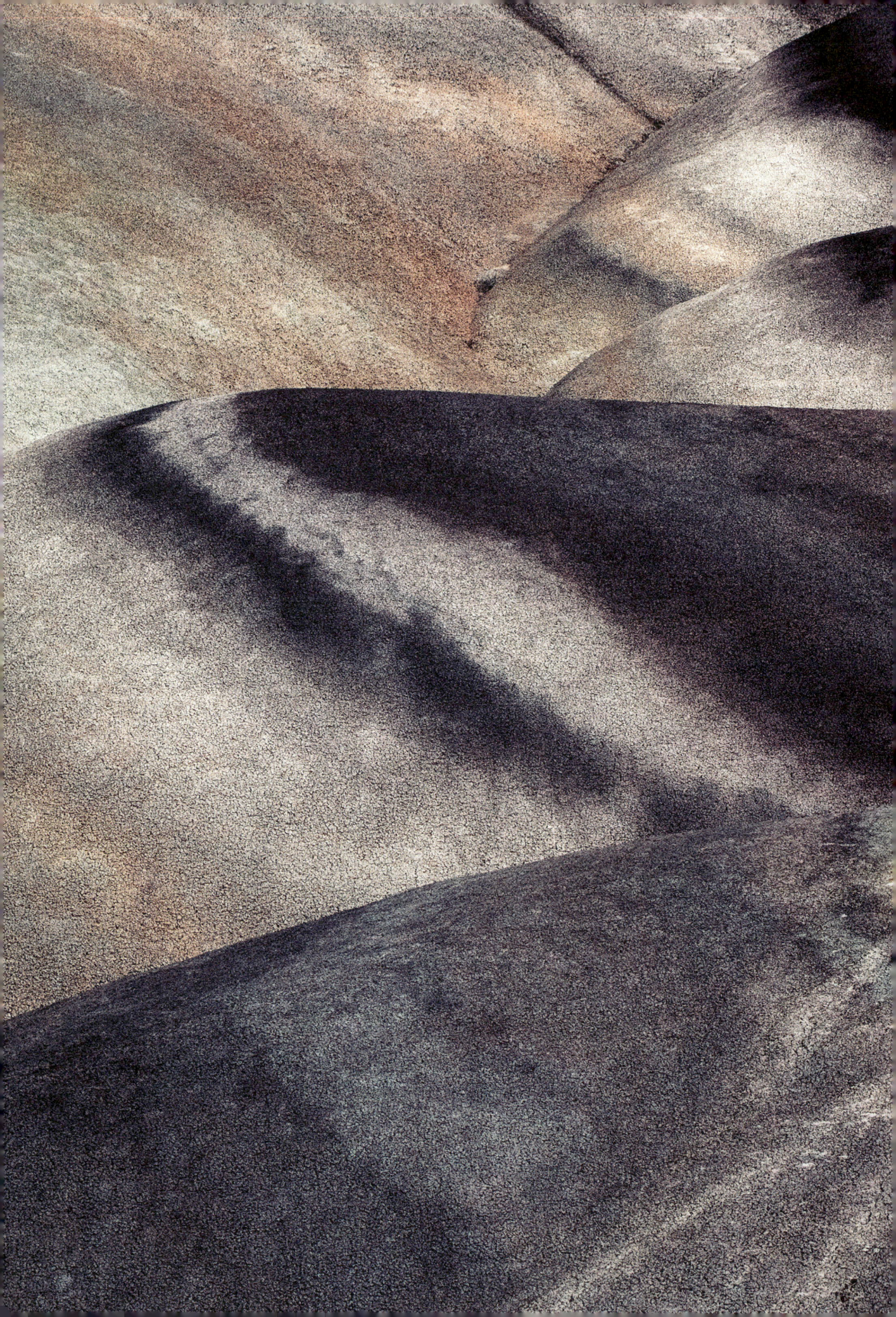

BLACK OCHRE

MAGNETITE

MAGNETIC IRON OXIDE

Etym. "way-stone" from *lode*, course, carry, burden, and *stone*, to stiffen, harden, "*burden stone*"

Fe_3O_4, $Fe^{2+}Fe^{3+}_2O_4$

PIGMENT SPECTRUM

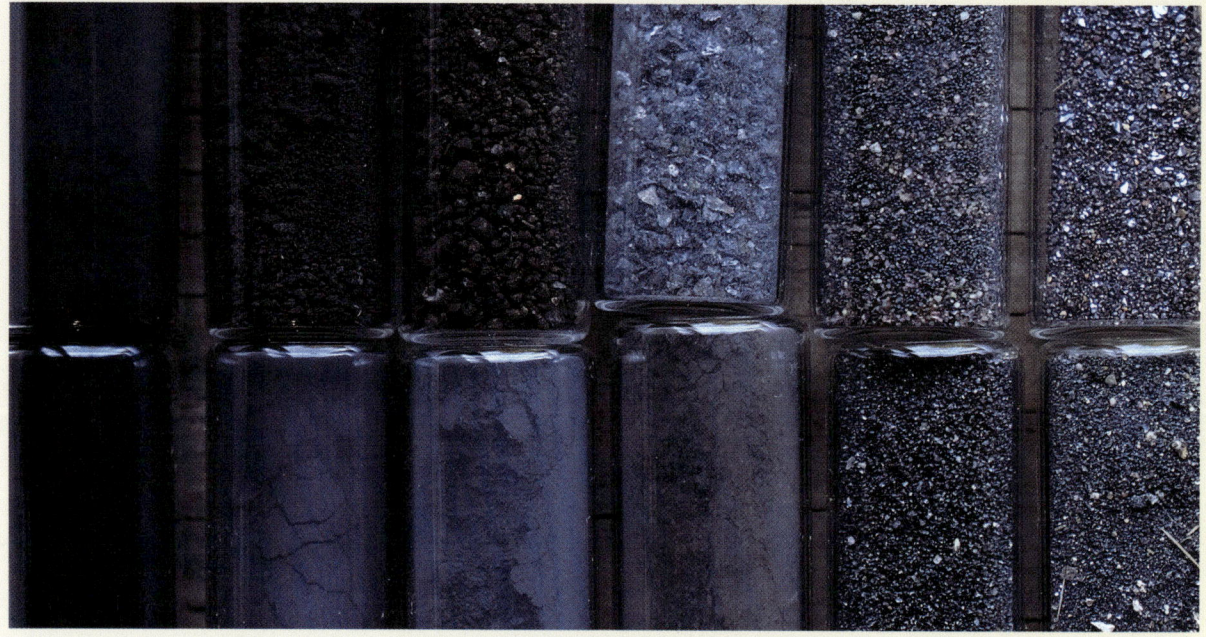

MATERIAL DESCRIPTION

M AGNETITE IS EASY TO IDENTIFY due to its magnetic properties, heavy downward weight, and blackish color, but it may prove difficult to find when simply walking around. Not to be confused with lightweight, carbon-rich blacks from organic matter and pervasive dusty black of any burnt forest, fire, or hearth (charcoal, coal, soot), magnetite is associated most often with cracks in the deep ocean floor, where the spreading seafloor releases iron particles from surrounding bedrock. Black sand radiates in stripes, as if a binary code storing the memory of magnetic true north. These attractive ancient seafloor zones are important ley lines for migrational species, who contain magnetite deposits within their bodies that function as an internal compass or orientation aid. Alongside hydrothermal or volcanic zones, conditions that form various magnetic iron-ore deposits include banded iron formations, magnetite skarns, magmatic fuel deposits with radical names like "magnetite-rich crystal mush in the magma mingling zone." For the average human wanderer, detrital weathering deposits of iron sand or black sand on beaches are easiest to locate.

Present in interplanetary dust particles and meteorites, magnetite uniquely contains both ferrous (Fe^{2+}) and ferric (Fe^{3+}) ions, implying yes, polarity, but more importantly synthesis and stability within an environment where oxidized and reduced states of iron are present and maintained.

NAMES

lodestone, way-stone, leading stone, compass ore, magnetic iron ore, ferrous ferrite, polestone, Hercules stone, iron tetroxide, black iron, brain iron, hammer's blow, black sand, iron sand, magnetic sand, ci shi (Chinese, "magnetite"), ferrofluid, Super Paramagnetic Iron Oxide Nanoparticle (SPION)

HABITAT

where mother ochre is — where iron is — near metal — seafloor spreading zones — beaches — brains — hearts — heads — spines — freaky places — Bermuda triangles — inside computers, credit cards, dollar bills, cell phones — old waves — ley lines — ways home

CHROMA

WAY BLACK	HARD SPARKLE
BLACKER THAN BLACK	HEAVY METAL
DEAD BLACK	THE WHOLE
DARK MIRROR	NIGHT SKY?
MAGNETIC NAUGHT	
WAVE LINES	
BETWEEN LEY LINES	

CRYSTALLOGRAPHIC SYSTEM

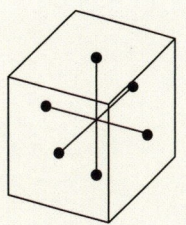

CUBIC

SECONDARY BLACK PIGMENTS

Iron Meteorites
(FeNi +/−), "sky iron"

Manganese Dioxide
(MnO_2) or "wads"

Carboniferous Matter
(charcoal, soot, coal, hydrocarbons, biochar, blackened bones)

BLACK OCHRE — Burden Stone

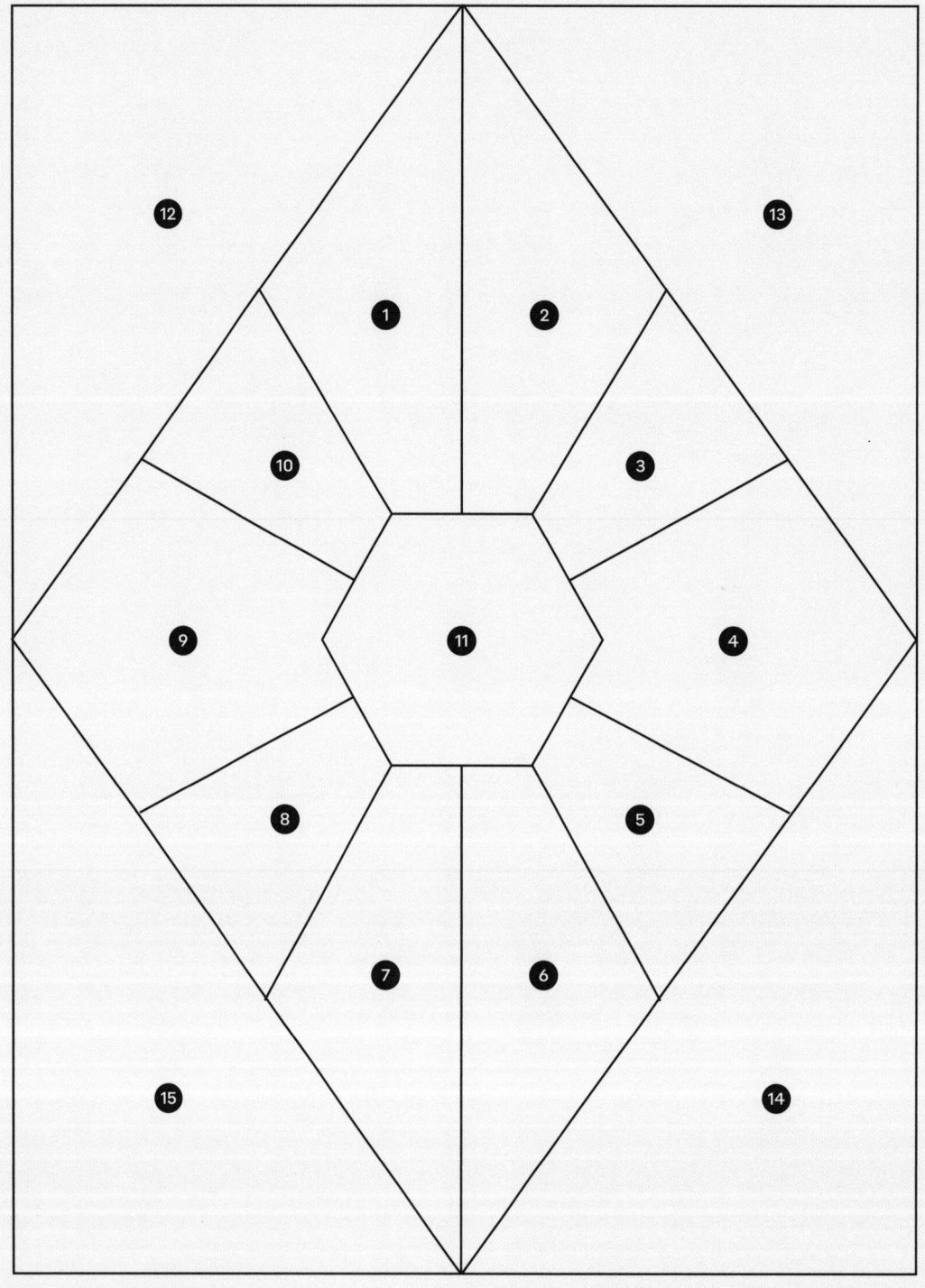

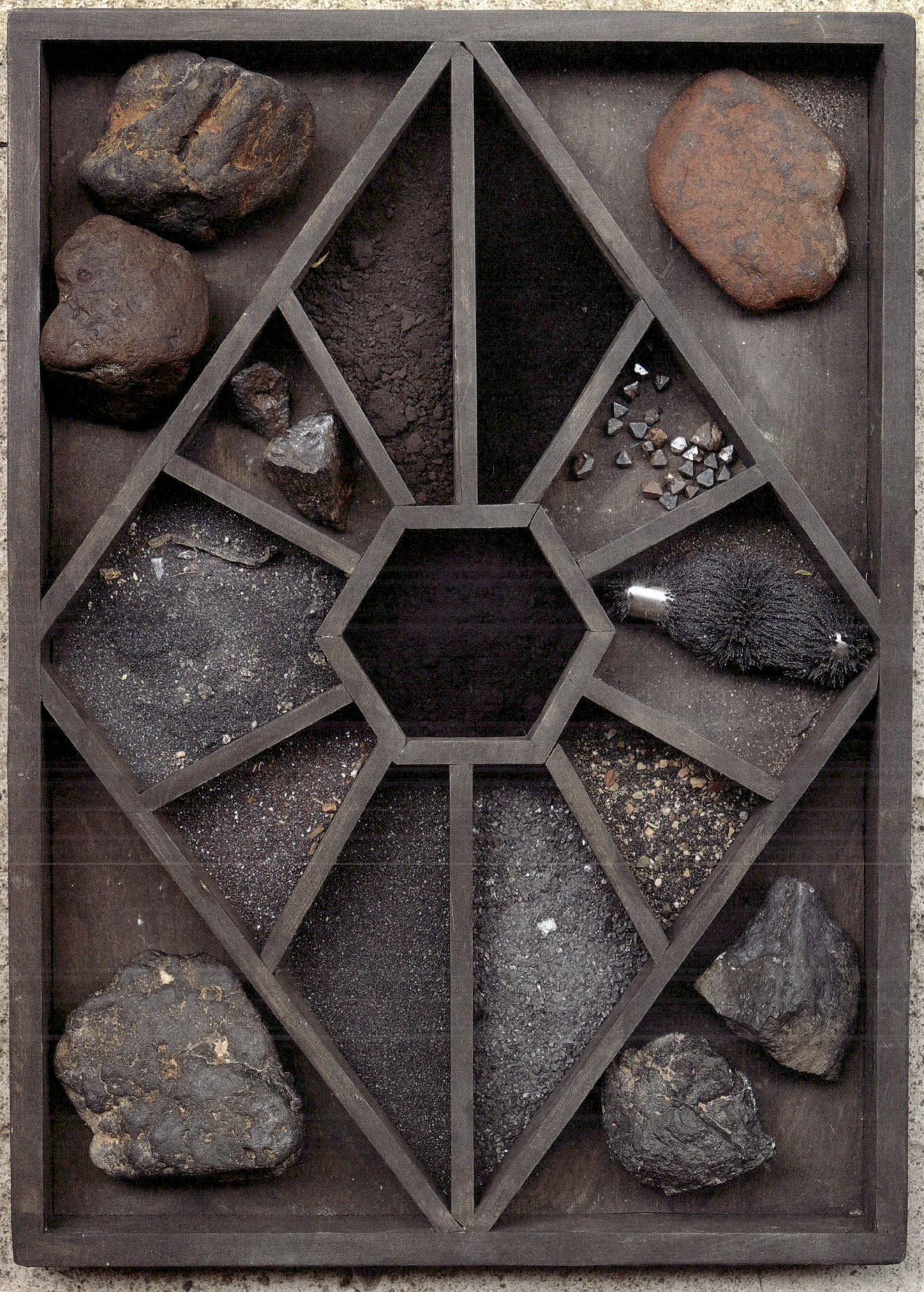

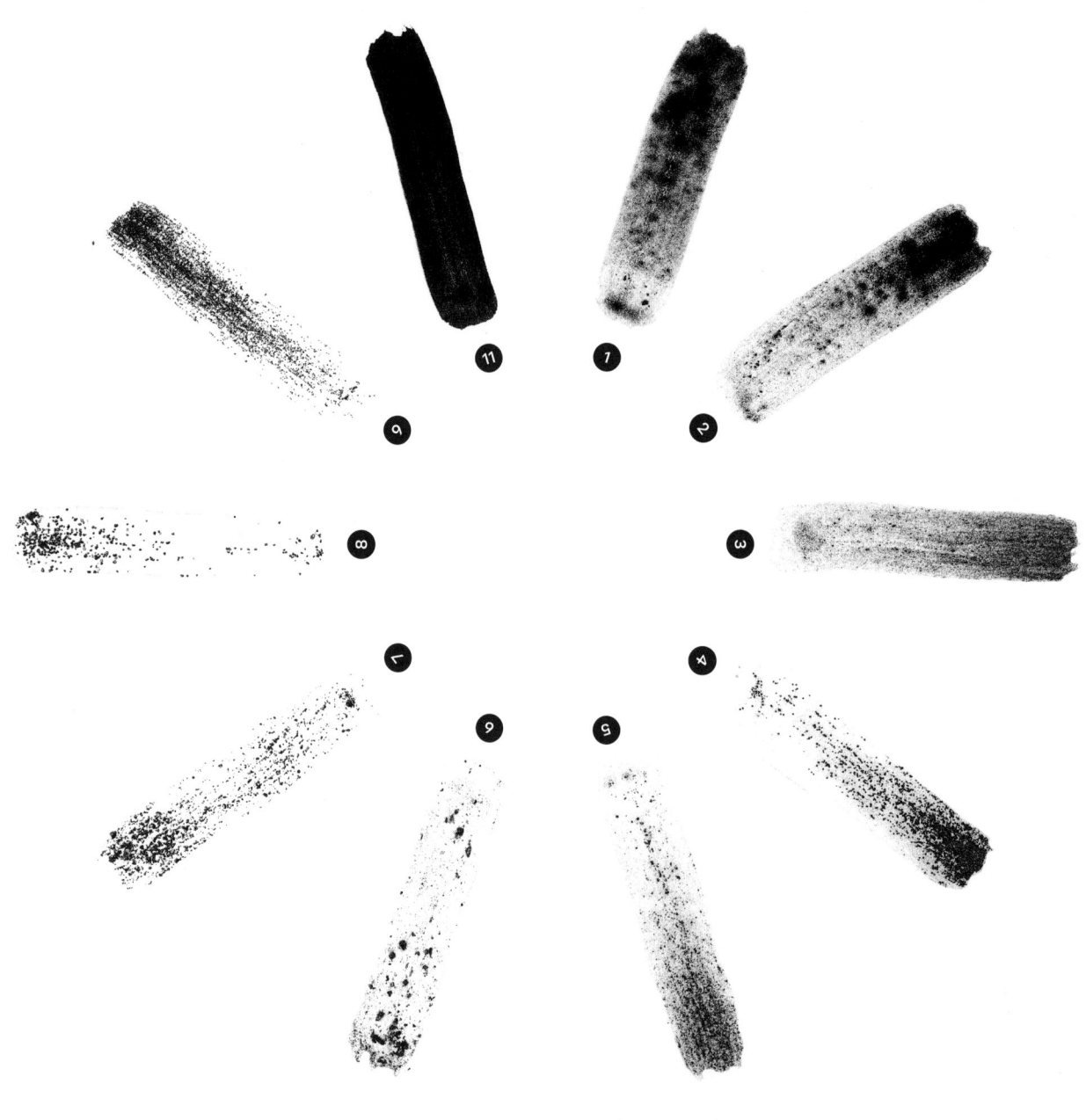

1. *Roman Black*. Undisclosed Italian origins, a manufactured blend of black iron oxide and other mineral earths.

2. *Dissolved Gun Lode*. Smells of far, far away and made from a dead and decomposed gun. Alchemized by the great Thomas Little using acidic wisdom.

3. *Heavy Metal Diamonds*. Technically called "octahedral magnetite crystals." I found these in a little bag on the attic shelves of a bookstore in Bellingham that was going out of business.

4. *Beach Sand*. I gathered it with a cow magnet on my grandparents' beach I grew up going to, on Whidbey Island. I didn't realize there was black ochre on this beach until after they both died and their spirits showed me how to look.

5. *Too Dark to Name*.

6. *Smashed Magnetite*. Not from the Kursk magnetic anomaly, but not that far away from there.

7. *Ironsands*. California.

8. *Black Sands*. Hormuz Island, Iran. If you dig your feet into the black sand, the red clay underneath comes up and turns the ocean into blood.

9. *Silver Sands*. Strait of Hormuz.

*10. *No-Man's Land*. I have zero idea where this magnetite was born because I found them alone in a junk pile in some vague borderland rock shop.

11. *Super Paramagnetic Iron Oxide Nano Particles*. Very difficult to track down the geologic origins. Work in progress. I try very hard not to breathe these because they could pass through my blood-brain barrier and accumulate.

*12. *Magnetite that Moves the Moon*. From Iron Mountain.

*13. *Faultline Lodestone*. Magnetite particles surrounded by hematite, exposed during floods in a mountain creek where I live.

*14. *Lodestone*. Ask the ferr-fairies.

*15. *Compass Sharpener*. From, actually, the middle of nowhere.

Swatch not pictured, see raw material/ color on page 143

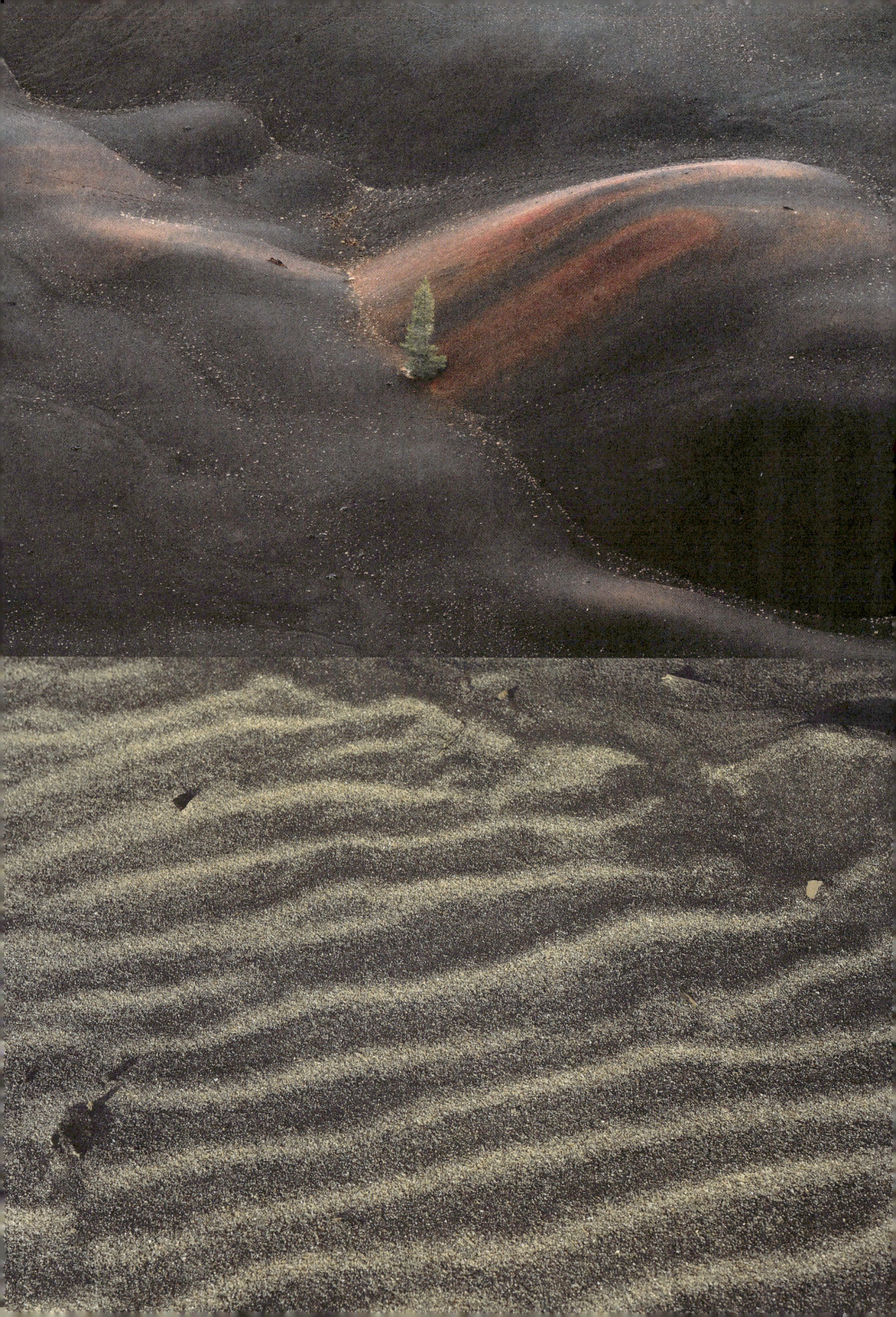

I don't remember her name, or what compelled me to forget
So drenched that night we all were from tough knowledge
Spilling out across the dark earth
In this vulnerable pulsing mother field.

—JOY HARJO, "MOTHER FIELD"

✳

Planets and stars, like glaciers and rivers and forests, are mortal.

—ROBERT BRINGHURST, "THE EDGE"

Lodestone

INDUSTRIAL IRON GIVES OFF AN evocative, intense scent. *"A curious black smell, the far, far distance, ash, rust, sweat . . . time buried in the black earth,"* as Russian writer Elena Dolgopyat describes it.[1] In her home country, the Kursk Magnetic Anomaly encompasses one of the largest iron ore deposits on Earth.[2] I've heard people estimate it accounts for about 50 percent of all of Earth's natal iron ore reserves. As a "superior iron formation" nobody seems to agree on how it formed or what formed it. People do agree it's likely cryptozoic, aka sometime between Earth's nebulous birth and the first formed seashell. The ore body contains several ochres, chiefly deep black magnetite or lodestone—way stone or burden stone.

I envision Kursk as a swarm of magnetic energies, located in the top of Earth's crustal skull. A soft spot disguised in a hard place. *Time buried in the black earth.* What's a magnetic anomaly? A strong local variation and amplification in Earth's magnetic field. When you get close to them, excess magnetism makes compasses dizzy and sometimes confuse north for south. Wayfaring's near impossible. Magnetic anomalies are vital energy spots of Earth's overall magnetic field. According to the *Eos* science news magazine article "The Herky Jerky Weirdness of Earth's Magnetic Field" author Jessica Duncombe, "The magnetic field protects Earth's atmosphere from harmful radiation emitted from the Sun."[3] Elsewhere in recent magnetic field news, reported by MIT, I read, "Not only does a magnetic field set the direction of our compass needles, it also acts as a shield of sorts, deflecting away solar wind that might otherwise eat away at the atmosphere."[4]

Without a regulated, magnificent shield, our planet becomes impotent, lifeless. According to the MIT scientists, our safe haven is directly powered "by the solidification of the planet's liquid iron core. The cooling and crystallization of the core stirs up the surrounding liquid iron, creating powerful electric currents that generate a magnetic field stretching far out into space."

Earth's core story lives in individual pieces of black iron ochre. "Iron-bearing minerals align to ambient magnetic field—they orient us to the current pulse/heartbeat of planet," notes genius geologist Marcia Bjornerud.[5] Magnetite knows deep time memory, acts as a lithic stethoscope listening by way of grains of black sand. In my land experience, lodestones are also unusually heavy to hold, as if always leading inward and downward, earthward.

Tiny magnetite deposits are found inside the bodies of diverse creatures: even dragons were rumored to have such "noble stones" in their brains that contained "marvelous properties."[6] Within wild rock dove's beaks, for example, magnetite particles are quite dense—the homing part of a homing pigeon. Each animal with an inner compass, a geomagnetic hotspot. A pigment deposit. Freshwater eels use their magnetic lateral lines to travel by sea at the end of their life, returning to their birth place in the Sargasso Sea, also known as the Bermuda Triangle—a well-fabled magnetic anomaly on its own terms—to breed for the first and last time.[7] When human scientists remove these sensitive organs—abdomens, eyes, noses, brains, lateral lines, nerves, cells, teeth, beaks—from animals, creatures tend to permanently lose their sense of home, direction, place: no longer in magnetic touch with Earth's heartbeat.

And, yes, crystalline ochre mines exist inside of you and me. All in our heads. Specifically concentrated deep in our brain stems, with a dose in the frontal lobe, or accumulated over time from breathing urban air. My own internal magnetic anomaly? What do I know about the iron mines in our brains?

First, iron flow and regulation are key to an active, learning human brain. Too little iron (Fe), and my neurotransmitters stop functioning. As major studies show, "Low Fe levels are typical of individuals suffering from severe depression and Fe supplementation improves cognitive functioning."[8] And several reviews[9] demonstrate that links between antisocial behavior and psychological instability and mineral deficiencies, especially iron, are unequivocal.

Iron aligns us to a social, even psychogenic, field.

Second, when *too much* iron accumulates, everything goes haywire. Magnetite uniquely corresponds with elevated iron deposits in the brain and pathological processes related to Alzheimer's, Parkinson's, multiple-system atrophy, and other brain-related dysfunctions.[10] The casual medical nickname for pathological magnetite concentrations is "senile plaques." In *We've Had a Hundred Years of Psychotherapy and the World's Getting Worse*, depth psychologist James Hillman puts it bluntly: "I've got rocks in my head. Just as you have physical scars, so

you have soul blemishes. And they're rocks. And they are what you are."

Let's imagine for a moment that "I've got rocks in my head" is a mythic definition. A truth that speaks to magnetite's affinity with accumulation, mental places, scarred realms, heavy stuff, and dysfunction—those places where we carry our burdens. Interestingly enough, in scientific medicine research, magnetite is an important material to assess and address body and brain problems. How? There are several ways, but each begins by smashing the dark wayward powder into very very fine particles, dispersing them in liquid, and calling them superparamagentic iron oxide nanoparticles (SPIONS). In general, people use magnetite to visualize inside our bodies, catalyze healing processes, and store memories we can't ordinarily retain. For example:

1. Black iron paint is injected into our bodies as "contrast agents" to make more precise MRIs, magnetic resonance images, of our bodies when we are hurt or slowly dying.

2. Black-ochre-impregnated particles go in our bodies as "delivering agents" carrying drugs across the blood-brain barrier to make us suffer less. Magnetite pigment is able to bind with drugs and then be released by altering a magnetic current.

3. Black ochre ink gets injected into our bodies and acts as an "electro-magnetic responder" that heats up on magnetic command and can locally destroy cancerous tumor cells. Magnetite pigment can be directed to a tumor spot using magnets and then using electric current, then pulsed to overheat the cells of a specific area.

4. Black ochre pigment bound with petroleum-derived binders creates magnetic tape. That's what allows for recording of sound, image, and binary data, and what creates a mechanized brain—memory extended outside of a human body—and also makes credit cards (those magnetic printed strips on the back) that remember who you are and where you go and what you do. And of course there's just plain old money and every single dollar that is printed with magnetic black ink.

vNow let's switch to a larger scale. In Western civilizations, we also use mined magnetic pigment to act as a *literal* catalyst: a catalyst that enacts a key part of synthetic hydrocarbon production.[11] Instead of magnetite nanoparticles, people use larger magnetite particles and bigger amounts of it, thrown into an enormous cauldron that spits out new, synthetic earth material—hydrocarbons. Hydrocarbons, of course, are the core material of plastic and oil and petroculture—the single largest influencer of climate chaos.

I find this fascinating and disturbing. We drill holes in Earth to make materials that greedily expand ourselves. At the same time, we're spewing tons of climate-altering materials into soil, air, and water—and the rocks in our head have catalyzed the sixth mass extinction. In the past, the climate-change-driven mass extinctions came from big volcanic eruptions spewing carbon into the air. Today, what starts by drilling holes in Earth's magnetic power spot ends with a kind of sprawling criminal blood splatter that goes *everywhere.*

I think of mass excavations in Kursk's Anomaly—irreparable punctures in Earth's skull. I think of the obscenely rich iron mining guys, wealthy internet gurus, deep-pocketed hydrocarbon billionaires. Worlds destroyed with money. I think of aging people suffering from slow brain disease. I think of what geologist Marcia Bjornerud said: "The psychological effects on humans left in the shadow of the decapitated mountains is beyond quantification . . . we've become agents of massive derangement."[12] A flash comes across my mind's eye as I remember a sculpture of large stone heads, carved by the Monte Alto people of Mesoamerica 3,000 years ago that were incredibly, and oddly, magnetized in certain anatomical areas, significantly the temples of the head. I think of Samuel Beckett's *Endgame*: "There's something dripping in my head. (Pause.) A heart in my head."

A heart in my head. I recall how heartmind (心 "xin") is one word in ancient Chinese Five Color medicine. And ci shi ("magnetite") is medicine. Ci shi, literally "the stone that nourishes," is indicated for use homeopathically for "shaking of the head," dizziness, and "usually indicated for unsteadiness of heartmind, insomnia, and fright palpitations due to fear leading to qi (vital energy) turbulence and failure of mind to keep to its abode."

I realize all these associational reveries, like a compass spinning wildly in my mind, do not yet seem to locate a true north. Intuitive material notions and images do not always need to complete themselves, back themselves up, have a goal or discovery. My writing here perhaps reflects the wound of my culture: *failure of mind to keep its abode.* There *are* rocks in my head. There *is* a hole in Earth's heartmind. With every giant hole opened, and dust wildly spread around the planet, I wonder: Are we creating a new magnetic anomaly in the middle of exactly nowhere—a placelessness, a displacement, that radically disorients humankind's sense of where home even is and confuses our empathic compass?

How do we carry the burdens, the "lode"? With lodestone in the palm of my hand, I try to feel and track what comes into the back of my mind. How do I find my inner compass? I move the lodestone between my hands, back and forth, like a worry stone. Like a clock ticking. As if building up and charging my magnetic body. Back and forth. Back and forth. Swaying and swirling, my mind starts to hum. I smell that curious black scent of iron . . . *hot metal, the far, far distance, ash, rust, sweat.* My hand feels like a compass. My fingers tingle and wiggle as if trying to grasp and tug on an invisible ley line. I call my brilliant alchemist friend Thomas Little, who still knows magical recipes; he knows how to extract magnetite buried within money's paper and a gun's metal body. He makes dollars and guns disappear at the same time as he recovers bits of dust from Earth's broken-open head and transforms them into a powerful, attractive sediment or very rich black ink.

Lodestone calls us to seek new bearings, to bring our knowledge to bear. My friend's gestures of dissolution help me because they are beautifully despairing; they feel honest to a polar darkness. Can we wonder and ask together: how do we follow the compass pointed toward not the "true" north but to abode in all the directions? How do I walk the right way, aligned to the shifting magnetic field surrounding me, the rocks in my heartmind, with my hands doing the homing?

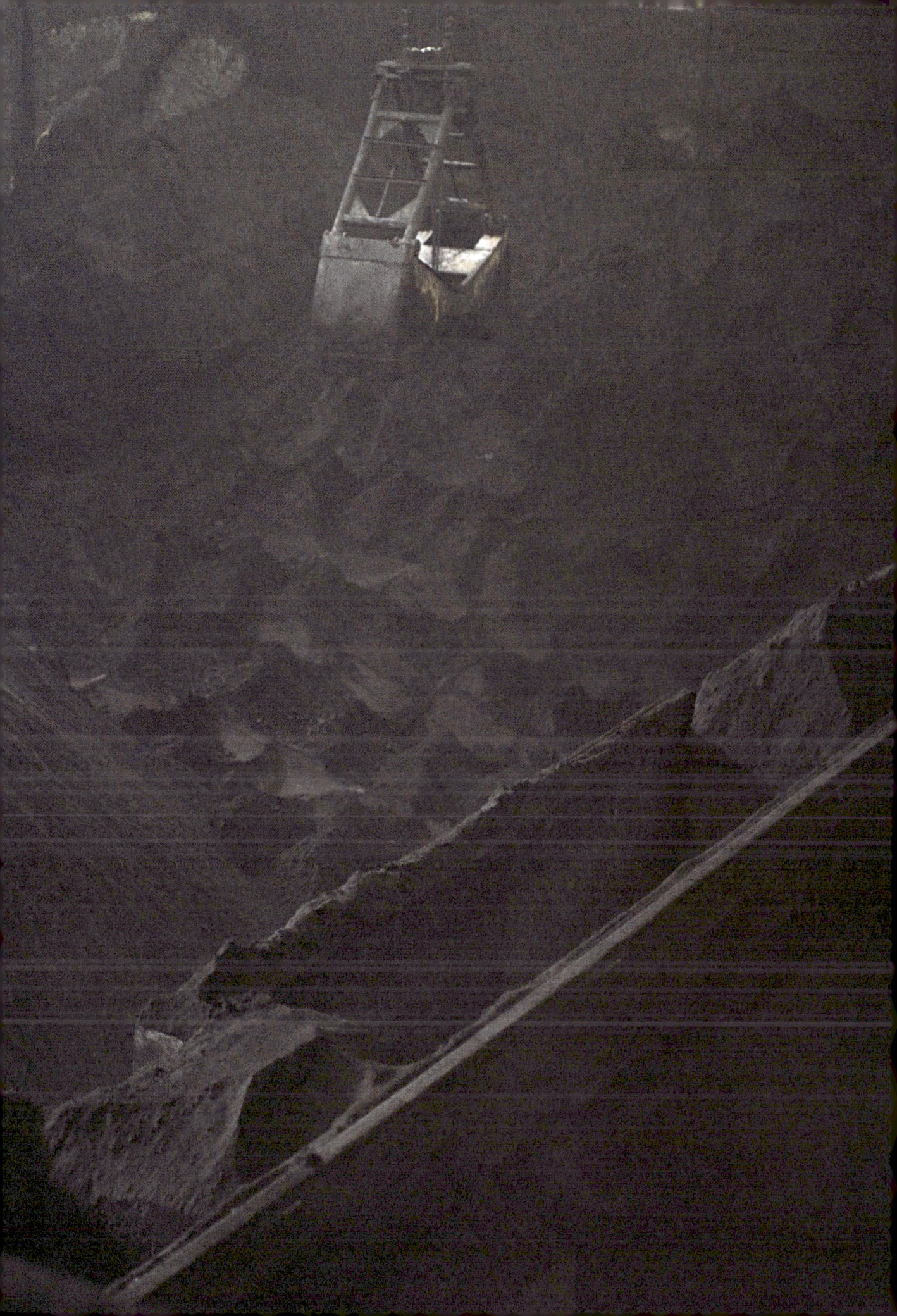

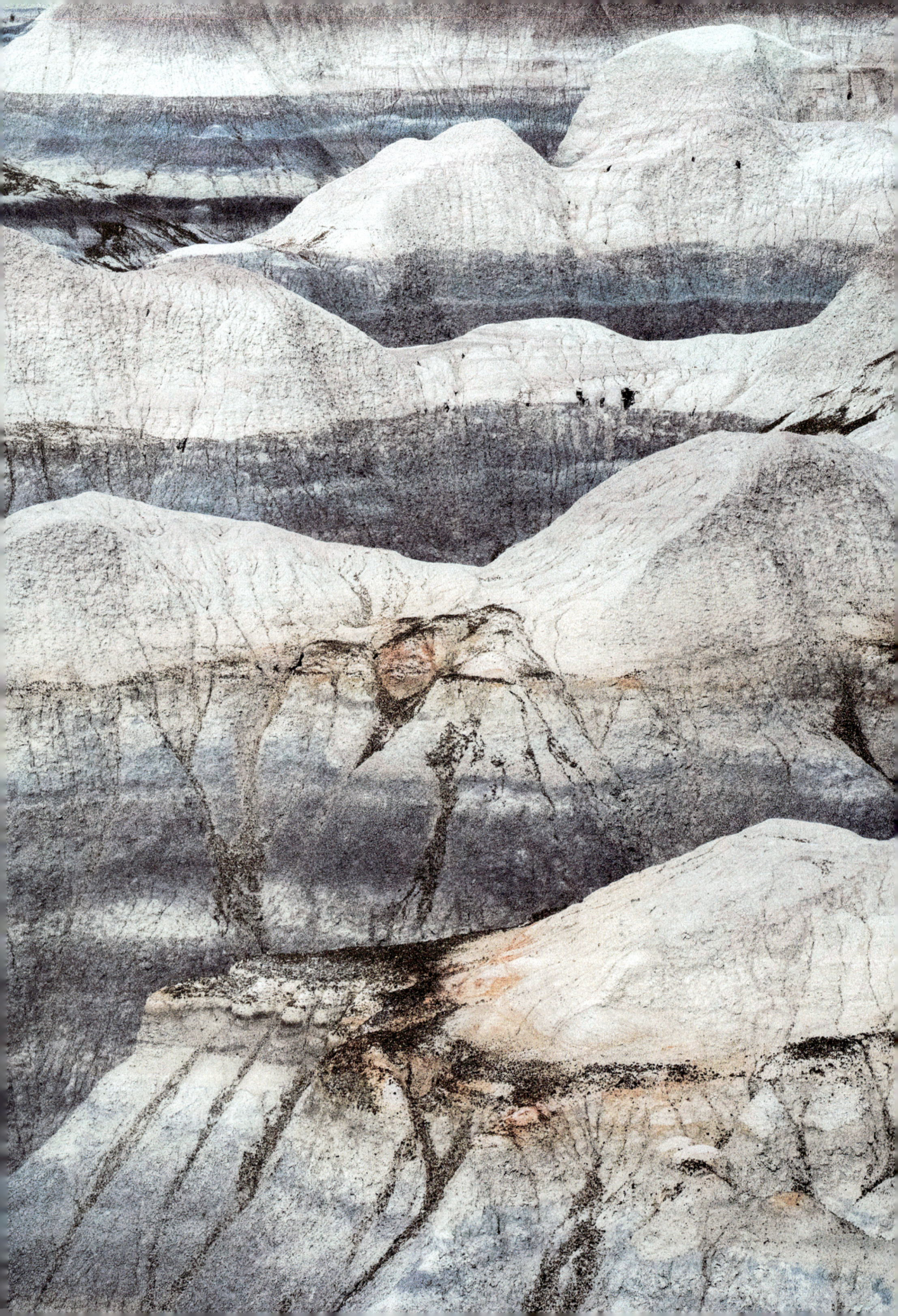

WHITE EARTHS

CHALK, LIMESTONE, SHELLS, BONES

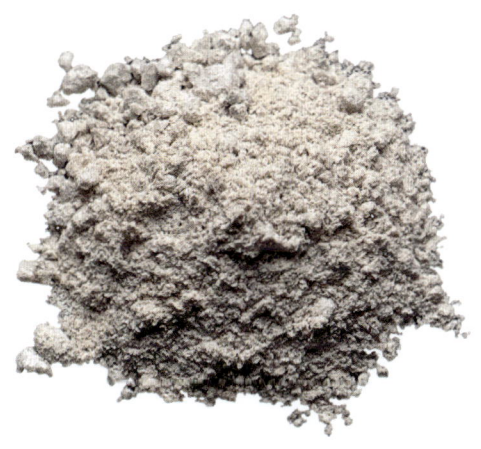

CALCIUM CARBONATES, CALCIUM HYDROXYAPATITE

Etym. from shell, *skaljo*, covering that splits off, a piece cut off and from bone,

ban, one of the parts which make up the skeleton, "one part split off"

$$CaCO_3, Ca_{10}(PO_4)_6(OH)_2$$

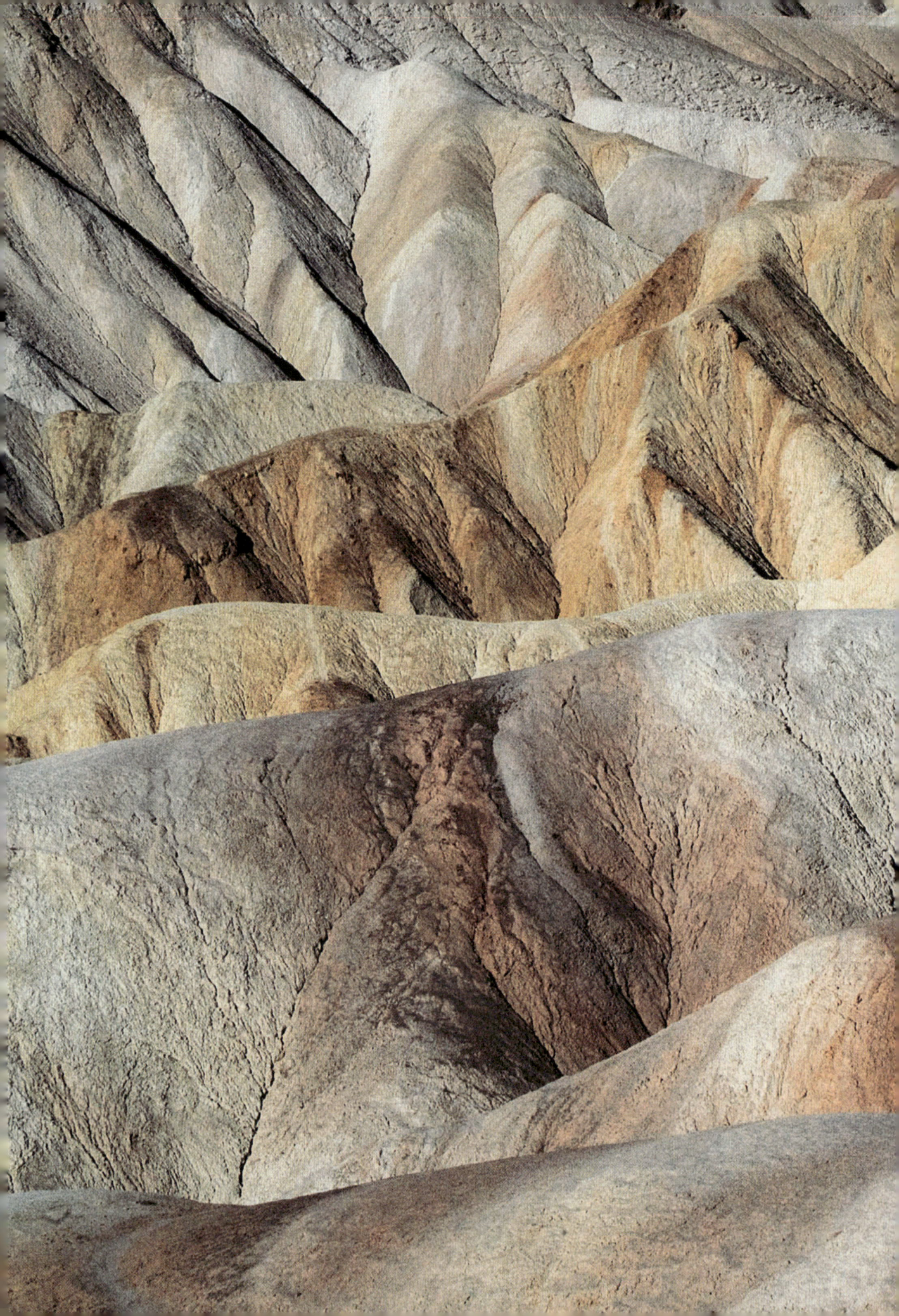

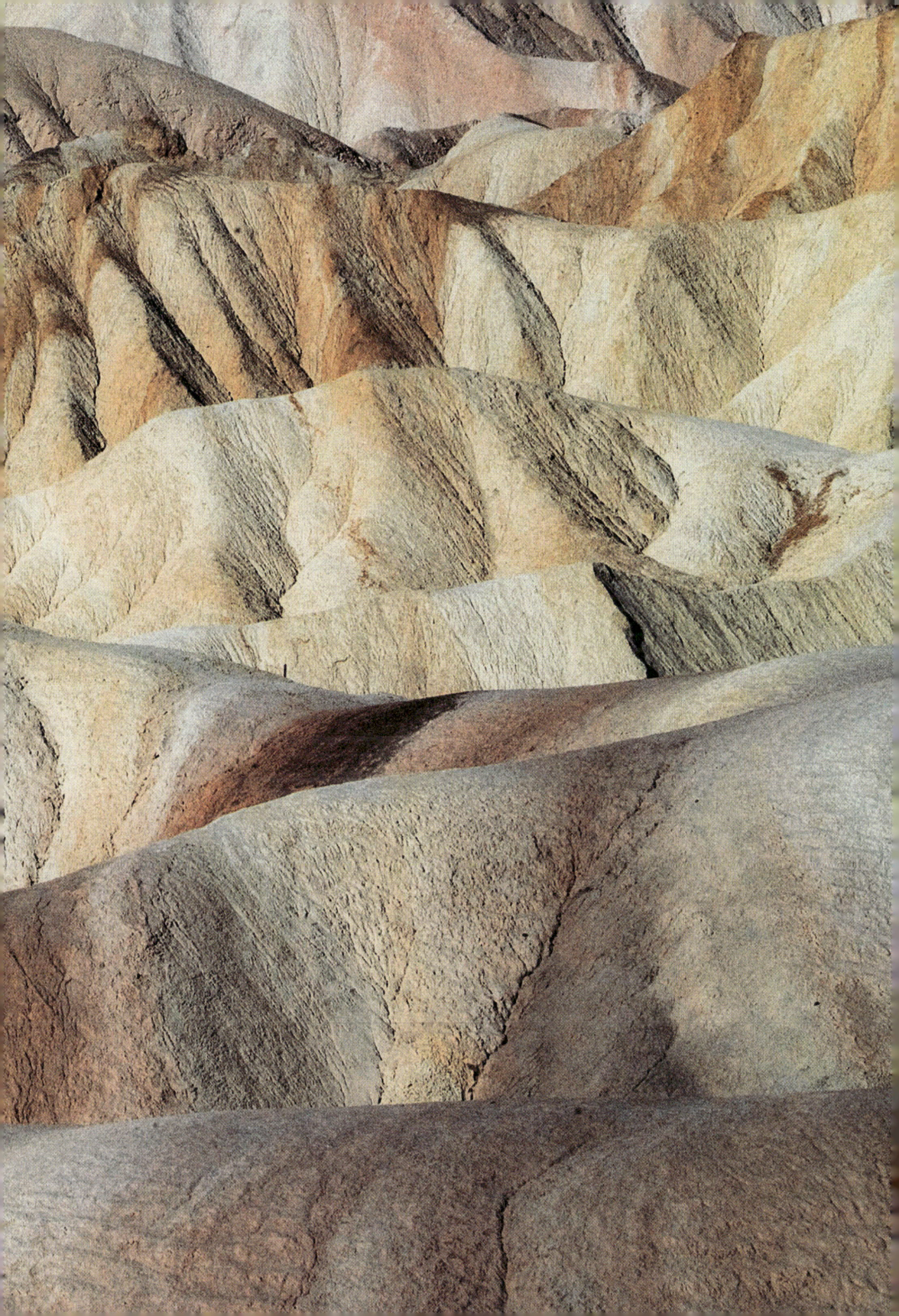

PIGMENT SPECTRUM

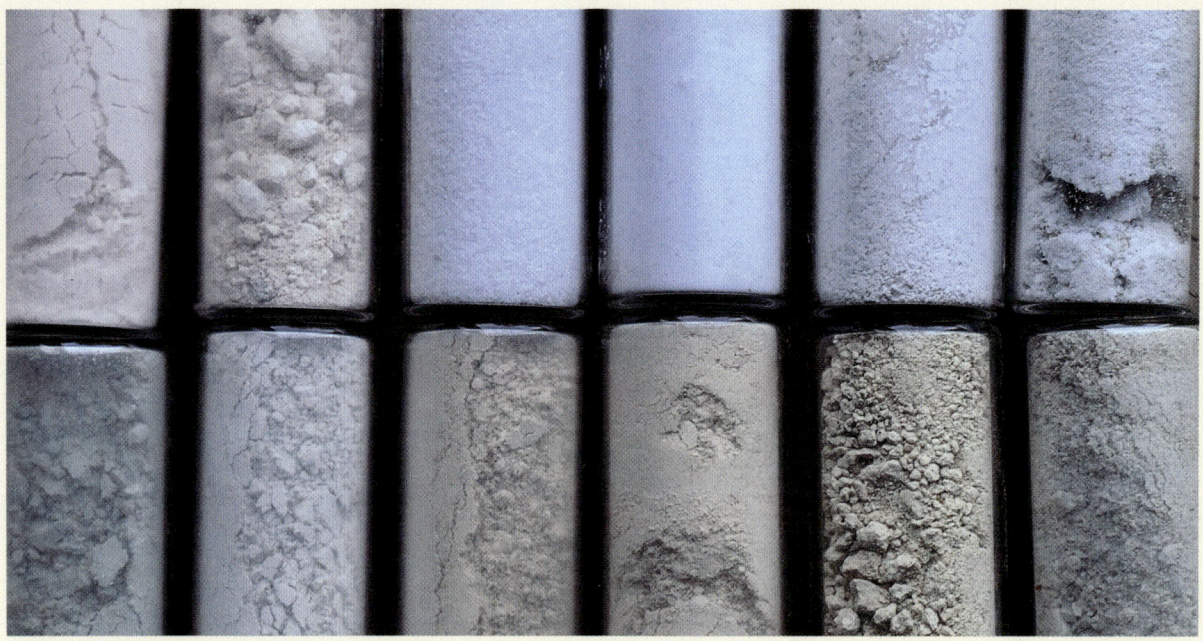

MATERIAL DESCRIPTION

EIGHT HUNDRED MILLION YEARS OF EVOLUTION in warm, shallow marine waters define white ochre pigment's skeletal material. Calcium carbonates are excreted from soft tissue of ancestral aquatic organisms to generate exoskeletons (i.e., shells). Their secretions were perhaps triggered as a protective effort against unpredictable ocean chemistry after major tectonic rifts and climate changes. Over time, several million creatures living and dying sink to the ocean floor as calcium carbonate. These skeletal remains compact into radical carbon sequestrations of marine fossils called limestone, marble, and other subsequently formed carbonate minerals.

Human inner skeletons (endoskeletons) evolved later, from a very similar cellular secretion process. Evolutionarily, bone anatomy is a kind of internalized shell. White earth minerals indicate a continuity between shell, bone, and rock.

Alongside shells, bones and their deep time communal forms as limestone, including diatomite (soft, biological, or fossilific limestone), marble (heated, pressurized limestone), and stalactites, stalagmites, and flowstone (evaporative cave limestones or moonmilk), there are other white earths and clays. These include non-limestone deposits such as volcanically derived kaolin clay, cave-loving gypsum (hydrous calcium sulfate), anhydrite ($CaSO_4$), and magnesium-rich huntite and dolomite.

Even though very little iron manifests in white earth pigments—thus technically they aren't ochres in my original definition—I consider them important ochre kin. And I want to pay homage to their traditional alchemical name, *flos ferri* ("flowers of iron"), describing coral-like calcium carbonate minerals that grow out of iron beds.

NAMES

white ochre, white clay, eggshell, shells, bone, coral, margaritae, pearl dust, pixie dust, rock flour, *flos ferri* ("flowers of iron"), moonmilk, limestone, chalk, creta, diatomite, fossilific limestone, marble, stalactites, stalagmites, flowstone, ground, lime, calx, calcium carbonate, kaolin, huntite, and what about owl pellets?

HABITAT

shallow marine water — tidal deposits — beaches — soils growing champagne and other fine wine vines — dead coral islands — heated shallow seas — beneath paintings — building dust — chunks of interior walls — below and around nests in a hatching time — graves — shell mounds — inside of you — around you — where life-forms are — where ancestors dwell

CHROMA

EGGSHELL	SHELL
SKY	MOON
BONE	
PEARL	
ASH	
DUST	

CRYSTALLOGRAPHIC SYSTEM

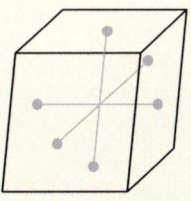

TRIGONAL

SECONDARY WHITE EARTHS

Huntite, $CaMg_3(CO_3)_4$

Kaolin, from Chinese, gāolǐng ("high hill"),
$Al_2Si_2O_5(OH)_4$

LOOK-ALIKES

Lead carbonate

Barite
(barium sulfate, $BaSO_4$)

Titanium oxide

Zinc oxide

WHITE OCHRE — One Part Split Off

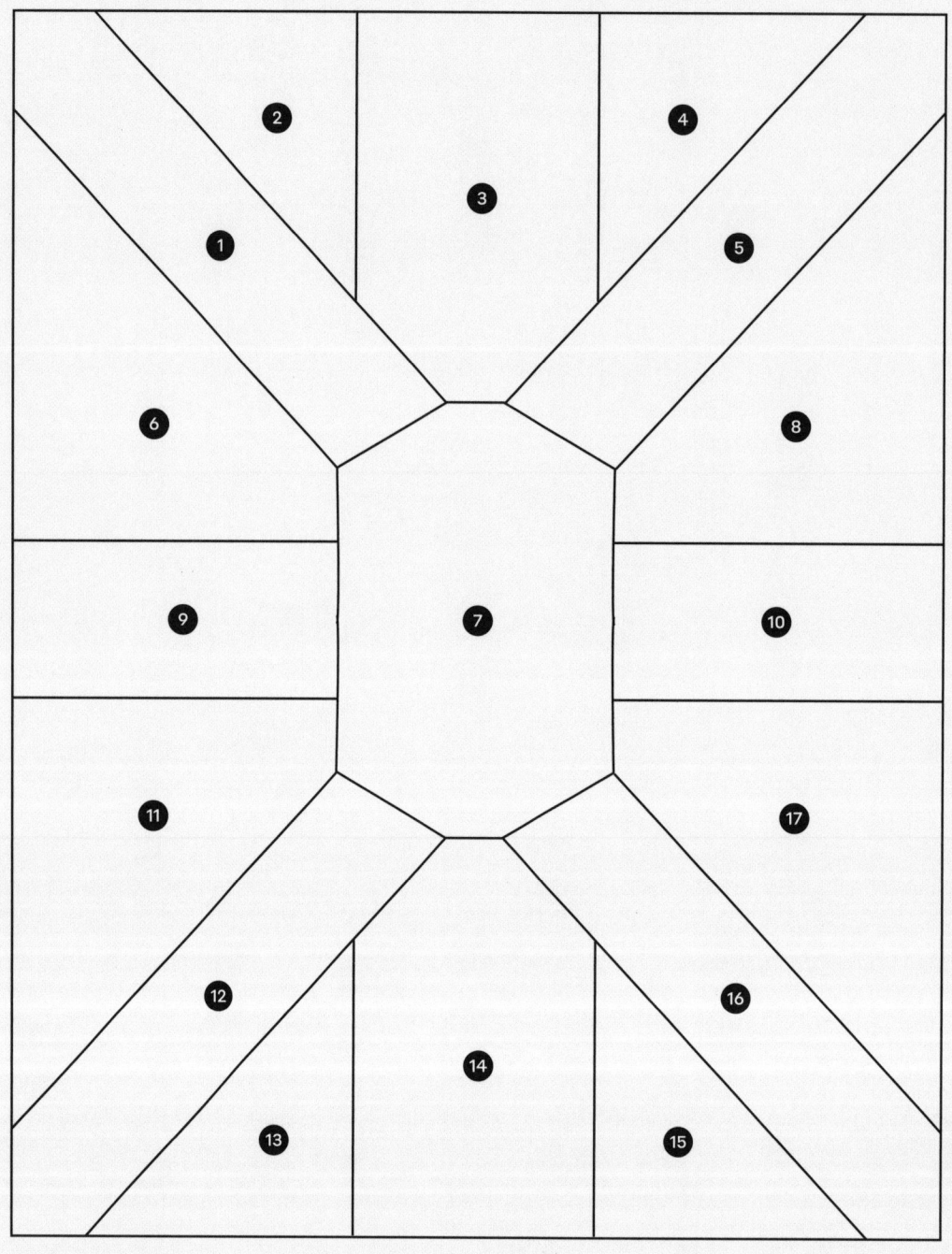

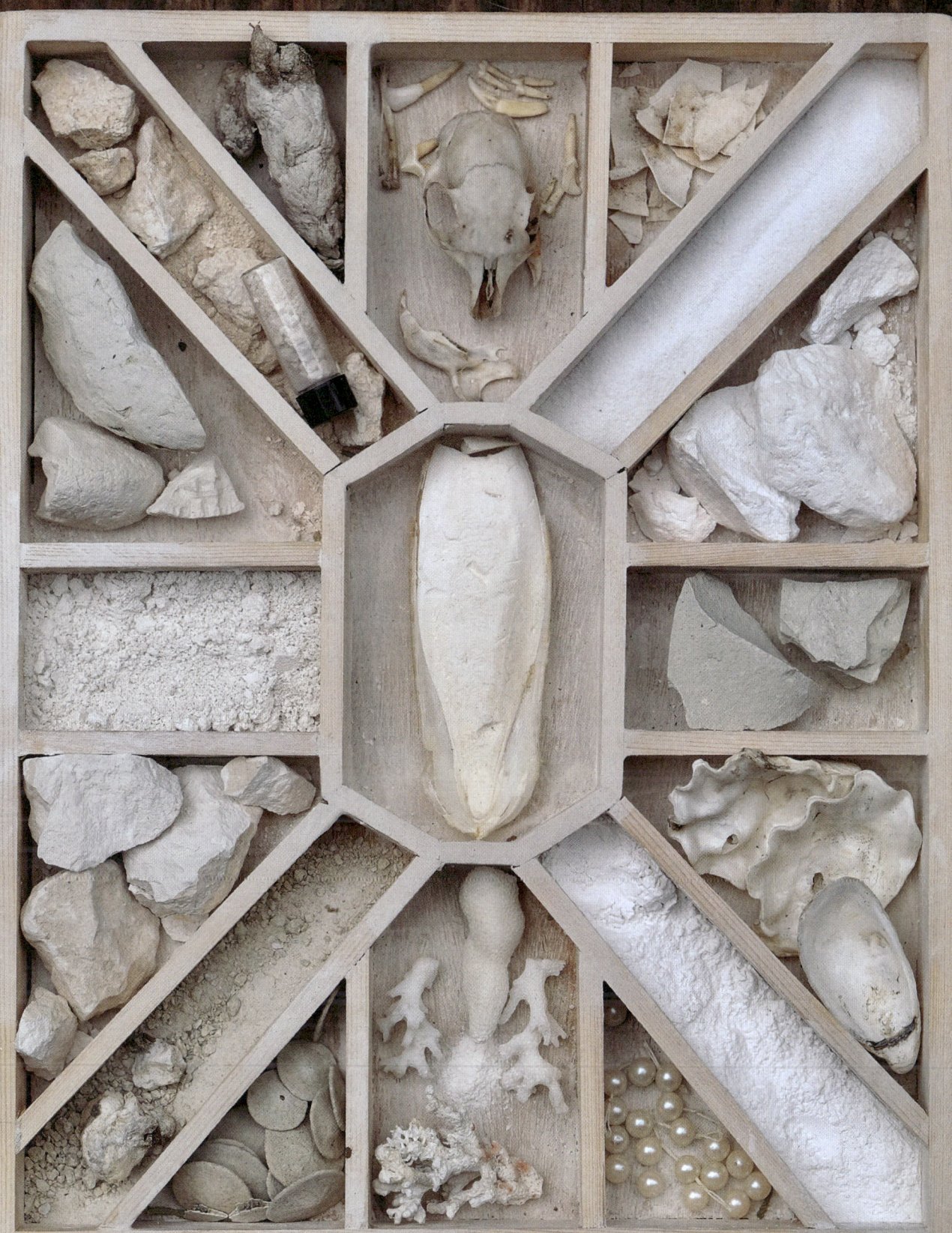

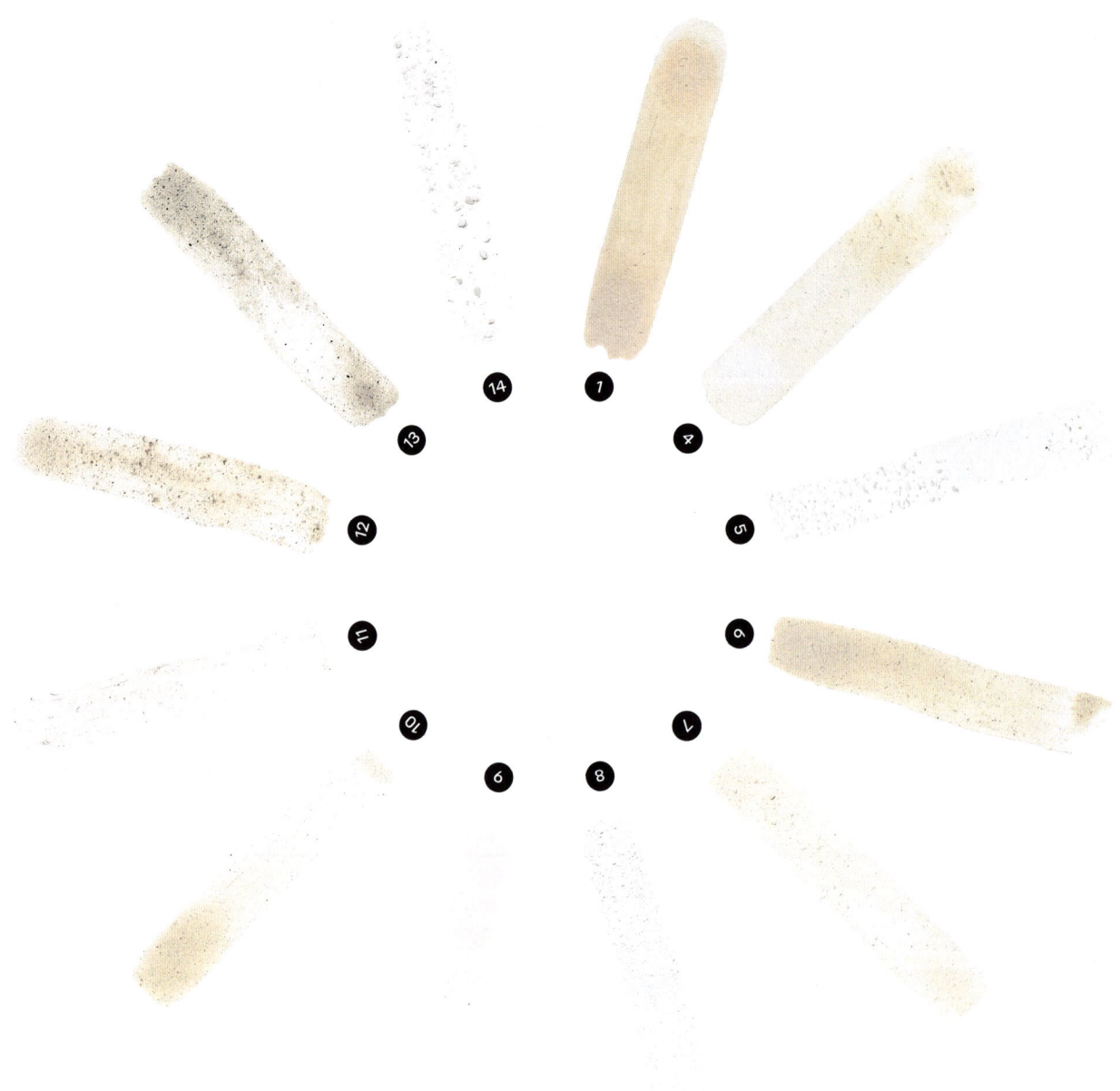

1. *Capital Caverns*. Deep karst and limestone dug up from under a superstore in Puerto Rico. There are more Walgreens and Walmarts per square mile in Puerto Rico than anywhere in the world. Rescued with love and gifted by local land caretakers Rosaura and Karla.

*2. *Old Poop*. Someone, probably a coyote, ate a lot of bones and left them beautifully digested and processed by their gut and the sun.

*3. *Skull and Teeth*. The shells within us; I found on walks in the woods.

4. *Eagle Eggshell, After a Hatchling*. Found broken open on the moss, on an island where the very rare mushrooms come up in the second growth forest, with Buck McAdoo.

5. *Marble Dust*. From Carrara, Italy, where iron veins grow in between long lost sea life.

6. *Beach Crayon*. Chalk pieces found washed up on the beach in Santa Cruz, California. Thanks to Shinehah Bigham.

7. *Cuddlefish Bone*. Not actually a bone. These are the remains of a cephalopod's internal shell, a pure calcium carbonate. Pet supply stores sell them to make "snowfall" in calcium-deficient aquarium habitats or for caged birds to play with and absorb as a dietary supplement.

8. *White Earth*. By way of ancient mines that are now "goat spas" (places where people can give water to their goats), in Milos, Greece, with the field support of Bucasso and Julie.

9. *Ooze*. Chalk from Champagne, France. Yes, it is the same geologic soil that grows the wine grapes.

10. *Weird Hydrophobic Clay from Desert Valley Hill Gap*.

11. *Root Chalk*. Gathered from the roots of a fallen tree in Great Ridge Wood, Wiltshire, England, in 2018 by Caroline Ross and friends in the Hunter Gatherer Challenge.

12. *Weathered Volcanic Pyrite*. Slowly turns to a kaolinish clay because of acidic water weathering.

13. *Baby Sand Dollars*. Gathered with artists during a jellyfish swarm on an island in the Salish Sea of Vashon.

14. *Coral Bones*. Rescued from an estate sale. I cannot tell where they are from.

*15. *Pearls*. Not from inside my heart, I am still working on that.

*16. *Limestone*. Common source of calcium carbonate artist's pigment. From Paleozoic extinction events over 252 million years ago, mined and extracted unsustainably in the Lucerne Valley, California, on ancestral Maara'yam (Serrano) land.

*17. *Oyster Shells*. Not from a shellmound. Found in discards from a modern oyster farm on the Salish Sea.

Swatch not pictured, see raw material/ color on page 159

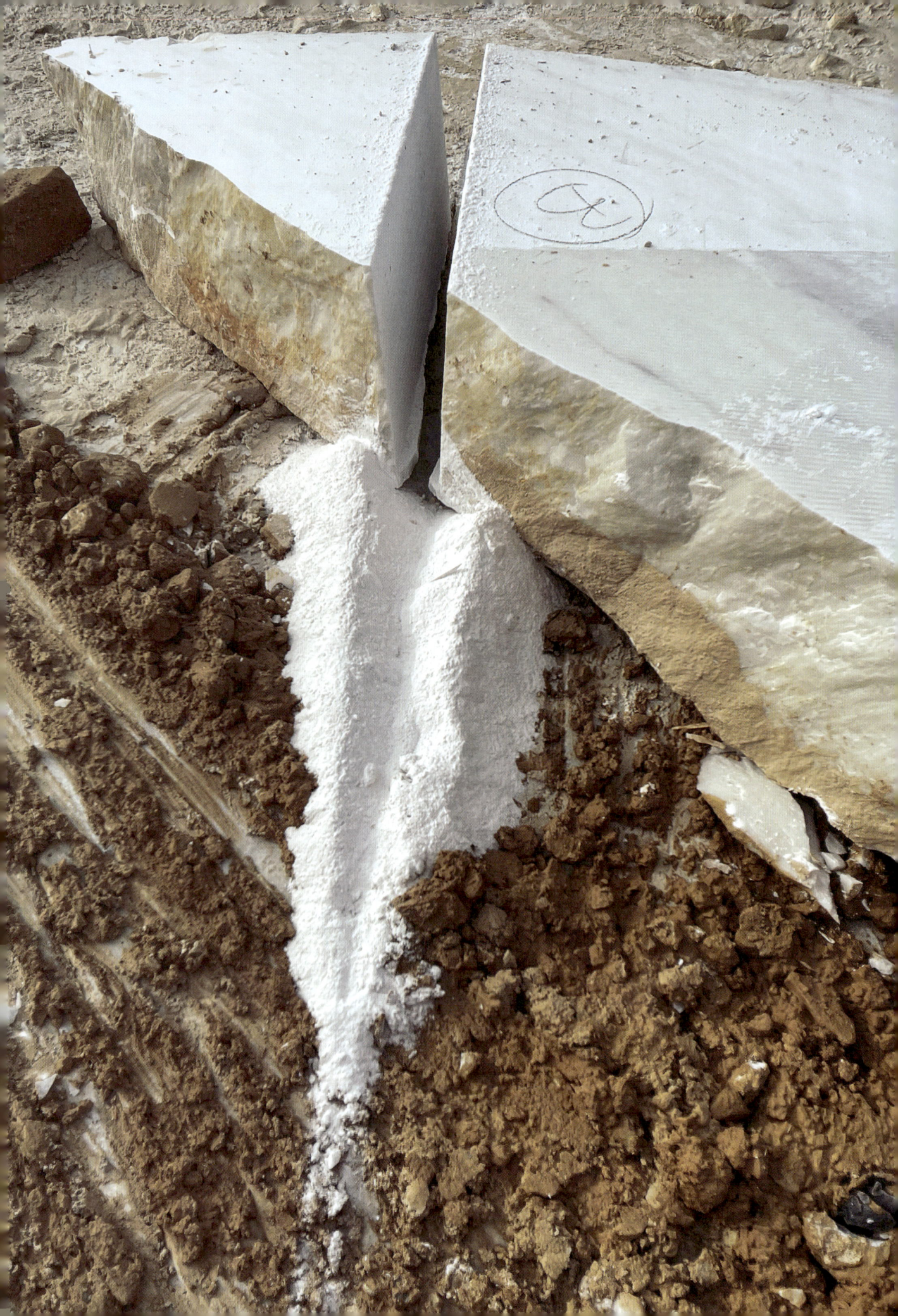

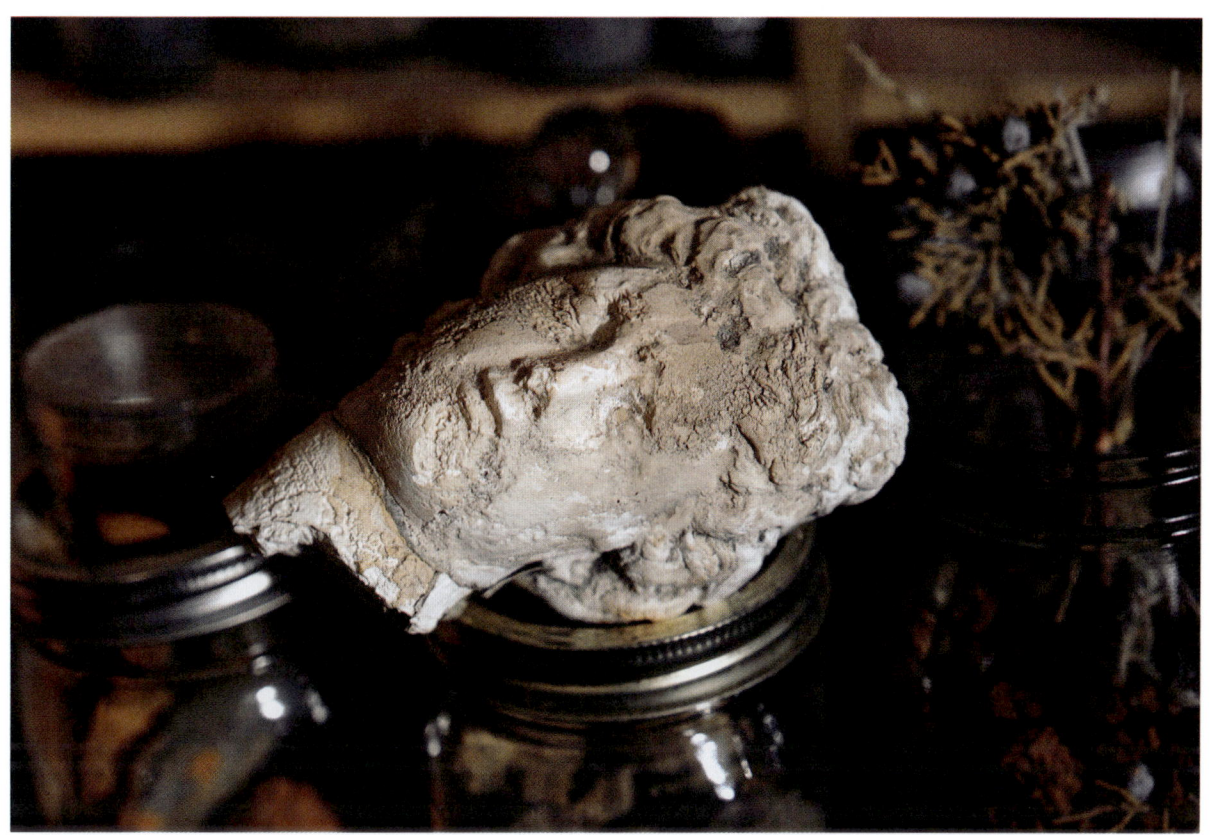

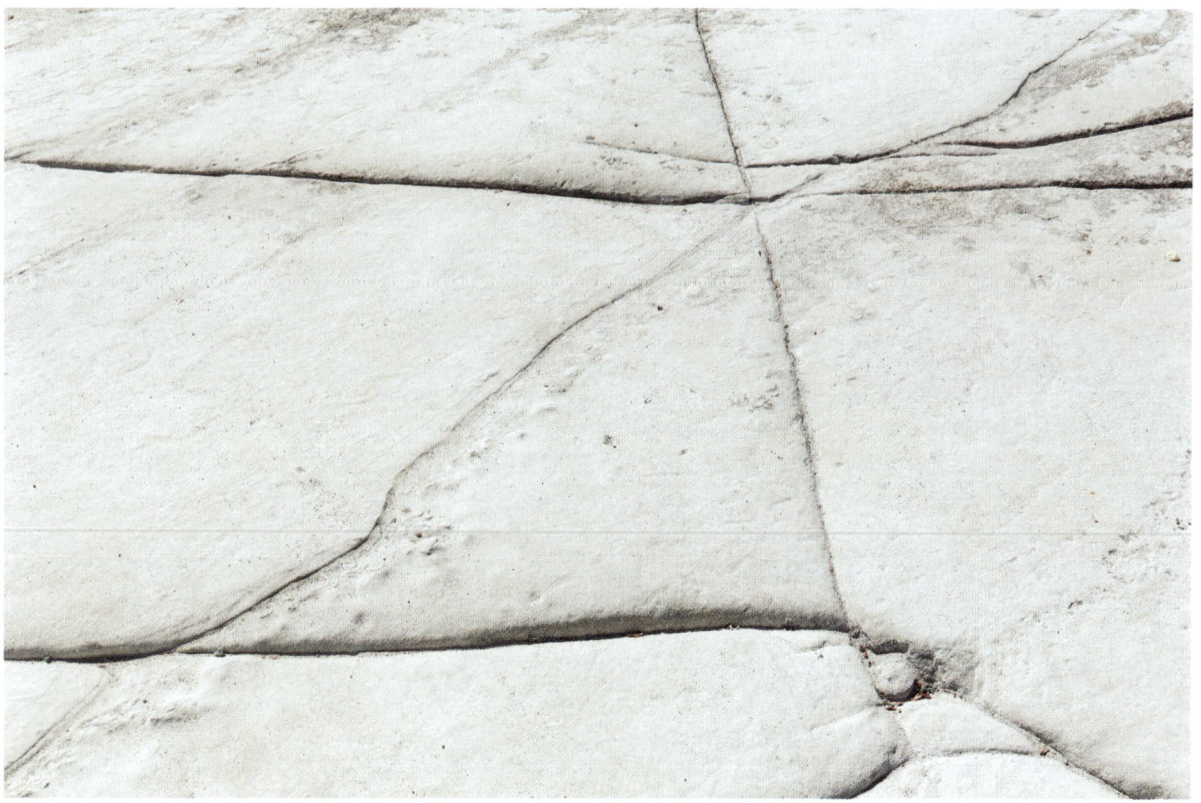

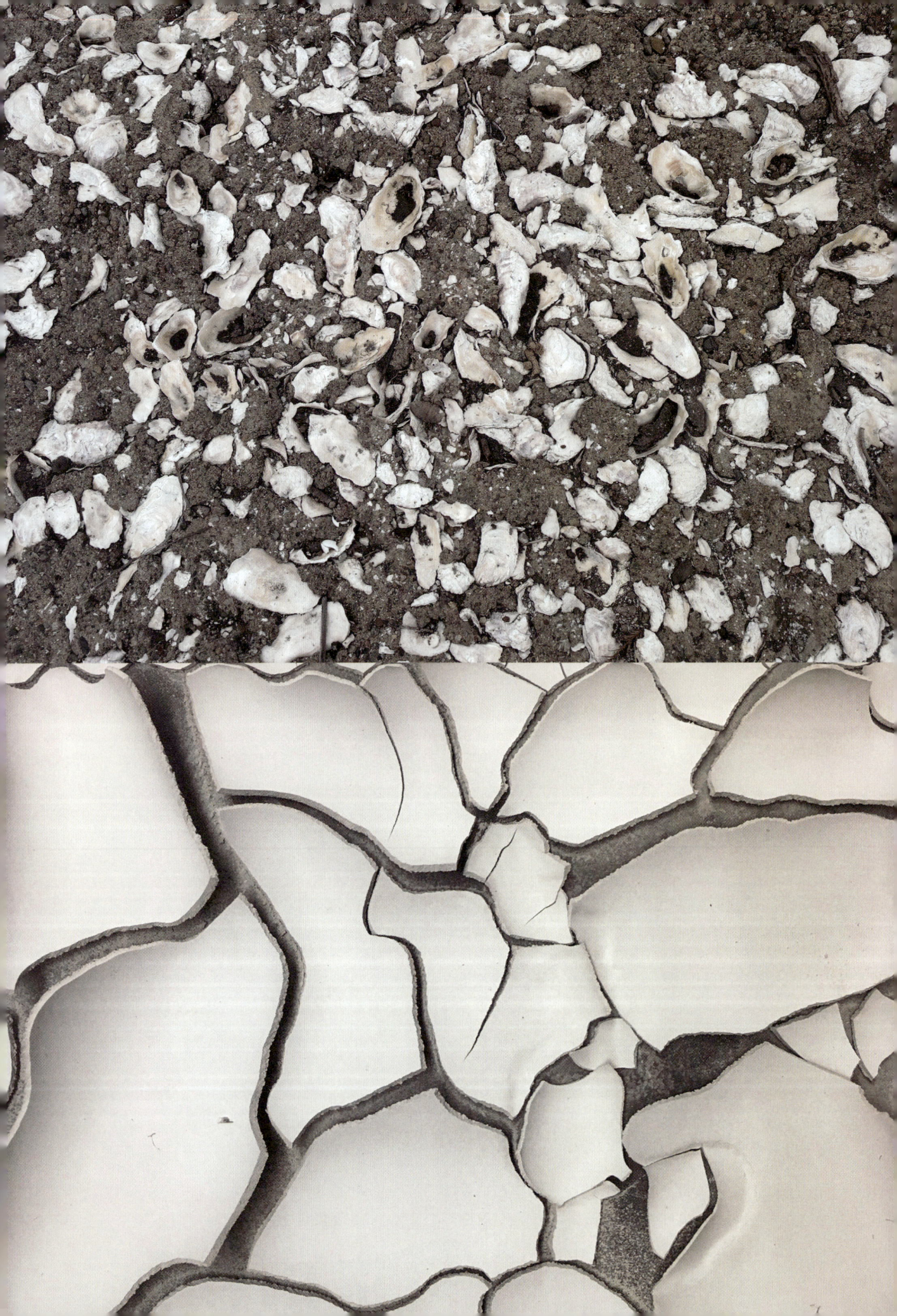

As if I had handled a stone which encloses the salt marshes of immemorial oceans or the ray of a star, I felt that I touched only the closed shell of a being who, on the inside, has access to the infinite.

—GEORGES BATAILLE, *INNER EXPERIENCE*

✸

Rocks make a deep mourning noise, not unlike the sounds of mining activity.

—MEGAN COPE, QUANDAMOOKA ARTIST

✸

One day their buildings will devour the sky.

—SACAJEWEA, *THE LOST JOURNALS OF SACAJEWEA* (TRANS. DEBRA MAGPIE EARLING)

Bone Dust

SOMETIMES I TRY TO IMAGINE when there were only very soft bodies in the sea. When rocks were rocks and creatures didn't make shells or contain skeletons. Early organisms didn't need mineral form, anatomy, bone, spine, or armor. Who knows why, but some say it was a lack of predation during the period known as the "boring billion" (boring because there's no fossil record, no evidence of mass extinctions or major traumas set in stone). I don't need a hard shell if nobody, nor the perinatal womb I develop in, threatens to harm me. Maybe I could remain soft gelatinous tissue in warm swaying water. Feels quite embryonic, like a dark peace.[1]

Our primordial grandparents had no mineral body parts, no organized skeleton, no inner geology. Biomineralization—the secretions of protective calcium carbonate minerals as extensions of body—begins around 800 million years ago, when changing chemicals altered the ocean. Mollusks and reef-building corals began gathering elements in saltwater and using them to secrete their exoskeletons. Probably in an anticipatory mode, aware of a changing ocean

condition that could dissolve their entire being. So soft creatures made seashells. Personal protection and domestic architecture emerges.

When sea animals and phytoplankton die, their geologic forms remain and sink to the seafloor. As their dust settles, a communal remainder forms ooze—*chalk*—and their artful exoskeletons compact together into *limestone*. Limestone metabolizes slowly into *marble*, creating a deep time process that assures carbon is drawn down and out of the atmosphere for millions of years. It's one of the largest ongoing carbon sequestration projects on Earth. This process secures survival of all the future animal ancestors of sea creatures (which arguably includes my own, at least for now).

Chalk, limestone, marble dust, shell, even eggshells all share the same germinal principle. These white pigments become the *imprimatura*, or base material upon which other things are laid.[2] Their material underlies plaster, cement, white office paper, gesso and primer (or ground) of paintings, paint, drawing chalk and pastels, makeup, toothpaste, ceremonial and protective body paint, champagne and other delicious wine grape soils, and several skeleton-supportive supplements, including prenatal chewable calcium. Each use of white earth pigment contains a trace of its formation: ingenious protective stuff born out of environmental climactic shift; a climax moment.

Even human bone evolved from shell. Why don't I remember learning that as a child? Calcium-binding cells that work to build marine exoskeletons are the same little workers who secrete endoskeletons in animals, us. In marine creatures, the secretory process creates more complex shell structures and even solid extremities and protection shields (picture the funny accordion-like shells of shrimp and lobster). However, researchers of anatomic evolution note, "rigid shells and shields did not allow much movement and locomotion; therefore, the next major change in the evolution of skeleton—dislocation of mineralized skeleton from the *outside* to the *inside* of animal bodies, proved to be a major adaptive advantage . . . the appearance of endoskeleton enabled the expansion of activity radius and habitation of entirely new environments."[3]

Shells *evolved inward*[4] around 400 million years ago, enchanting brave new worlds full of "sunken exoskeletons." So seashell metamorphoses into skull and teeth and a spine and, like magic, animals begin to swim. Millions of years later, us land creatures still contain a shellskull protecting our ancient primordial soft tissue brain, throat, mouth, and locus of tightly packed sense organs and fragile apertures. Semi-rigid crustacean shells now proudly form a flexible spinal column. Rock takes exquisite shape in us.

No matter if you are an oyster, ray-finned fish, plant, vulture, or human, our skeletons become an important type of carbon photograph, a "precise climatic moment fixed in the body," as structural geologist Marcia Bjornerud puts it.[5] Both during our lives and in our death, we carry our environment internalized. For example, in *Angels and Saints*, essayist Eliot Weinberger notes a precious detail about blind, hunchback saint Margaret of Città di Castello: When she died, "they dissected her heart and found three pearls."[6] As if her heart was *still* part oyster! We each

accumulate our experience—where we go; what we breathe, eat, and drink; and how we pray—this account remains legible in our bones and dust. Life decays to soil, soil feeds and harkens rock. Bodies become bone, bone transforms into a part of an enduring fossil fixative, an image suspended in land, a fossil record. Our mortal coil clings to prehistoric reef.

Today, our shells and skeletons and architecture still evolve and respond to a unique climactic moment. I see this, for example, in how ultraviolet, colorful living coral become bleached and bonelike. The swaying, swimming, colorful part of them retreats back to hardened rock, a fossil, a fixed image. I think of a similar elemental transition that occurs when water becomes ice. A huge transformation in mobility and flow occurs with only one single degree of change, 1°C to 0°C. In one moment, what was fluid becomes solid, hard, dense. Likewise, we could imagine how today, a single degree of ocean warming, like a single degree of water cooling, initiates a material phase change.

Our uses of ochre and earth pigment (especially in our creative expressions), unconsciously or not, remains symbiotically allied with their deep time and mineral origin stories, formation, process, and habits across an ongoing geologic continuum. Ochre follows *geologic*—Earth's logic. In white earth pigment's case, the logic is defined on *skeletal, protective*, and *ancestral* terms. I think it is meaningful to imagine—in a simple, daily act of painting gesso on a canvas, applying makeup, brushing your teeth, or reading text on a white page—that it is still possible to feel and sense and respond to the sacrifice of a gazillion sea stars, oysters, and kin, whose incredible, compassionate offering gives us shelter, joy, and creative momentum, movement, mobility.

I'm not alone in this quest to track the deep time of shell material. Manuel DeLanda, a Mexican American philosopher, took up a similar task, tracing the long shell-skeleton-limestone story into the present. He notes, as I and many others agree, that "the human endoskeleton was one of the many products of ancient mineralization," but goes further to say, "Only about eight thousand years ago, human populations began mineralizing *again* when they developed *an urban exoskeleton*. . . . This exoskeleton

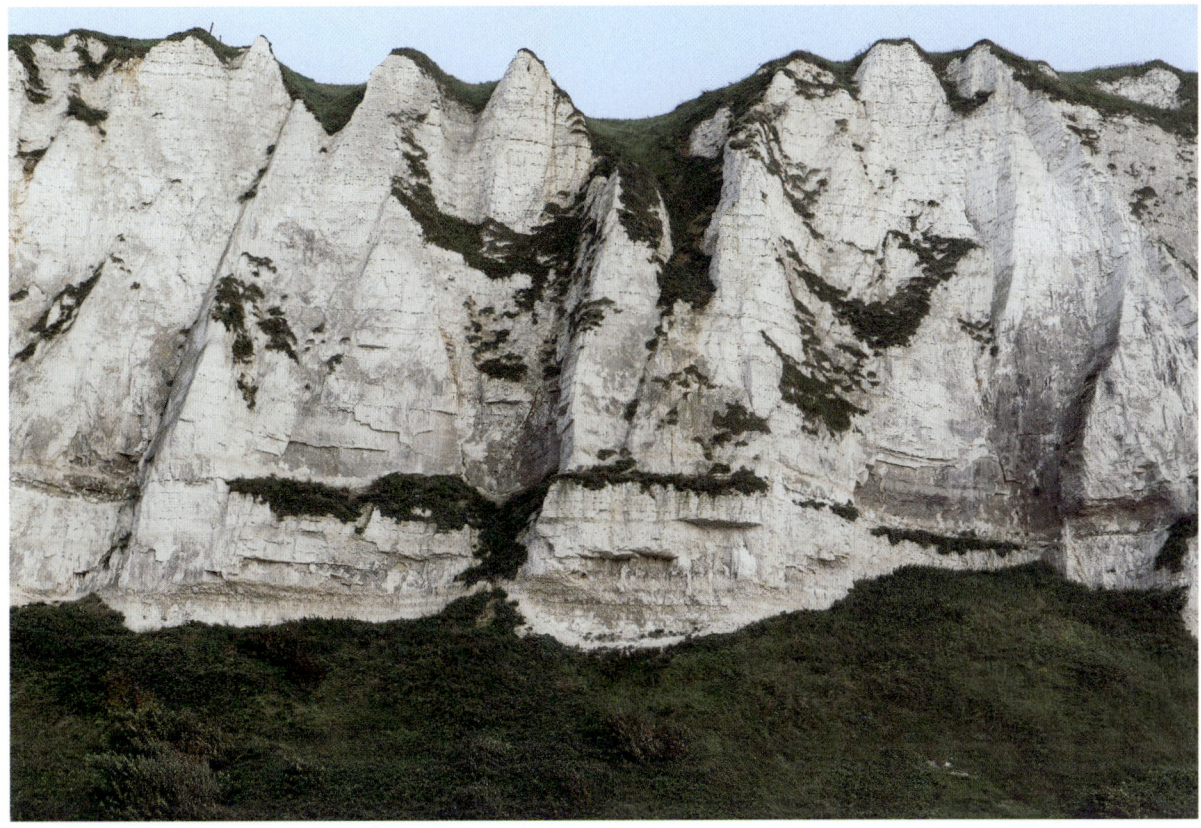

served a purpose similar to its internal counterpart: to control the movement of human flesh in and out of a town's walls. The urban exoskeleton also regulated the motion of many other things: luxury objects, news, and food. . . . Thus, the urban infrastructure may be said to perform, for tightly packed populations of humans, the same function of motion control that our bones do in relation to our fleshy parts."[7]

This really hit home for me, especially in my early ochre refuge years, when I lived in the Bay Area of California. I was very often tired, haunted, and depressed. I would walk the northern shorelines to try to get a grip. There was a particular landfill peninsula with thousands of pale ceramic extrusions, like piles of little hardened porcelain fingerbones. A former industrial ceramic company dumped their waste and trash on the beach. Next to these ceramic bones lay bodies of dead stingrays or rotting fish bones eaten or thrown back by the local fishermen. Farther down the beach was an estuary mouth where five creeks flowed together around incredible mounds or "middens" of

fused oyster shells, now a protected habitat for migratory shorebirds. There's no modern signs or designations of how these large heaps of shell were made or what they were for.

In later years, I began to read archival research and articles and heard public conversations of Bay Area Indigenous leaders and activists. I did not know that by naming these places I walked as landfill or midden "refuse places," there was an ongoing and strategic erasure of sacred Indigenous land and cultural memory. I learned, from reading the reports by Sacred Land Film project director Toby McLeod, that before colonial occupation, there were over "four hundred shell mounds clustered along the estuaries and inlets of the region, some acres wide and several stories high. The shell mounds were left by peoples of the Ohlone," which were created over time because "the activities of daily life—eating shellfish, making tools, cooking, butchering animals, building shelters—led to the accumulation and compaction of tons of shells and other material in sloping mounds of rich soil. Generations of coastal dwellers returned to these shell mounds again and again, using the sites to bury ancestors—a way to intertwine their daily lives with the afterlife."[8] I was struck by the strong material connection between shells, ancestral burials, animal bones, and architectural structure.

In only the last few hundred years, these ancient shell mounds were mostly flattened and looted by colonial occupants. Tons and tons of oyster shells were dug up and pulverized, along with memory, bodies, kin. Why? In part because of the calcium carbonate. White earth pigment. The shells were often collected, crushed for landfill, or burnt in limekilns to make lime, the key ingredient in cement and concrete, which build the foundations, sidewalks, walls, columns, vaults, basements, buildings of colonial culture—controlling the movements of densely populated humans, as Manuel DeLanda might say. Technically the process of burning shells into cement and into colonial buildings and parking lots can only happen if the carbon gets burned off. In that sense, people today are still *secreting* carbon, like nervous sea creatures, following an ingrained millions-of-years-old defensive habit evolved from oysters themselves.

The oldest surviving native Ohlone shell mound, over 3,500 years old, was designated one of America's 11 Most Endangered Places in 2020 by the National Trust for Historic Preservation. Most folks would not recognize this land as a sacred place by looking at it. Compacted shells, human bones, and few remaining cultural objects are cemented over and fused beneath what is now some of the most expensive real estate in the United States, a few hundred square feet of parking lot valued at several million dollars. Underneath the cement lot is where the outflow of an enormous creek estuary should flow. Over the past several years, Ohlone leaders, including Lisjan Ohlone tribal spokesperson Corrina Gould, have been trying to protect this ancestral homeland against massive development through multiple pathways, including legal, financial, and political.

Standing next to the cement-covered shell mound of her ancestors, Corrina once said, "This one piece of ground, this one place that doesn't have building, this one place that is open to the sky." Her words mirror a haunting phrase—"one day their buildings will devour the sky"—said by Sacajewea (as poetically imagined by Bitterroot Salish writer Debra Magpie Earling). I read these words in an obscure and stunning book, *The Lost Journals of Sacajewea*,[9] published by an arthouse letterpress in a building a few blocks away from this very parking lot, also built atop the extensive buried shell mound village area.

Across the great Pacific Ocean, in Australia, Lutruwita (Tasmania), and Aotearoa (New Zealand), similar stories of extracting limestone from shell mounds are well known. I've heard about how shell mounds, which are also community villages, burial sites, and intentional aquaculture-seeded sustainable oyster reefs, were pillaged and mined out to build sidewalks, cities, and other colonial structures, like the infamous Sydney Opera House. "From first arrival until the late nineteenth century, reefs were dredged and middens burnt for lime, which was used in mortar for colonial buildings," as contemporary Quandamooka artist Megan Cope,[10] whose ancestral shell mounds were destroyed in this way, describes. Megan makes artwork re-envisioning decimated middens, noting "the burning of middens for building materials destroyed key markers of Quandamooka occupancy and reinforced the myth of *terra nullius* [British Latin law term meaning 'land belonging to no one']." She hand-casts hundreds of concrete oyster shells (reclaiming the shells and ancestors hidden inside the concrete), layers them over copper slag (from a mine that sits on parts of her land), and creates a new shell mound, one that resists colonial occupation and "re-imagines what a young midden might look like and reinstates this symbol of Aboriginal people's continuous habitation."[11]

The poignant act of breaking down concrete to breathe life back into land inspires me beyond measure. All the while, in my own cultural backyard, I feel the sky becoming hard and rigid like a new carbon shell, like water to ice, or coral to bone. I feel my body failing to absorb and do what our soft ancestral kin did: adapt to an elemental change in atmosphere. White ochres, as a habitual ground and architecture of our souls (and reefs, paintings, sidewalks), offer one very humble, very simple teaching: The sky is still the limit.

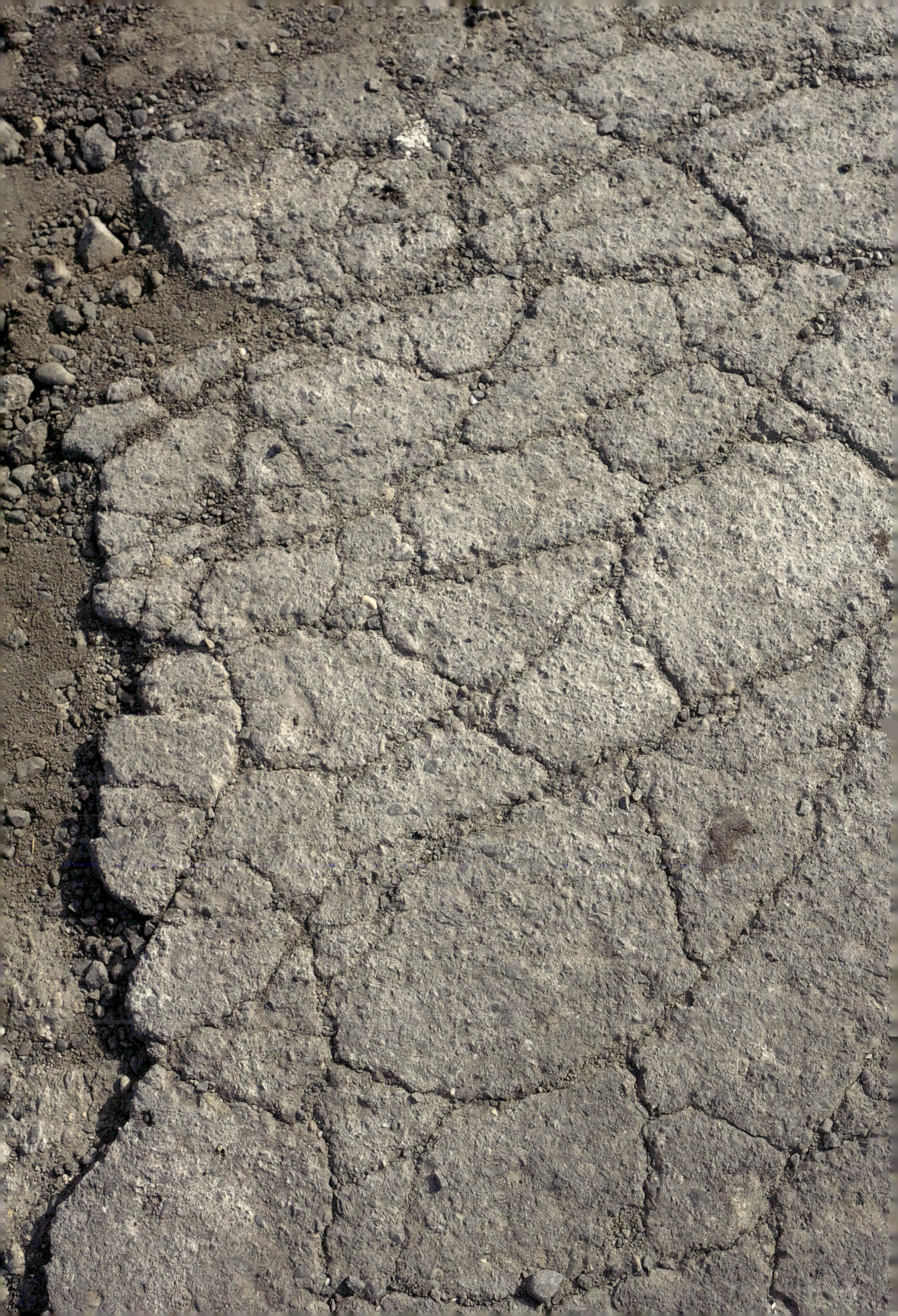

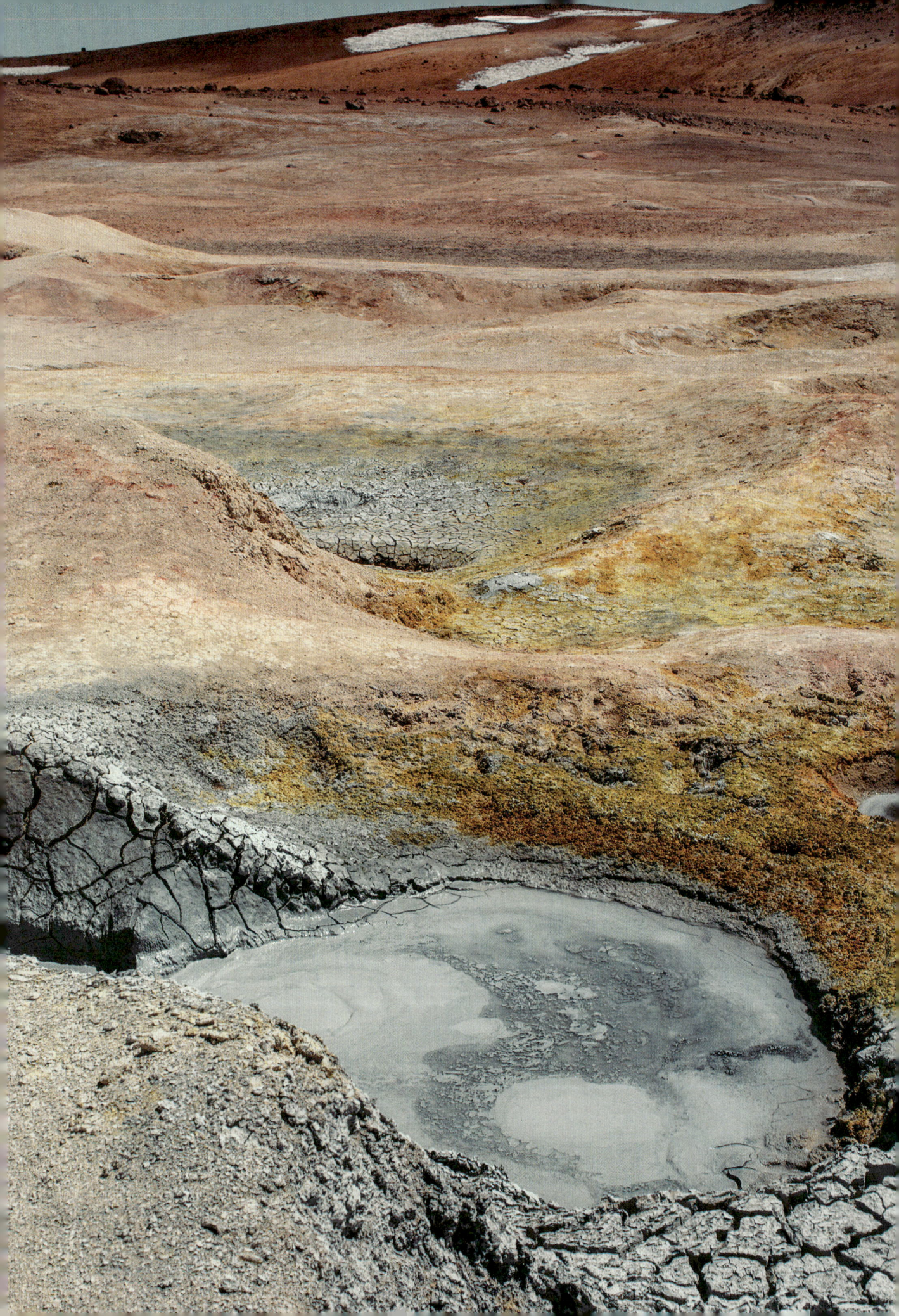

A NOTE ON SOIL

He hokinga mate, he hokinga kāinga, he hokinga oneone.
A return to suffering, a return to home, a return to soil.

—HONE HEKE, NGĀPUHI IWI RANGATIRA (CHIEF) [12]

Admire the riot of colors, the magnificence of the material itself.

—ROBERT CAILLOIS, *THE WRITING OF STONE*

Undersoil, subsoil, laterite, paleosols, ferralsols, loess, "loose dust deposits," clay, silt, sand, pigment, crumbling earth—these are soil types, soils who are the primal incubator and inheritor of our lives. All earth pigments are soils. "Soul and soil are not separate," as Terry Tempest Williams says beautifully. "Neither are wild and spirit, nor water and tears. We are eroding and evolving at once, like the red rock landscape before me."[13]

Making color could be considered the evolution of rock into splendor. Or else just plain old erosion into dust. Either way, rocks cannot writhe, they cannot scream. Something precious, or semi-precious, is formally lost forever when artistic pigment is made. An ancient part of earth becomes soil. That is a huge risk. On the other hand, there is a chance that a stone bound to mark eternity *wants* to break down and "return to soil"—to turn soft, mobile, writhe, bleed, hold, speak, and be *re*-membered.

What I do know is that making pigment expedites a natural geologic process called soil forming.[14] Soil forms generally from the weathering of rocks, broken-down land from parent material over time by natural forces like tides, wind, water, fire, organisms, etc. When I make color, I shapeshift into a weather system, a wondrous whirling dervish changing a part of planet Earth. When I make pigment, and paint, *I am tending soil.* "All art is ecological," as the book title by philosopher Timothy Morton boldly affirms.[15] Color is soil, and paint is part of "the biologically excited layer of the Earth's crust" (one of the more fun scientific definitions of soil).

And this fact that *pigment is soil* means, to me, that anyone who makes, uses, plays, or paints with color becomes an integral part of not only a complex ecological system—a mooring and an anchorage to nourish life-forms—but a transformative cultural complex.

Perhaps this is why color-making was (and can still be) a specialized profession reserved for those willing to risk wading through a lot of muck: realm-hopping medicine people, artists, iron-workers, ritual and spiritual leaders, alchemists, pharmacists, farmers, remote hermits, children, meditation masters, and any person who remains in the churning vortex where spirit, soul, land, creature, bodies cycle together and alter Earth.

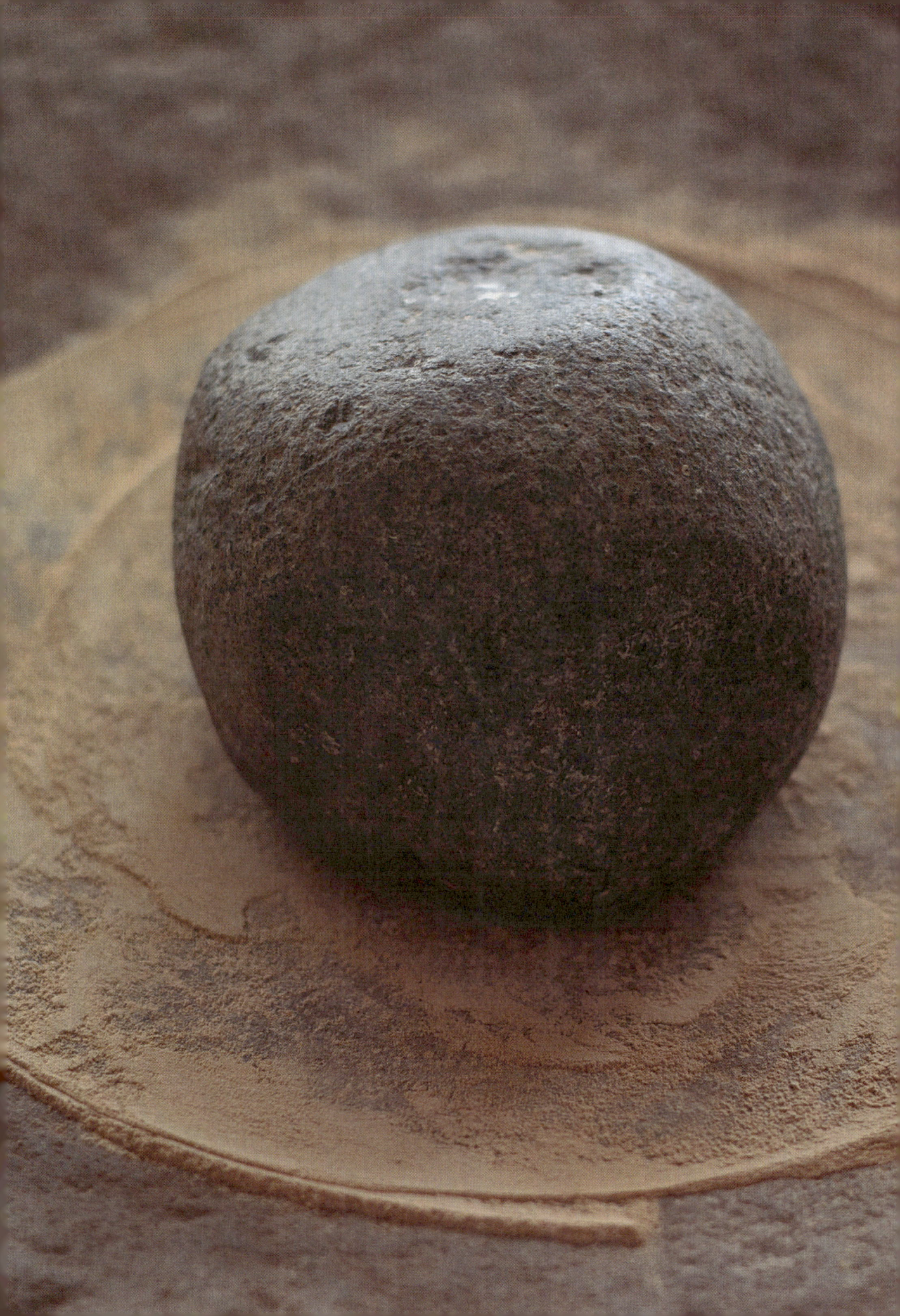

FROM
ROCK
TO
PAINT

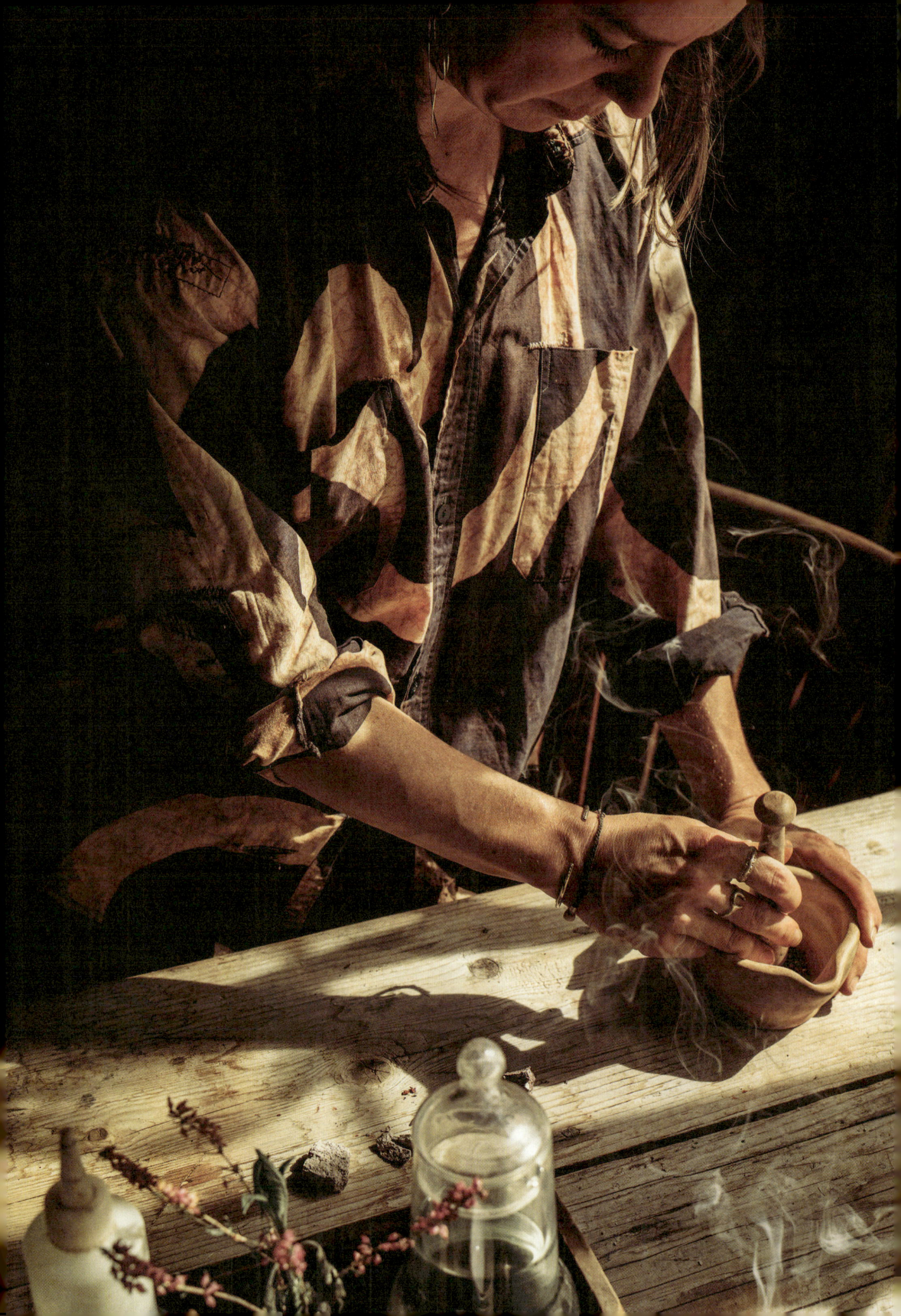

Listen until you can hear the dreams of dust that
settles on your head.

—BARRY LOPEZ, *DESERT NOTES*

✳

There was a word inside a stone.

—URSULA K. LE GUIN, "THE MARROW"

KEY PROCESSES

Making rock into pigment is simple and primal and can suddenly—often without warning churn the depths of your being. Humbling questions and creative capacities emerge from following that inward and earthward momentum. There are many ways of processing ore into dust—into paint and beyond. I celebrate the old, raw, open instructions involving the least amount of tools and machine processing as possible.

Rocks come from somewhere, have something(s) to say, and contain layers of meaning, of matter. If I pick up a rock and make paint, I try to stay curious about basic reality: land and water are dynamic, mortal creatures who are humongous. You and I are Earth. How do you honor all of that? Here is what I've heard: Learn about the history of land and inhabitants across deep time through to the present day. Support Indigenous and Native land stewardship and sovereignty, and honor local protocols and ways of learning. Ask permission from appropriate beings and caretakers before you engage, gather, photograph. Listen for a response and be grateful for what the Earth is willing to give you; listen for what the pebble and bedrock need. Gather only what you are offered; take only what you are asked to handle and carry. Respect intuitive feelings, emotions, spirits, and ancestors, and that which is not meant for you to know or engage. Be generous in giving back. Make offerings. Proceed with joy, eros, risk, and caution. And, maybe sometimes, just let the earth hold you?

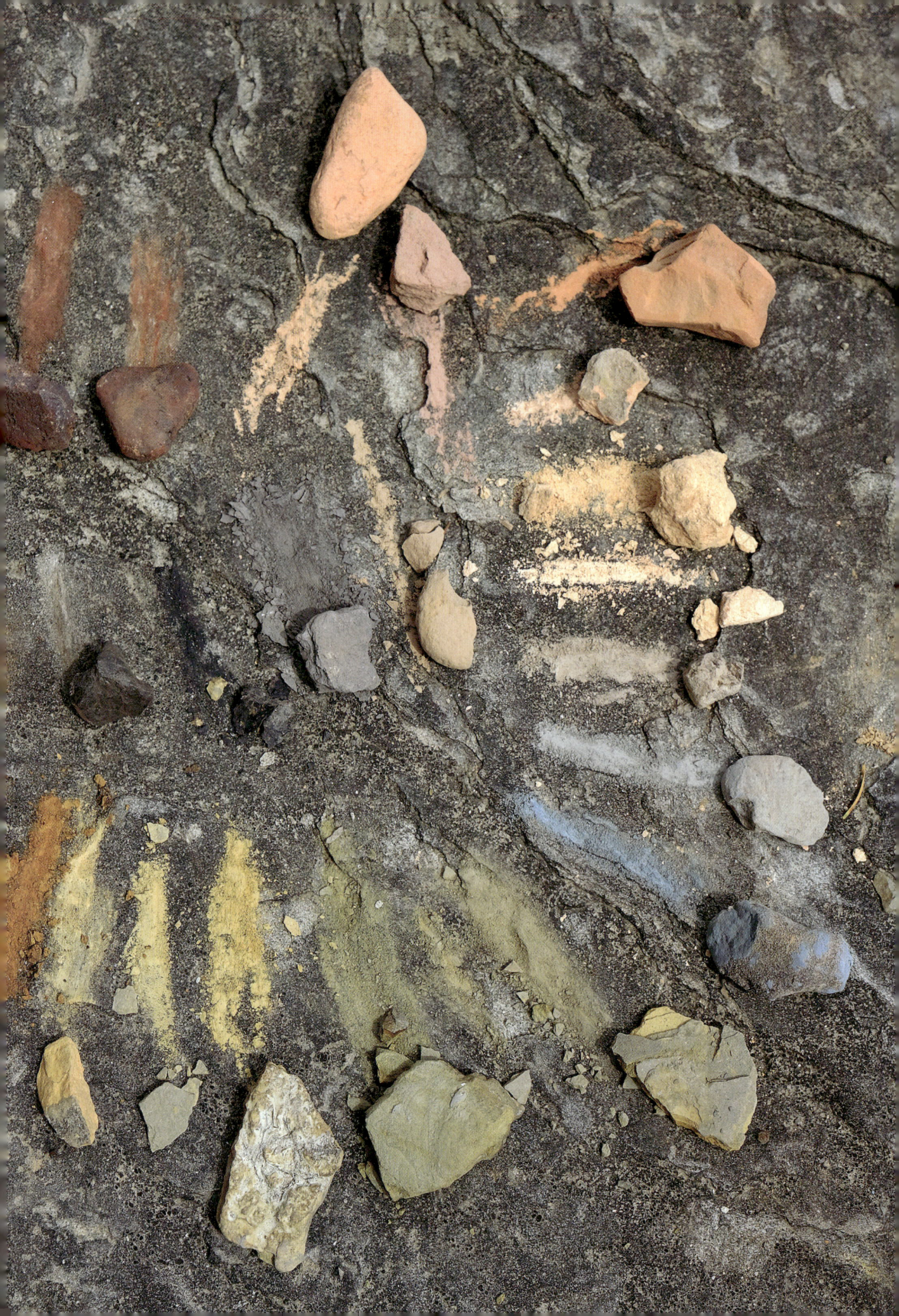

RUB

- Rub a stone on another stone. Is there color? That's pigment.

- Add liquid. That's paint.

- This works especially well when using softer stones against large, hard, flat, smooth river rocks.

RUB & GRIND

- Take two stones, a flat one and some other one.

- Put the colorful earth in between them and crush until soft and fine.

GRIND

- Use a large stone mortar and pestle (or smaller ceramic, agate, or marble works well too). Get to know your rock. What do they feel like? Are they heavy, cold, crumbly, soft? Does pigment stain your fingers? Make sure the rock likes the idea of being crushed. How you figure this out may take years!

- With a large rock, break it into smaller pieces and remove larger chunks for later. (I always save a piece of rock so I remember whose pigment is whose.) Only pound and grind what you need. At this stage, you can add water if you want to help lessen dust. Keep pounding until it becomes powder (or smooth slurry if using water)—watch for color changes as the particles get smaller. Feel with your fingers along the way for texture. Take pinches out and test with a drop of your preferred binder, if desired.

Things to consider

- What direction are you grinding? Clockwise or counterclockwise? Is this with or against the natural spin of Earth?[1]

- Is there a rhythm to your grinding? A song, a beat, a sound?

- Do you want variable textures? Maybe keep some pigment at different granularity or particle sizes, so you can experiment.

- Several nonhuman species move earth around and refine geologic matter. Take notes from them about how to grind pigment. I've seen, for example, creatures that move dirt with their bodies to make their homes or nests—and as a result of their embodied care and work, create tender, fine, soft soil that a human's refining can't obtain. Consider the earthworm: They process pigment and soil by taking it through their digestive systems.

- How long should you grind? Some pigment masters recommend grinding certain materials for twenty-four hours. (Don't worry, most masters live in caves and have lots of time on their hands.)

- In the last few hundred years, the development of industrial steam-powered ball mills and mechanical grinders gave rise to production of larger quantities of pigment, homogenizing particle size (i.e., fineness) and color uniformity. I think of it as akin to making a monocrop. This type of process, which I don't discuss here, could be very good for super smooth, controllable paint. Kremer Pigments is an excellent source of information on more commercial-oriented process.

Safety note: Mineral dust isn't good for your lungs (and not always great for your skin, either). Wearing a high-quality mask (and clothes and gloves) can help protect your lungs and skin from damage. Working outside is best, or inside with windows open and proper ventilation. Fine pigments can accumulate in your body (e.g., those magnetic senile plaques!), and working with pigments is risky business. For more in-depth guidance, see Wild Pigment Project's Safety Guidelines at wildpigmentproject.org /safety-guidelines.

REFINE

- Once pigment becomes powder, you can filter or sieve it further to help make the particles uniform. But you don't have to; it depends what you want to do with the pigment (or what it wants to do with you). I refine pigments using mesh sieves (see *A Note on Scale*, page 117, for a mesh and particle size chart). I recommend mesh #100 or #200 for granulating paint pigment. I use a hand-sized sieve.

- Pour some dust into the sieve. Shake it until the finest particles pass through the mesh. Be sure to catch what you grind. The remaining chunkier pigment can be reground or offered back to the soil or saved for later.

Things to consider

• Don't have access to various soil sieves? Paper coffee filters or a cotton T-shirt are good alternatives.

• Can you cover your sieve with a lid or sock to prevent dust?

WASH OR LEVIGATE

- *Water* is very good. You can use water to clean, refine, and/or mix with pigment.

- *Wash.* Use water to *clean* pigment and remove organic matter. Simply pour water over pigment in a jar or bowl. Strain off organic stuff that floats to the top. Let the pigment settle. Remove water using a baster (if you are in a rush) or let evaporate over several days.

- *Levigate.* Use water to sort pigment particle sizes and further refine it. Put pigment in a jar or bowl. Pour lots of water over it and swish around. For most ochres, smaller, brighter particles of pigment will stay suspended in water longer than larger, heavier, or sandier particles, which sink to the bottom. Wait 60 to 90 seconds. Pour the colored water into another container (some people pour the water through a coffee filter at this stage and let dry flat.) This will be your more

"refined" or "levigated" pigment. Wait for particles to sink to the bottom (this can take hours or days). Pour off the clean water. Let dry. Sometimes it's best to spread the pigment out on paper or a baking tray evenly, in the shade or a spot that won't have uneven light or heat.

Things to consider

• Try washing multiple times to get different tones and textures from the same rough pigment.

ALTER COLOR USING HEAT

- Using *fire* can be important for ochres. Almost all ochres will change to other colors (even slightly) with heat. Bio, yellow, green, and blue ochres are especially sensitive to heat. I find it very fun to throw a rock into a night fire and come back and see what happened the next morning.

- Try putting a rock or pigment in the oven at the highest temperature for 5 to 15 minutes. Use a kiln, wood-fired oven, or backyard grill (or other very hot or enclosed environment, like a hole covered by coals) with more specific and hotter temperatures to test when exactly a specific pigment will shift valences in color or mood.

PALETTES OF PLACE

- Every ochre stone makes a different pigment. Even ten similar ochres found near each other may display a wild myriad of hues and behaviors depending on how you treat the rock. Hum, sing, dance? Grind, wetness, fineness, shape of particles, geomorphology (the story of how the rock formed), season, moon phase, sunlight, organic materials (did this rock support something growing out of it, like moss, plants, people?), geologic age, and inclusion of trace minerals create a spectral and elemental fingerprint unique to each ochre—a fingerprint that can be forensically traced back to its creative source and geologic place.

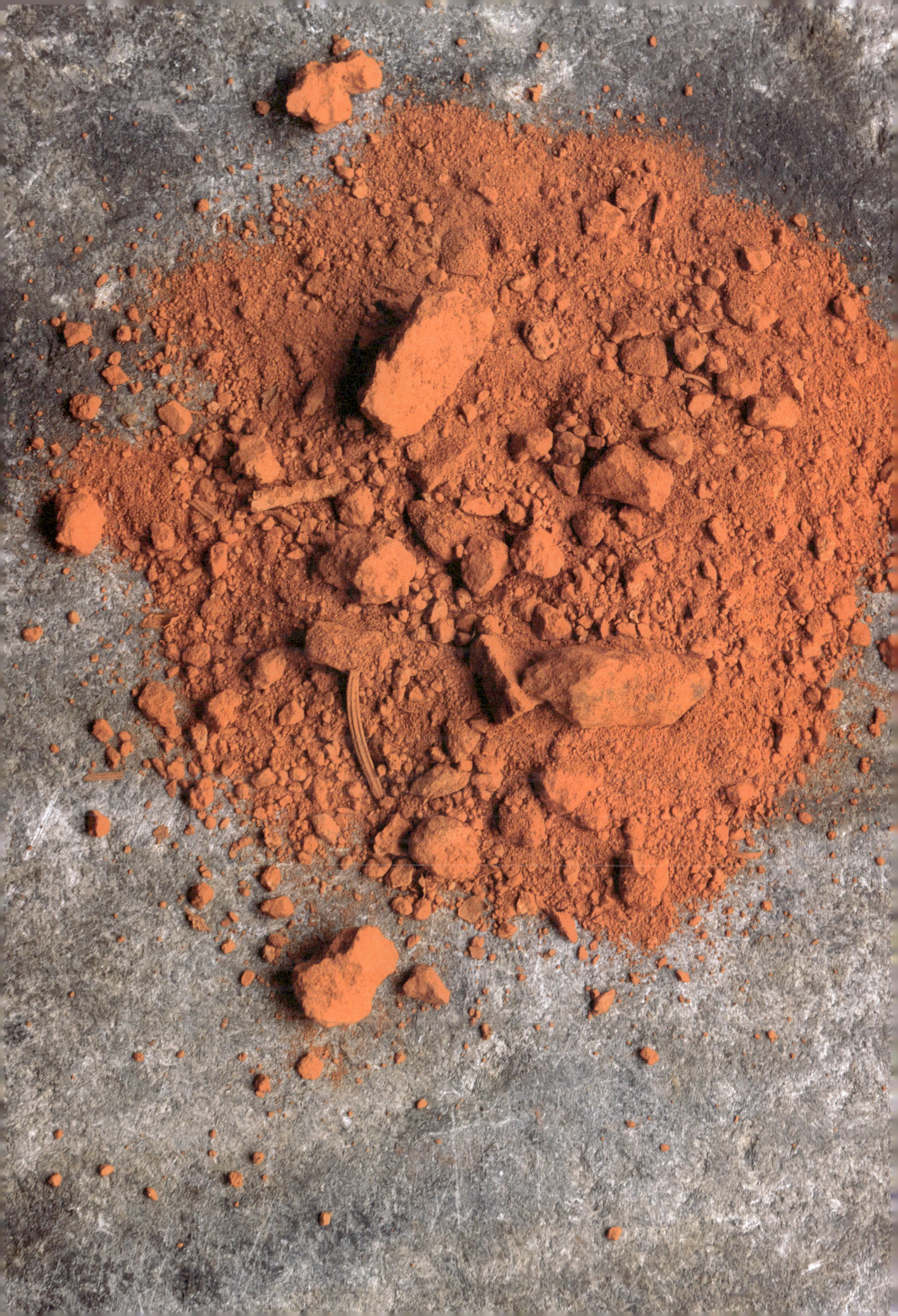

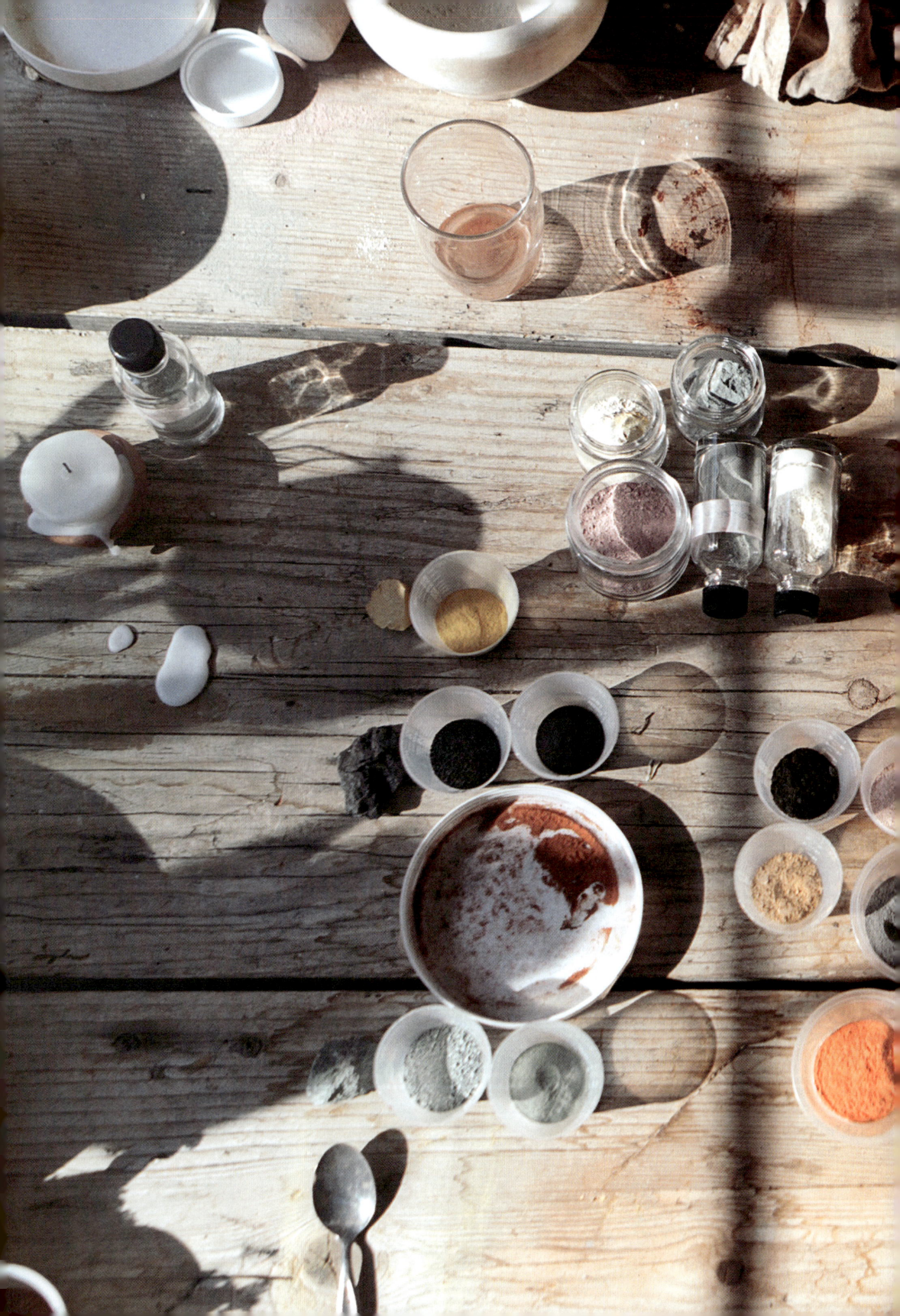

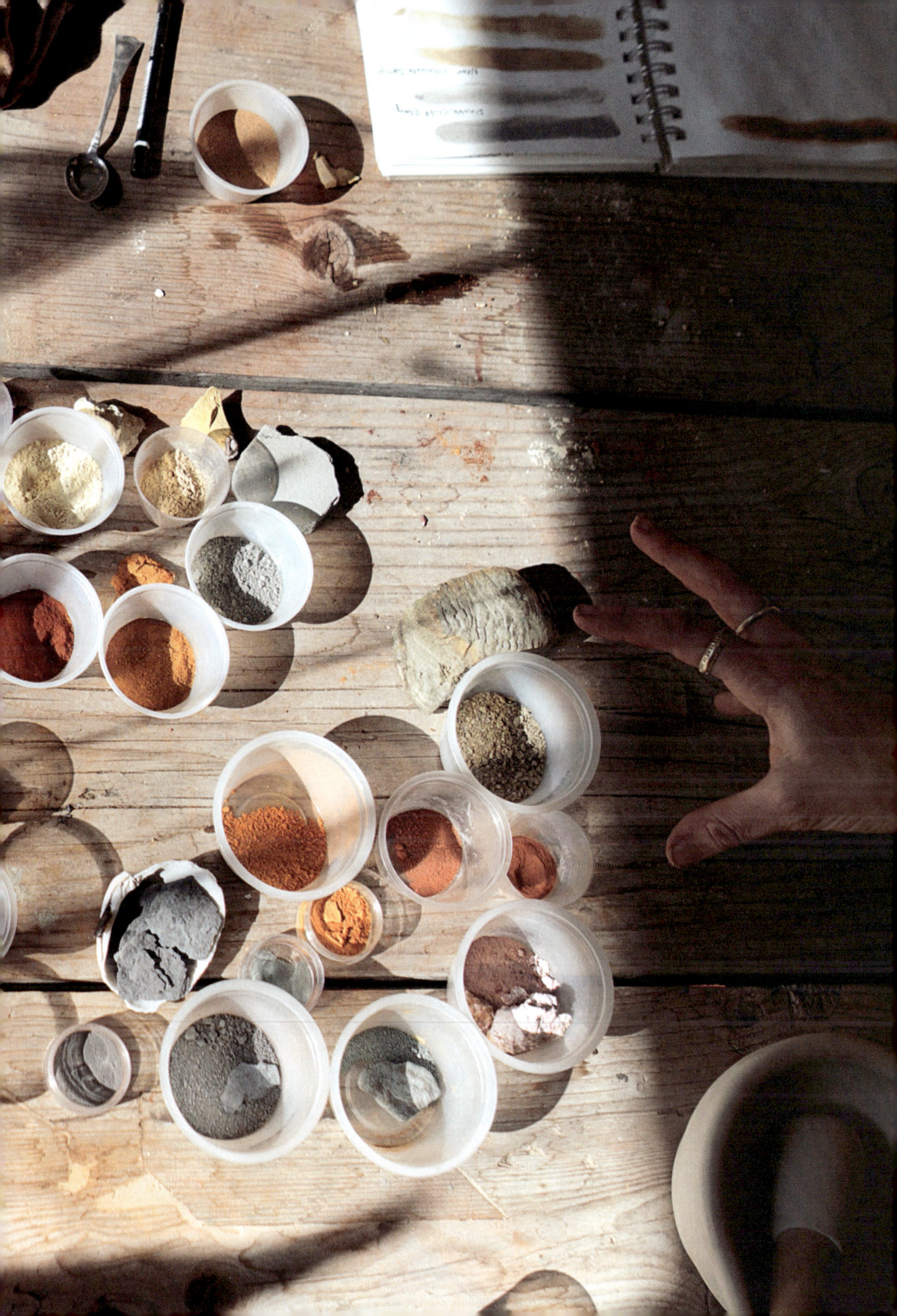

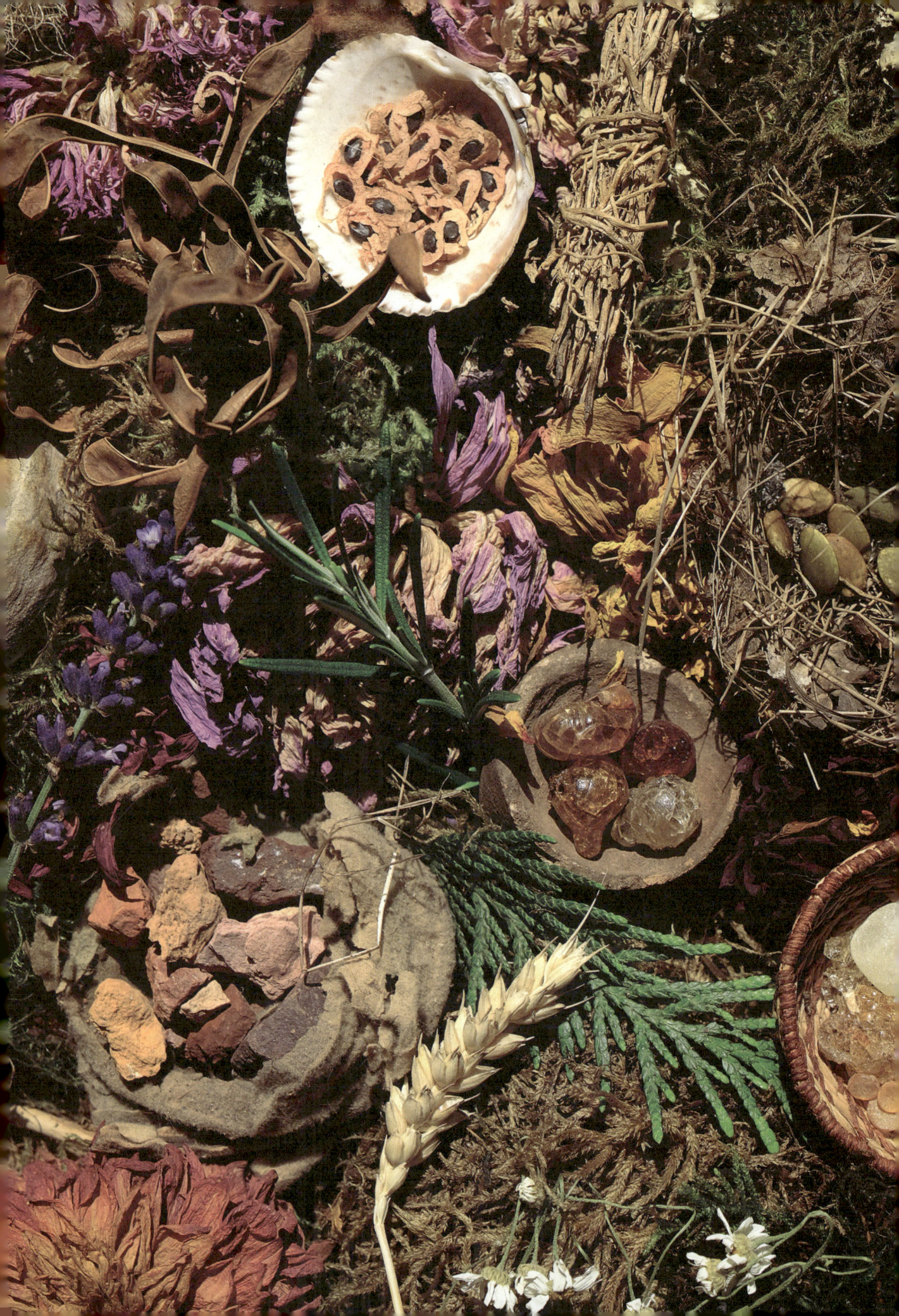

A NOTE ON BINDERS

(aka Fluids, Vehicles, Carriers, Mediums, Fixatives, Mordents, and other Viscous Substances)

Paint is water and stone, and it is also liquid thought.

—JAMES ELKINS, *WHAT PAINTING IS*

What a
rock thought to do
was rain and it
rained.

—CHRISTOPHER PATTON, *DUMUZI*

I'm in a dream place where the rocks are myriad colors. A rainbow place where I feel at home and where my ochre guide walks beside me. As we walk, she points to the plants' songs and trees' tears. *What roots in ochre knows how to bind with ochre.*

I'm in a place where the soil is glacial clay. A plain place where I also feel at home; these are places I went to as a child, and where my ochre guide walks beside me. As we walk, she points to the mugwort and yarrow and cherry tree sap. *What roots in clay knows how to bind with clay.*

I'm in a fallow field where the river delta ground is super fine and pale and cracked. My rural neighborhood where I wander and where my ochre guide walks inside me. As we walk, she points to the old dry wheat stalks and flowering chamomile growing there. *What roots where you live knows how to bind with you.*

Wheat and chamomile and groundwater are now my home binders. When I want to make paint, I pound wheat before rock to get a fine flour. Chamomile soaks in hot water, deep well water that's from the aquifer beneath my house, to make a tea. Wheat flour stirs and stirs into hot flower water, and I keep it all warm until it forms a thick goo. Like a thick fat. I pour the fat into a bowl and put it outside in the cedar woods to cool. (Or in my fridge if it is raining.) After glutinous soul of wheat cools, I can mix it with any grit pigment I want and paint it on anything and it will harden and stay there. Somewhere official, it is said that wheat paste (also called flour paste) is used to bind paper inside books, to glue and to paint interior walls naturally. I just call it my inner house paint. Also, if someone becomes hungry later, I can just scrape it back off and feed them color.

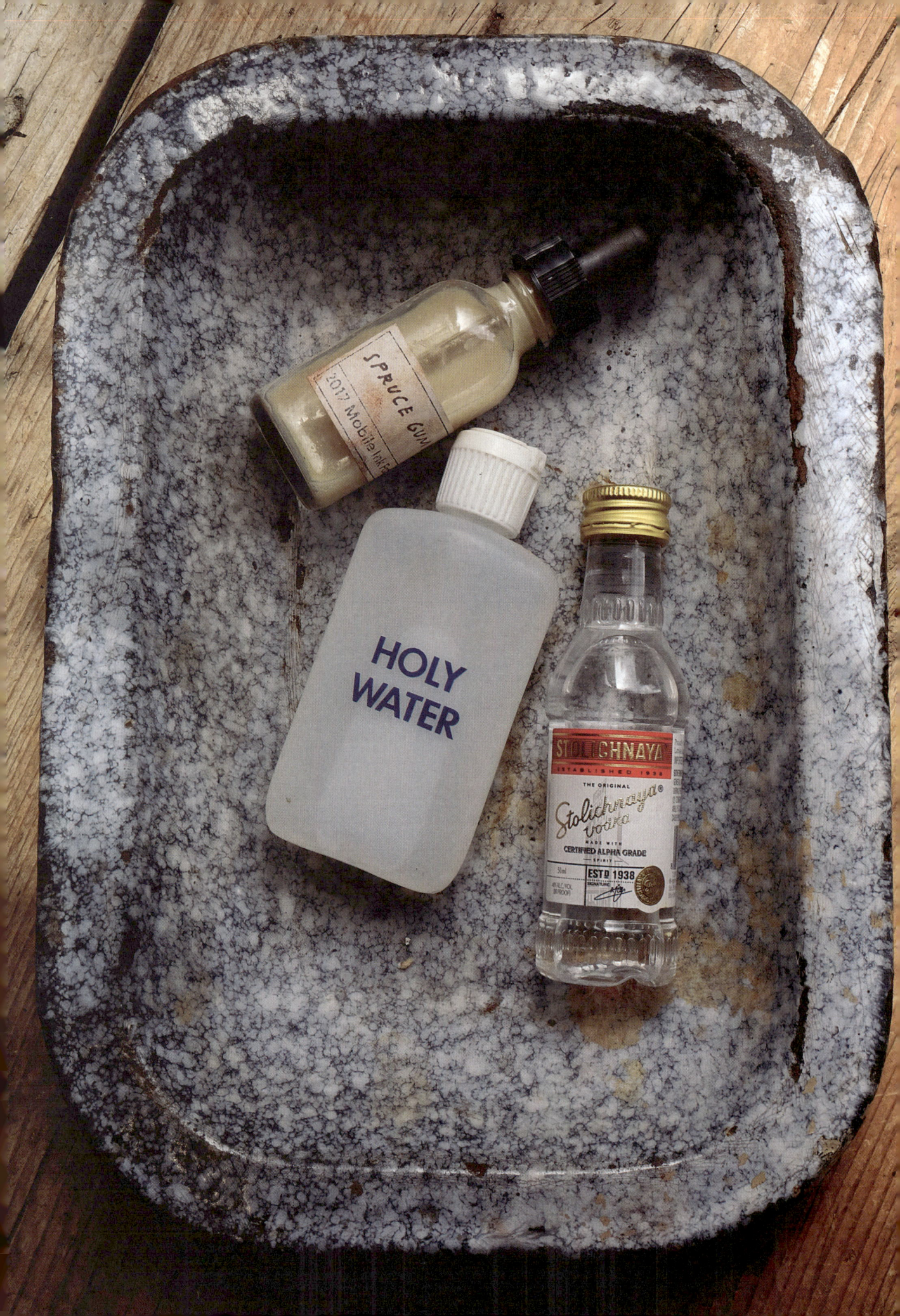

WATERS

rain

floodwater

groundwater or aquifer

surface water, river, lake, stream, ponds

waterways

glacial & snow melt

atmosphere

icicles

dissolving ice

evaporations

distilled water

processed or municipal

gray water

wastewater

ever-weeping mountains

celestial water

murkiness

ocean

seawater

sugar water

thick water

plant-infused water (like rosewater)

water stored inside plants such as horsetail and desert succulents

hot spring water

holy spring water

holy water

dew

fog

Note: Water loves earth pigments. Pigment mixed with water prefers to be applied to surfaces that have an innate ability to absorb minerals and dust, such as *tender, porous materials* like your own skin, or wood, and stone—humid cave walls of limestone, gypsum, or sandstone—and of course, directly down on the ground and field and in other soil. If applied to natural fibers (i.e., raw canvas and textiles), pigment water could stain but may only temporarily adhere and may not technically bind or soak in, thus could wear off with movement, more water, touch, or time.

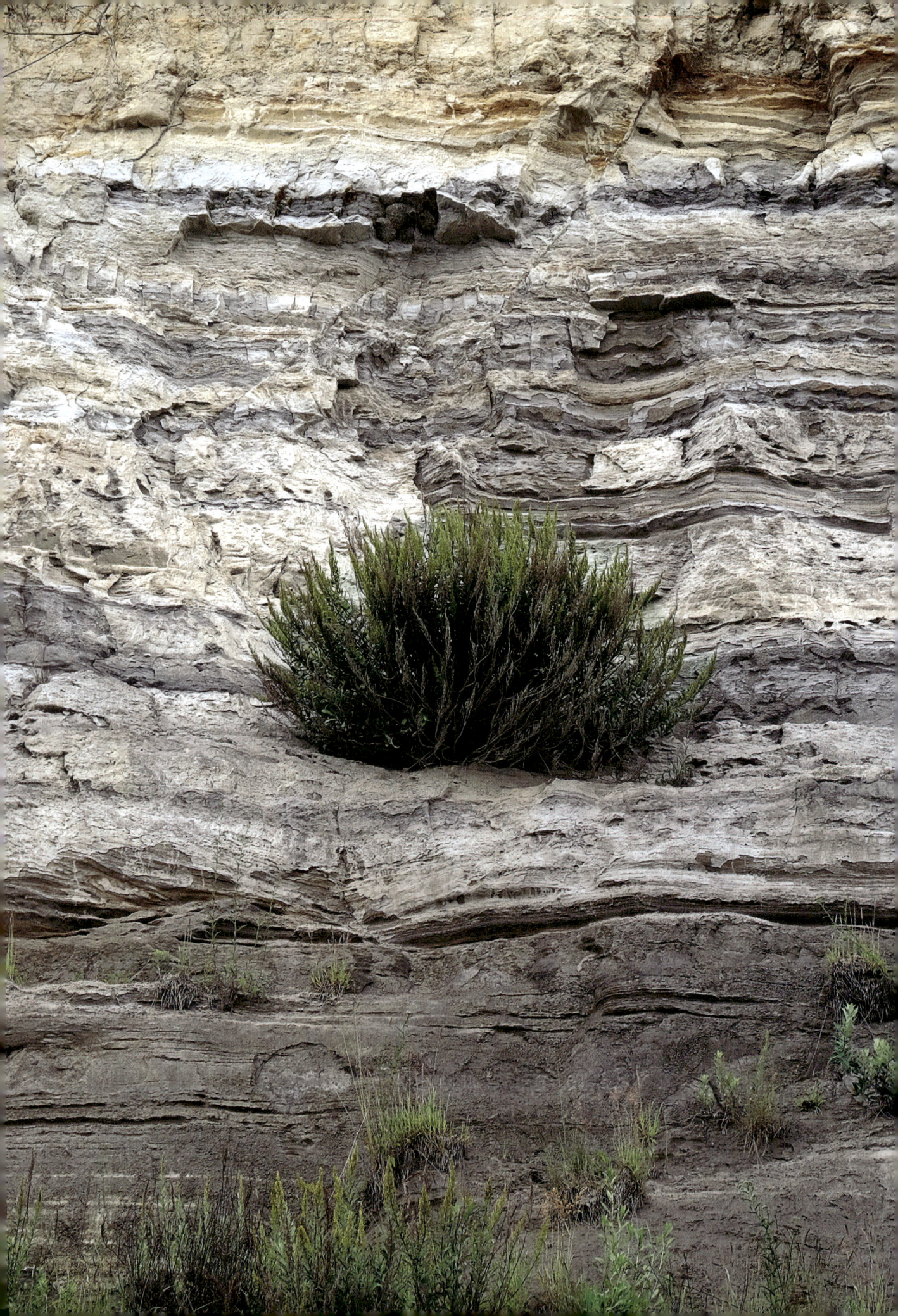

FLORA / PLANTS

tears of tree

gum of conifers and fruit trees

maritime pine and balsam pitch

sap

hardwood gum

melaleuca gum

acacia gum

piñon gum

mastika gum

dammar resins

Dragonsblood
(dracaena cinnabari sap)

fossil resins of amber or copal

nut oils of acorn, walnut,
and others

swamp root oil

yucca sap

wild orchid juice

green plum resin

eucalyptus resin

bloodwood and paperbark resins

red camwood bark

forest mango or hog plum bark

juice of flowering plants
(fruits, vegetables)

mezcal or mesquite soaked
for sugar water

previous season garlic juice

milky juice of
young sprouts of fig

milky juice of milkweed,
dandelion, and others

fatty oils of rapeseed,
hemp, cottonseed

white poppy, flaxseed
(i.e., linseed) and other seed oils

sun-bleached or
sun-thickened linseed

sunflower oil

absinthe (wormwood) oil

vinegar

bark (often boiled, often birch)

roots (often boiled or soaked)

leaves

whole seed, crushed

chewed wild vine fruit
(e.g., squash, melon, cucumber)
or cotton seed

wheat and grains, warmed
with water to fatten

flours and starches of rice,
wheat, rye, and other grasses

plants of the sea, algae, funori

aromatic herbs in teas
or squished

oils of flowering rosemary,
lavender, thyme, marjoram

alcohols

ouzo

juniper gin and tears

vodka

neat mezcal

moonshine

beer

evaporated beer

Note: Living creatures contain, and excrete, binders from their bodies. You contain, like a barley stalk, an intimate and rather erotic supply. Spit, readily available and renewable, is a special solvent of a sort (a function simultaneously important to break down food for digestion). Gums are sugar, soluble. Resin is not. Sweat's salt. As a general principle, plant and animal binders work best applied to themselves, and with things they like and like to take in or eat: *body, skin, hide, bone, feather, shell, and plants* (textiles, paper).

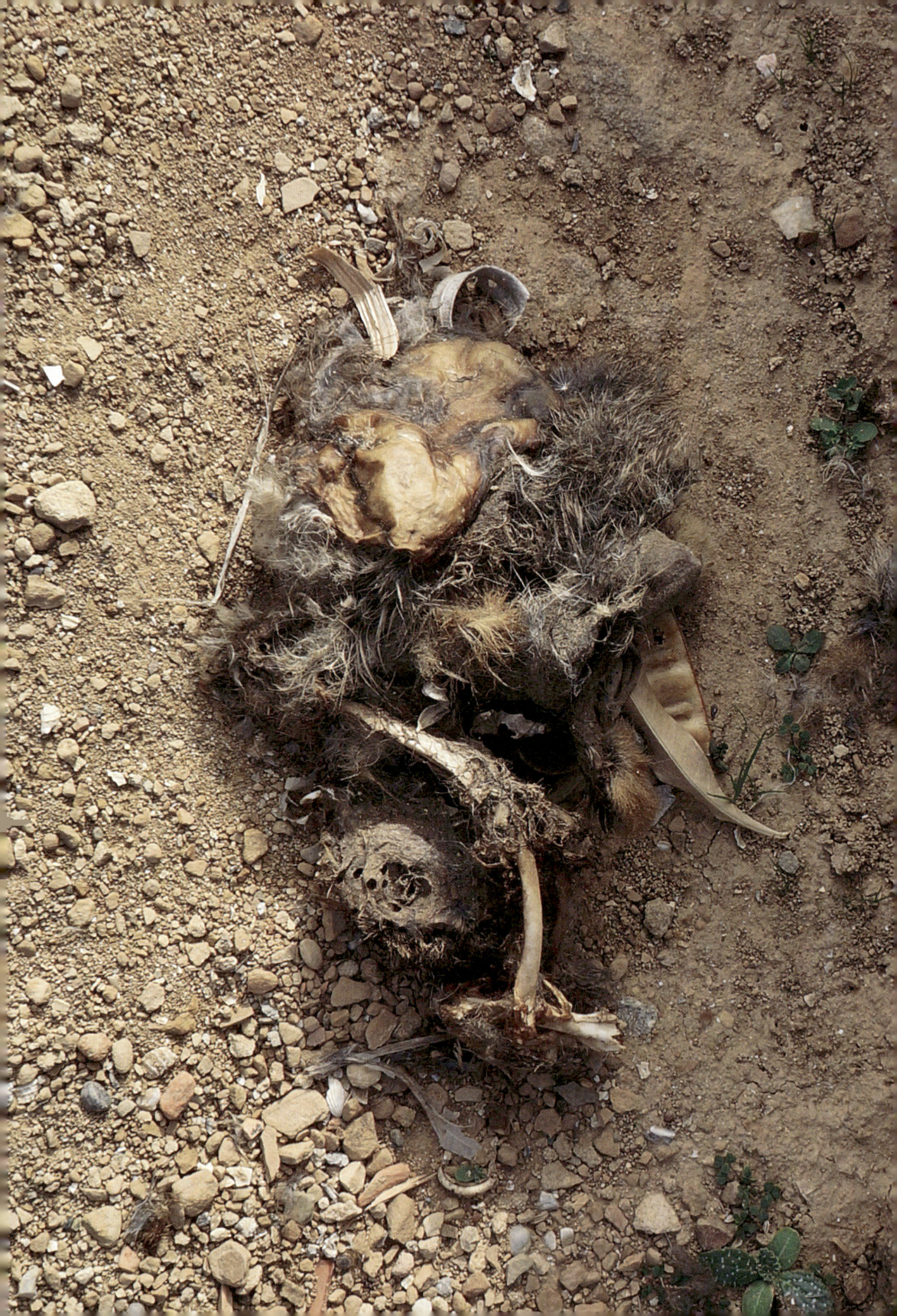

FAUNA / CREATURES

marrow

blood

blood of freshly killed eland,
auroch, equid, saiga antelope,
reindeer, horned animals

dried cattle heart

boil'd hooves

mammal fats and oil (but only
during the right season)

masticated organs (animal glue)

liver and kidney fat oil of
shark and ray

seal, dolphin, whale oil

cod and fish oil

bear grease

deer tallow

melted suet

tit milk (casein)

fat-free white cheese (quark)

skimmed milk

gull egg

avian egg yolk free of its
membrane

albumen of eagle eggs

not duck eggs—save those
for baking

sea turtle egg yolk

egg white

wild honey and propolis

beeswax

beetle juice

oil of scorpion

skin, especially of rabbits
(rabbit-skin glue)

boiled bones, especially of
hoofed mammals

urine

urine of hyrax

urine of rock rabbit

hyraxeum

whole body of termites

feces

raptor poop

oil obtained from different
body parts of mammals

glands, eggs, and fats of fish

gnawed salmon eggs

gnawed termite

octopus ink

sturgeon bladders

whale ambergris

sperm oil and secretion
from the head of whales

fish fawn

constrictor milk

snake oil

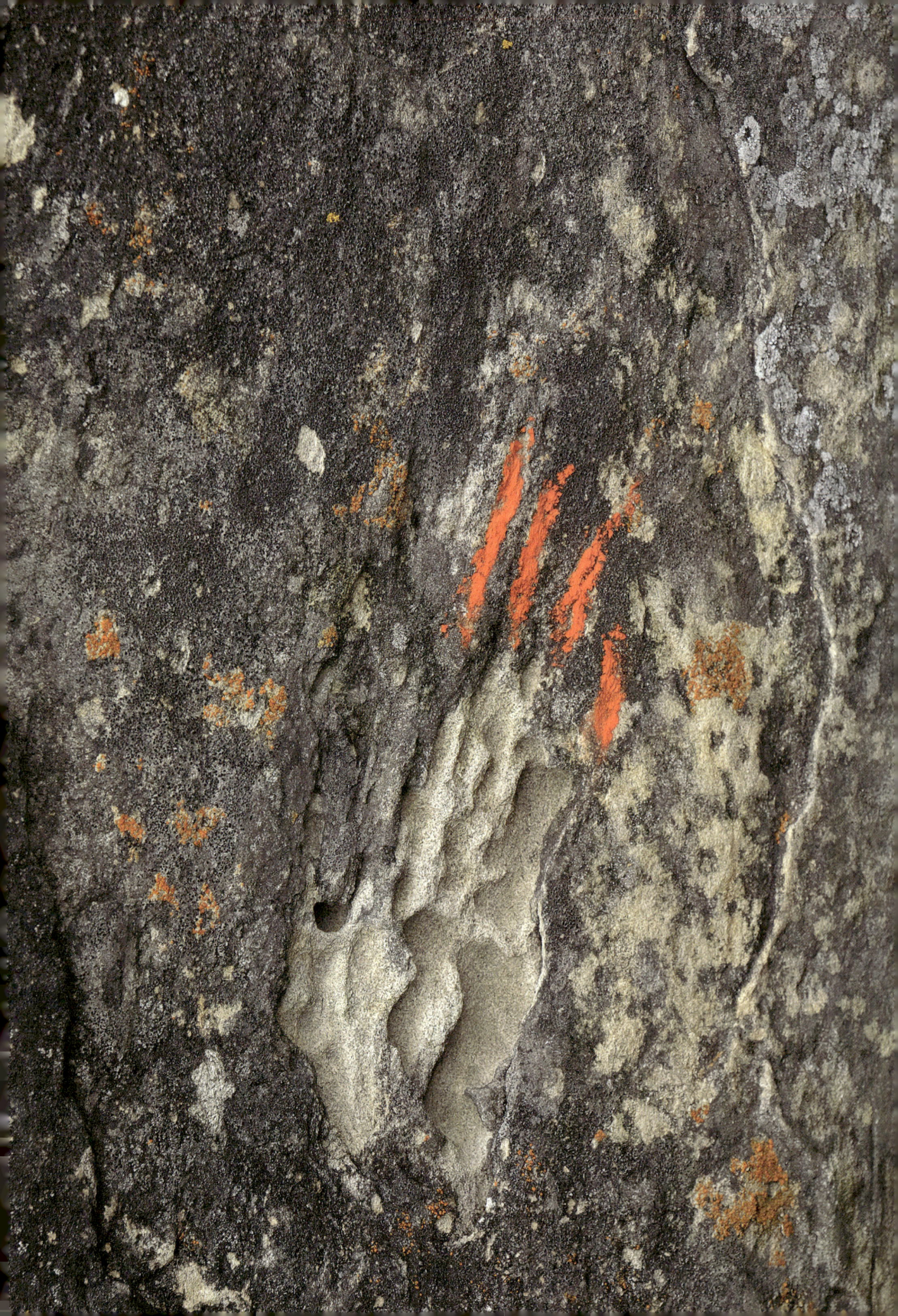

HUMANS

saliva, spit, and drool

saliva after chewing pumpkin
or other seeds

breastmilk

vein blood

menstrual, placental,
or umbilical blood

sweat from movement

sweat from fear

sweat from sex

eye water

tears from pain

tears from grief

tears from love

cum or semen

skin

and bones

and my own fat,

and all anthropogenic excrement
and petrochemical invention

an interstellar medium

dust to dust

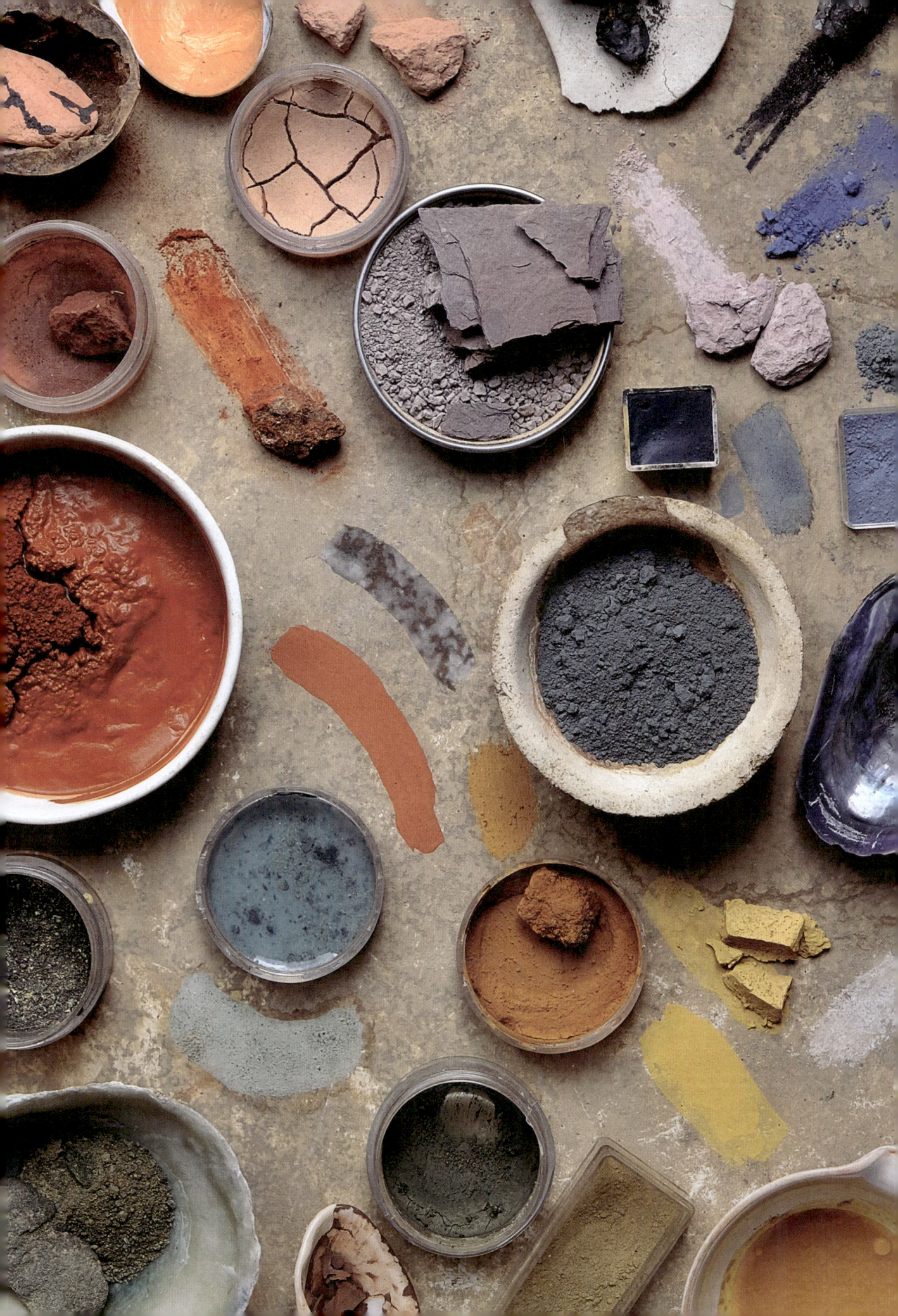

SEED RECIPES

Our planet is not so *blah*. Color is not just symbolic. Ochre affects our brain, bodies, and our soul. Material choices reveal intimate knowledge, kinships, access (whether consciously or not), and much more. Parking lot rock, cheap glue stick, rusty nail, Wet 'n' Wild lipstick, gum from a felled cherry tree: Our marks and art tell a story about who and where you are, and the deep memory, dreams, and lifeforce gathered and shared together for hundreds of thousands of years.

   Some questions I ask myself in my own creative process: How do my materials invite empathy and intimacy with Earth, with other creatures and nonhumans? Could they renew and reconcile ecological connection? What is around where I live? What is available? What's being offered? Are there invasive trees who need pruning, whose ooze-filled wound might be of creative benefit? Is there a dead animal on the road whose body could be joyfully honored? How am I bound? Am I listening? Am I being a good guest? Am I honoring my specific culture and engaging my ancestral burden? Does this work require yolk of egg or tear of tree? Does my work need spit or crushed bone? Are there wastewaters that need some tending? Acid-mine

rivers that could heal from extracting the ochre? Is it raining in my cup? Is the snow melting? Is there a river flowing under the cement? What's in my local landfill? Growing in the cracks of my street? Can I scrape fresh pigment off that crumbling white building wall? Gather brick dust? Old makeup? How do I extract pearl or rust color from my own heart and spear? Are you acknowledging the rocks in your head? Let earth remain earth? Let this book be a rock?

   I find experimental process both freeing and simple. But don't get me wrong, I realize that mainstream culture often errs on the side of detailed precision, and I, too, deeply value knowledge traditions as important springboards. I am offering recipes and examples here as a jumping-off point—they are imaginal portals and germinal seeds. I encourage you to adapt and adjust these recipes relevant to your own creative needs, culture, and traditions, perhaps with the guidance of a living teacher who has regional knowledge.

Note: Recipes are designed for immediate experimentation, not a commercial shelf life.

Land Color

1 part soil (i.e., pigment)

- Rub this anywhere you want raw color.

Personal Pigment

1 part pigment from one place

1 part another pigment or two or three or more from many different places

+/- fluid, time, soul

Art Water

(e.g., color, ceramic stain, wall paint)

1 part pigment

2 parts fluid

+/- extra drops of liquid until consistency desired

+/- essential plant oils

Note: Art water is paint or mud dye that will stain but may or may not leave a permanent mark on the surface it is applied to (as there is no technical binder or adhesive element).

Edible Ferment

(fermented seafood like Iranian suragh)

1 part pure ochre soil

2 parts seawater (or 1 part water and a bunch of sea salt)

1 part plant or creature (little fish, for example)

Mythic Blood

1 part red ochre pigment

10 part grain-based alcohol (beer, wine)

Safety note: See page 197.

Example:

FRESH WATERCOLOR PAINT WITH PLUM GUM

- Find a tree that's been trimmed or wounded and has healing gum ooze or sap (plum, cultivated cherry, invasive acacia, almond, and many others make suitable gum). If there isn't one locally, try to source ecologically restorative and fair-trade acacia gum (hashab) from suppliers protecting against desertification in the African Sahel region.

- Take a chunk. Dissolve in hot water (try 4 parts water to 1 part gum).

- I mix small amounts of paint at a time using 1 teaspoon of pigment to 1 dropper full of gum liquid. Add more gum or pigment as needed for desired consistency, as each pigment will absorb gum water differently. Add a drop of honey and a drop of antifungal oil if desired to prevent mold and/or to add other qualities (like aromatherapy). The texture of the paint is largely determined by the quality and grind of the pigment. The flow of the paint is influenced by the type and amount of gum and water and honey used.

Tree Water

(e.g., watercolor, ink)

1 part pigment

1 part dissolved tree gum (often gum arabic from Acacia trees)

some water

dash of honey

dash of antifungal plant oil (thyme, lavender, clove, etc.)

Example:

WALNUT OIL PAINT
WITH LAVENDER

(Pigment plus tree nut oil mixed with low-altitude broad-leaf lavender was sometimes used by Leonardo da Vinci.)

- Put 2 tablespoons of pigment on a smooth surface (glass, marble, smooth stone).

- Make a crater in the middle, like a volcano (if you've ever made fresh pasta, this will be familiar).

- Add 1 or 2 drops walnut oil in the crater.

- Blend together with a palette knife.

- Add more oil as needed to make a creamy paste.

- Every pigment is unique in how much oil it will need to absorb.

- If desired, add a drop of lavender spike oil (note the broadleaf variety is different from true lavender), which will thin the paint and add a delicate aroma.

Seed Paint

(e.g., oil paint, ink, drinks)

1 part pigment

1 part juice or oil of seed or plant (e.g., flaxseed oil for modern oil paint, plant alcohols and vinegars for ink)

Example:

TRADITIONAL
WOOD PAINT

Often used on furniture,
but used on wood paneling or walls as well.

- Pour a quart of skim milk in a bowl with juice from one lemon or a ¼ cup (60 ml) vinegar. (This is a good way to save and use old milk, also.) Let sit for a day or overnight, to curdle. Strain through cheesecloth and keep the curd. Mix with 2 to 4 tablespoons pigment (add more or less for desired tint strength).

- Some folks add a little bit of lime (chemical lime from limestone or calcium carbonate, not the fruit!).

- Paint on wood! The paint will smell like old milk, but not forever.

Milk Paint

3 parts curdled fresh non-fat skim milk (curd) or
dry milk (nonfat, casein, or otherwise)

1 lemon

1 part pigment

Example:

HEIDI'S INNER HOUSE PAINT

- Prepare the flour paste: Get local wheat and grind it into fine flour. Or get flour that's already ground. (I prefer flour that's been grown and milled from land/soil I live near.) Whisk together 1 cup (240 ml) water and 3 tablespoons flour. Heat on the stove over medium heat, whisking constantly (activating the "glue" part of the flour), making sure there are no lumps. Once the liquid begins to thicken into gravy or paste, pour into a bowl to cool (it will continue to thicken). Fresh paste can last a few days stored in the fridge or a cool place.

- Make paint: Stir a spoonful of cooled flour paste with a spoonful of pigment. If you want, add a dash of pigment/soil from the field where the wheat is grown. Add drops of water or anti-fungal oil for consistency as desired.

- This paint can be applied directly to many surfaces of your home (inner or outer): stone, natural plaster, wood, paper, skin, etc.

Grain Paint

1 part flour paste of wheat, barley, rice, or other grain
(see below)

1 part pigment

+/- water as needed
(for consistency)

+/- antifungal oil if desired
(to inhibit mold, e.g. thyme, clove, lavender, wintergreen, oregano, lemongrass, clary sage, eucalyptus, etc.)

Note: Although it binds strongly and easily, grain paint is not permanent and can be washed off or removed with water.

Example:

EGG TEMPERA PAINT

- Crack an egg and carefully move the yolk between your hands to remove the clear white.

- Take a needle and poke the yolk, gently squeezing the pure yolk into a cup (being careful not to mix in the slightly thicker membrane that surrounds it).

- Mix a teaspoon of pigment into the yolk with a brush (or spoon is fine too).

- Add drops of water (or wine or vinegar) until desired consistency.

- Add drop of antifungal oil (if desired).

- Add more pigment, if needed.

- This paint is best layered in thin layers (works well on prepared wood panel or canvas, but also works on paper and several other surfaces).

- Use immediately and make fresh every time you want to use.

Embryo Paint

(e.g., egg tempera paint, icon paint, egg face mask)

1 bit of pigment

1 egg

*for "egg tempera paint" use yolk only,
free of thin surrounding membrane and whites*

*for "glair" (used for glazing or sizing) use beaten and
settled egg whites only*

some water

+/- vinegar or dry white wine or other alcohol
(as preservative)

dash of essential oil (for scent) or antifungal oil
(to inhibit mold)

Example:

SOFT PASTEL OR DRAWING STICK

- Put 1 teaspoon powdered gum tragacanth and a generous ½ cup (120 ml) water in a closed jar to dissolve over two days. When ready to use, warm up on the stove over medium heat. Add beeswax (if desired) into the liquid, simmering until beeswax is melted. Some people will strain through a cheesecloth (if the gum doesn't dissolve all the way).

- Mix some pigment (try starting with 2 tablespoons) with a dash of chalk or calcium carbonate or pumice for texture (play around with this!).

- Form the pigment into a small mound.

- Add a few drops of liquid gum and a few drops of water to the pigment. Mix with your hands until the consistency forms a dough. Roll into snakes and cut into cylinders.

- Every pigment will absorb binder and work differently. Experimenting with the amount of gum and water, and adding a little oil or beeswax, can be a fun way to experiment. Personally, I just use a little gum water and pigment to form a consistency I like.

Color Stick

(e.g., pastels, crayon, pencils)

4 parts pigment

1 part water (+ more drops as needed until rollable dough)

+/- gum tragacanth (a legume) or preferred binder

+/- a little chalk (for smoothness)

+/- a little melted beeswax (for waxier texture; optional)

+/- pumice (for grip and grit)

Example:

LIP & CHEEK BALM WITH YARROW

- Take 1 tablespoon shea butter or other local plant butter or premade balm of oil and beeswax.

- Add a dash of your favorite color pigment.

- Add a drop of yarrow (or other plant essence that you love).

- Mix together until smooth. Apply!

Rouge

(e.g., lipstick, face balm, sunscreen)

1 part pigment

1 part fat (of plant or animal or self)—add more until consistency desired

dash of scent or essential oil, if necessary

Safety note: Natural pigments can contain trace elements of heavy metals and other cancer-causing elements. Thus, for anything that may be absorbed through the skin, accidentally swallowed (as in lipstick), or purposefully eaten, I officially must recommend you only use pigments you've had tested in a local soil or chemistry lab for full elemental analysis.

When in doubt, keep it simple:

Break stone, mix dust with waters.

(There are more or less laborious ways to do that.)

Play with color somewhere.

(There's a gazillion choices of technique and style and surface.)

Ask for diverse, expressive life to remain on Earth.

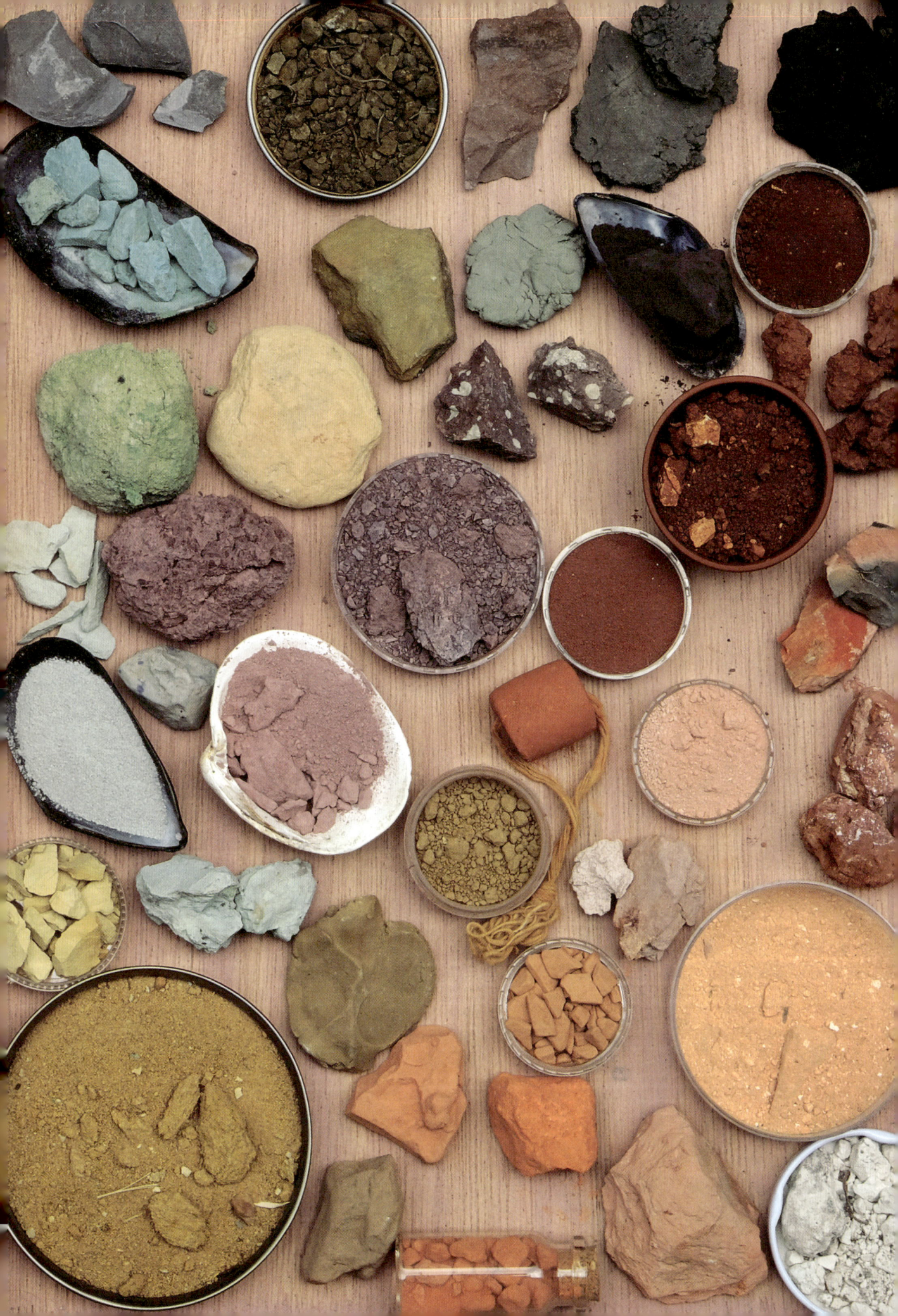

A QUICK REFERENCE CHART

---

earth pigment

+ nothing = **colorful dirt or medicine**

+ nothing, calcium carbonate, talc +/- others = **cosmetics**

+ water or booze = **faux blood, iron-enriched alcohol**

+ water and/or fat = **body paint, sunscreen (if worn in the sunlight)**

+ water and/or fresh soy milk = **textile paint**

+ water and straw (and clay-rich ochre) = **adobe bricks**

+ gum resin, water, honey, and/or antifungal oils = **watercolor paint**

+ acacia gum = **adhesive (or paint)**

+ flaxseed oil (i.e., linseed) = **oil paint (or stomach pain relief)**

+ egg yolk (sac removed) and/or water = **egg tempera paint (or healing clay face mask)**

+ calcium carbonate (i.e., chalk), water, and/or soap = **pastels, pencils**

+ wheat paste (or rice paste) = **plaster paint (wall paste, papier-mâché, and bookbinding glue)**

+ hot or cold beeswax = **encaustic**

+ beeswax and olive or other oil or butter = **tinted balm, lipstick**

+ pine tar or pitch = **wood stain (or barn and exterior paint)**

+ wood ash and/or water = **ceramic glaze**

+ lime (burnt limestone) and water and/or sand = **plaster and cement**

+ shea butter or other fats/lipids = **cosmetic rouge, lipstick**

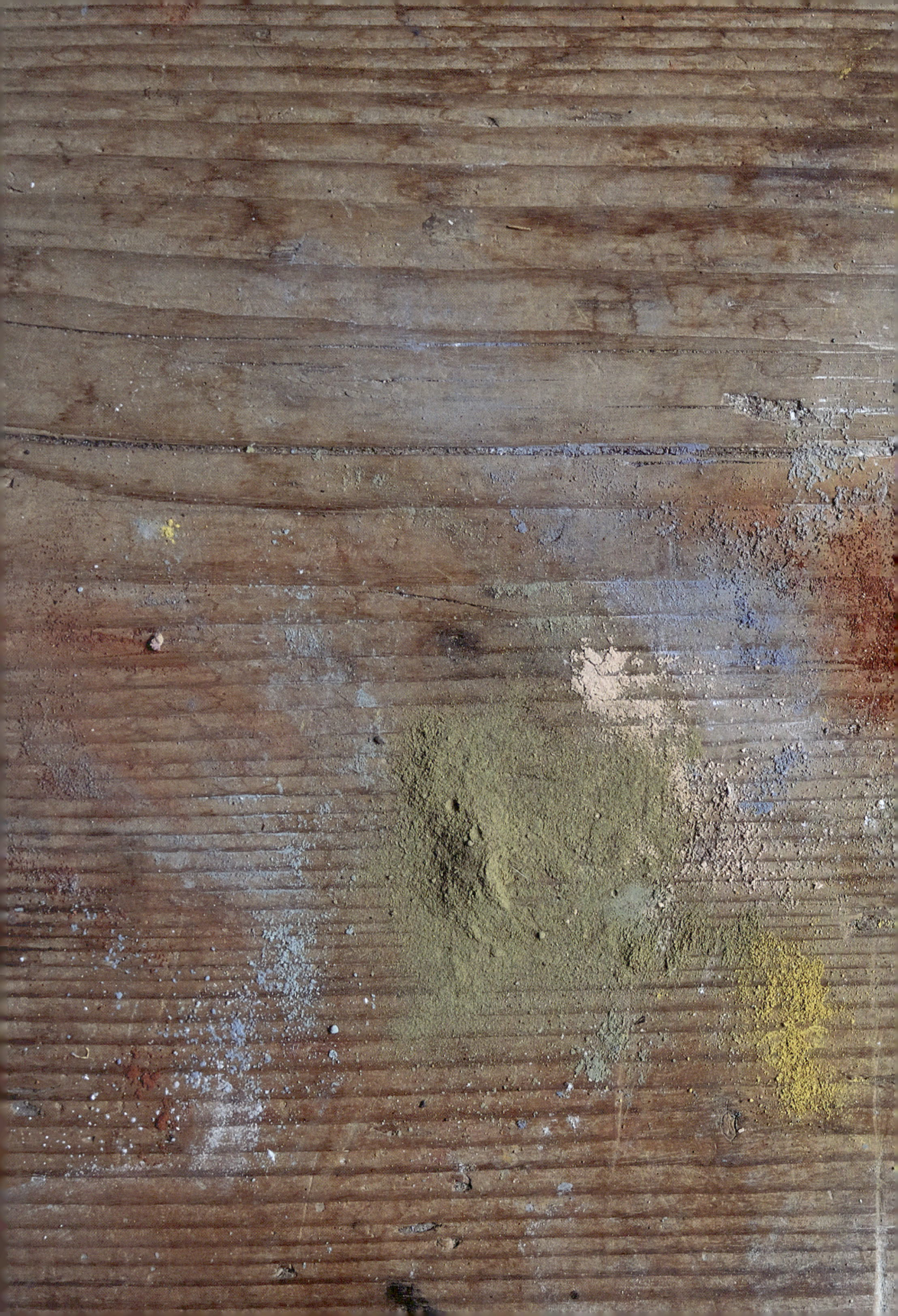

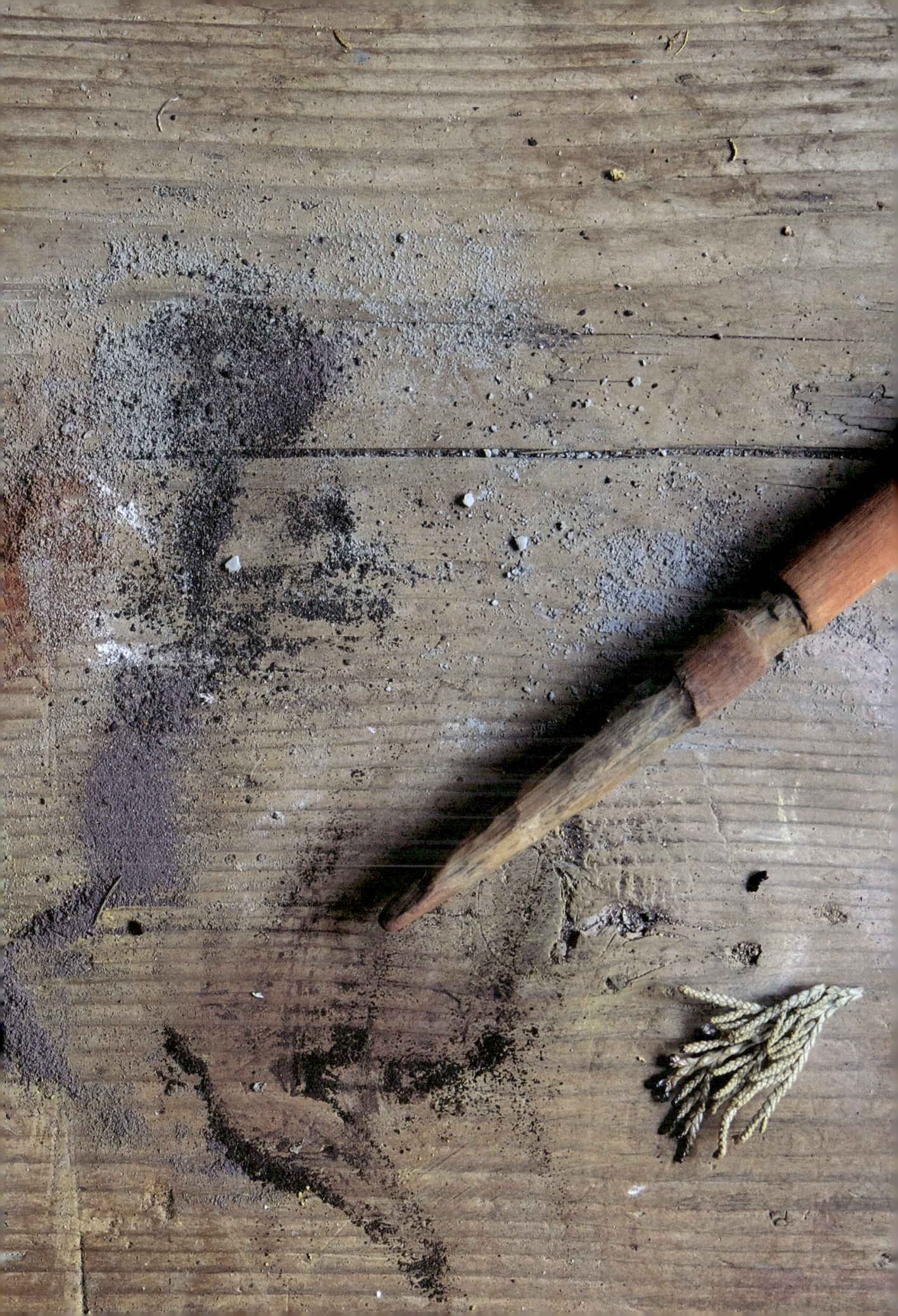

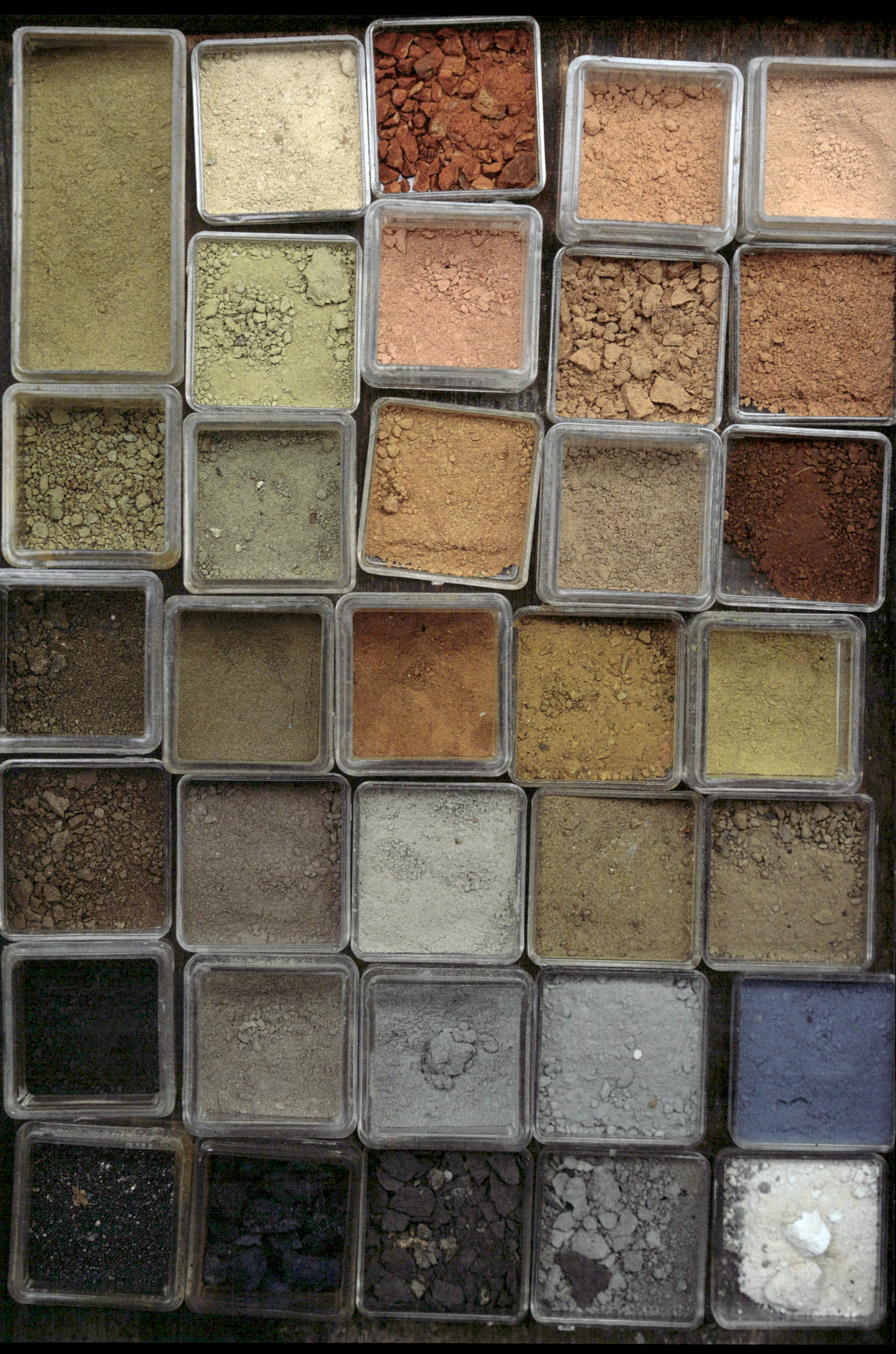

NOTES

1. Sigo, *Guard the Mysteries*, 40.

2. Steven's teaching and scholarship is also influential; including Goodman, *The Buddhist Psychology of Awakening*.

PREFACE

1. Kimmerer, *Braiding Sweetgrass*.

ON OCHRE

1. Inspired by Povinelli, *Geontologies*.

2. Elizabeth Kolbert, *The Sixth Extinction: An Unnatural History*.

3. As said on *The Tonight Show Starring Jimmy Fallon*, April 2018, https://www.youtube.com/watch?v=YplKPH_qcRw.

4. Introduced in Popelka-Filcoff and Zipkin, "The Archaeometry of Ochre Sensu Lato."

5. Personal communication from Elizabeth Velliky.

6. For an online summary of Iron Oxide futures see: https://www.futuremarketinsights.com/reports/iron-oxide-market.

7. Velliky et al., "The Ochre Experience Model (OEM): towards a deep-time perspective on the earth material heritage of ochre."

8. Finlay, *Color*, 25.

9. From Taungurung elder and artist Uncle Roy Patterson (1940–2017), as quoted by Annette Sax (@wa_ring_) on Instagram.

10. Faivre, *Iron Oxides*, 10.

11. Foraged from a reference in Palmer, *In the Aura of a Hole*, 90.

12. Known as the Chandrasekhar limit. For more on the topic, visit sciencedirect.com/topics/physics-and-astronomy/chandrasekhar-limit.

13. Thanks to astrophysicist Reed Garber for the scientific review on this point.

14. As Jamie Fox puts it, "Others, however, think iron is incidental to blood's colour, and that redness is actually created by the ligands (bonds) between the haem [blood] and oxygen." From Fox, *The World According to Colour*.

15. Evidence, albeit somewhat sporadic and speculative, is cited in various anthropological studies, such as Hodgskiss, "Ochre Use in the Middle Stone Age," and Watts et al., "Early Evidence for Brilliant Ritualized Display."

16. Roebroeks et al., "Use of Red Ochre by Early Neandertals."

17. Rifkin, Riaan, "Ethnographic Insight into the Prehistoric Significance of Red Ochre."

18. For a talk summarizing the history of ochre, watch Dr. Tammy Hodgskiss, youtube.com/watch?v=24COZZTF2ys. Excellent review of red ochre use in the human cultural record can be read in Watts, "Red ochre, body painting, and language: interpreting the Blombos ochre."

19. For more on prehistoric cosmic depictions, see Marchant, *The Human Cosmos*.

BIOGENIC OCHRE

1. Several species and their lifestyles are discussed in Chan et al., "The Architecture of Iron Microbial Mats."

2. Not only bacteria, but all organisms from bacteria to birds to humans require iron with only one exception (lactic streptococci), as noted in Holland and Turekian, *Treatise on Geochemistry*.

3. Captive breeding parents are observed going immediately to their nest after having had a bath in ochre water and rubbing it on their young, as documented in Arlettaz et al., "Deliberate Rusty Staining of Plumage in the Bearded Vulture: Does Function Precede Art?"

4. Specifics of ochre plumage protection further explored in Tributsch, "Ochre Bathing of the Bearded Vulture."

5. Beautifully documented in MacDonald et al., "Hunter-Gatherers Harvested and Heated Microbial Biogenic Iron Oxides to Produce Rock Art Pigment."

6. From Collins, *A History of the Animal World in the Ancient Near East*.

7. Margalida and Villalba, "The Importance of the Nutritive Value of Old Bones in the Diet of Bearded Vultures (*Gypaetus Barbatus*)."

8. "The Lammergeier," Tibetpedia, April 28, 2017, accessed July 27, 2022, https://tibetpedia.com/lifestyle/the-lammergeier.

9. Observations of vultures storing dried bones for safekeeping (as if a high mountaintop acts like a

modern refrigerator) by Houston and Copsey, "Bone Digestion and Intenstinal Morphology of the Bearded Vulture."

10. Hai, Van Cam, and Lauren Shapiro (trans.), "The Last Bird Burial Master," *91st Meridian* 6:1 (Spring 2008), https://iwp.uiowa.edu/91st/vol6-num1/the-last-bird-burial-master.

11. Sky burial practice isn't limited to Vajrayana Buddhists of the Himalayas. A similar practice is found in several cultures worldwide. Most closely to my own home in North America, the Cherokee, Sioux, and Lakota cultures are documented doing a similar sky or "air burial," practice, placing the dead on offering platforms and in trees. The Cherokee word for *vulture* is said to translate into English as something like "peace eagles."

12. Hai and Shapiro, "The Last Bird Burial Master."

13. As noted by German chemist, Andreas Libavius in *Alchymia*, 1606.

14. Egan, Timothy, "The Death of a River Looms Over Choice for Interior Post," *New York Times* (Jan. 7 2001), accessed September 26 2022, https://nytimes.com/2001/01/07/us/the-death-of-a-river-looms-over-choice-for-interior-post.html.

15. A quick news search will reveal several global examples: Brazil's Tapajas River, Paraopeba River and Doce River disasters, Slovakia's Slana River, Colorado's Animas River, Spain's Rio Tinto River, China's Ting River basin, Russia's Ural River, South Africa's Witwatersrand regional streams in the Vaal and Limpopo River watersheds. Denmark even introduced a legal framework, "The Ochre Act," designed to mitigate toxic iron oxide drainage in their wetlands, rivers, and streams.

16. To learn more about how not just vultures, but artists are transforming toxic mine waste into useable paint today, see John Sabraw's work, available online at johnsabraw.com.

17. In this passage the vulture goddess Anat is speaking a message to storm-god Baal. I translated the original Ugaritic passage from cuneiform tablet facsimile printed in Bordreuil, Pierre, and Dennis Pardee, *A Manual of Ugaritic* (Winona Lake, IN: Eisenbrauns, 2009): 164–167. Standard translation available in Smith, Mark S., and Wayne T. Pitard, *The Ugaritic Baal Cycle. Volume II* (Leiden: Brill, 2009): 202–203.

18. Great discussion in Kerényi, *The Gods of the Greeks*, 156.

19. Kerényi, *The Gods of the Greeks*, 158.

20. Beautiful account of these skills and their relevance in Stein, "Hephaistos."

21. Detienne and Vernant, *Cunning Intelligence in Greek Culture and Society*, 194.

22. Wonderful in-depth account of Hephaestus's embodied capacities in Hall, "Hephaestus the Hobbling Humorist."

23. I am guided here by Nor Hall's musing, that "Hephaistos as 'midwife of Earth's birth'" in a footnote in Hall, *Irons in the Fire*, 50.

24. Eliade, *The Forge and the Crucible*, 95.

25. According to Aeschylus, *The Libation Bearers*, 631–34, a Lemnian deed is a crime considered to be the greatest of all transgressions.

26. Also known as Lemnian *sphragis* (Greek, "seal"), and other far-famed names: Lemnian *miltos* (Greek, "red earth"), Lemnian *rubrica* (Latin, "special red"),

terra Lemnia (Latin, "earth of Lemnos"), Lemnian *terra sigillata* (medieval Latin, "sealed earth"), clay bearing little images, or stamped earth, *axungia solis* (Paracelsian term for "sun-grease"), and *tin-mathtum* (Arabic, "sealed earth metal").

27. "Shade of the Sun" is also one of the many secret alchemical names of the philosopher's stone, according to Ruland, *A Lexicon of Alchemy*.

28. Galen describes the ritual in 166 AD: "The priestess collects [the earth], to the accompaniment of some local ceremony, no animal being sacrificed, but wheat and barley being given back to the land in exchange. She then takes it to the city, mixes it with water so as to make moist mud, shakes this violently and then allows it to stand. . . . She takes small portions and imprints upon them the seal of Artemis [the goat]; then again she dries these in the shade till they are absolutely free from moisture. . . . This then becomes what all physicians know as the Lemnian Seal," as quoted in Dasen, "Magic and Medicine: The Power of Seals."

29. Joos van Ghistele, a Dutchman who visited Lemnos in 1485, wrote, "Terra Sigillata is produced in a pool which dries up every summer and is full of water in winter. When this pool begins to dry up, a thick scum, variegated in colour, forms on its surface. This is skimmed off and laid on clean planks as required, according to the method in use locally. When dry, it is made up into round pellets or flat cakes." As mentioned in Photos-Jones et al., "Archaeological Medicinal Earths as Antibacterial Agents: The Case of the Basel Lemnian Sphragides."

30. A donkey is also the animal Hephaistos rides in several Greek images, vases, and described in mythic portrayals, thus the

presence of a donkey-cart likely signals Hephaistos's presence in the processing of the ochre.

31. Further reasoning in Hall and Photos-Jones, "Accessing Past Beliefs and Practices: The Case of Lemnian Earth."

32. Eliade, *Forge and the Crucible*, footnote on p. 70. For a deeper investigation on iron and bones of the "Earth-Mother," see pp. 43–70.

33. Schweitzer et al., "A Role for Iron and Oxygen Chemistry in Preserving Soft Tissues, Cells and Molecules from Deep Time."

34. Photos-Jones et al., "Archaeological Medicinal Earths as Antibacterial Agents: The Case of the Basel Lemnian Sphragides."

35. Athanassakis and Wolkow, *The Orphic Hymns*.

36. For more on Hephaistos epithets and embodied association, see Detienne and Vernant, *Cunning Intelligence in Greek Culture and Society,* specifically chapter nine, "The Feet of Hephaistos," 259.

37. MacGregor, "Medicinal *terra sigillata*: a historical, geographical and typological review," 115.

38. As well as controlled animal trials in which there are several cockerel survivors of viper bites after being fed Lemnian seals, as noted in Retsas, "Geotherapeutics."

39. Photos-Jones notes that a sacrifice of wheat and barley would be equivalent to "the acknowledgement that Lemnian Earth is a gift from the earth and therefore an indication of its real medicinal value" in "Accessing Past Beliefs and Practices: The Case of Lemnian Earth."

40. Based on Photos-Jones's research, the historical recipe likely contained: absorbing clay of layered aluminum silicates (montmorillonite, bentonite, illite), quartz, iron oxide of hematite and/or goethite, some dissolved alum (from hydrothermal waters); and fungal bioxanthracenes (likely from mixed in barley and wheat "blessings") and iron-loving microbial species native to the holy spring itself.

41. McClain, J. B., and K. Cooney, *The Daily Offering Meal in the Ritual of Amenhotep I.*

RED OCHRE

1. These categories are inspired by Popelka-Filcoff and Zipkin, "The Archaeometry of Ochre Sensu Lato."

2. Siddall, "The Origin of Ochres: Interbasaltic Beds."

3. Johnson, Sarah Stewart, *The Sirens of Mars: Searching for Life on Another World*, New York: Crown Publishing Group, 2020, quoted in Steve Squyres, *Roving Mars: Spirit, Opportunity and the Exploration of the Red Planet*, New York: Hyperion, 2005.

4. A longer exploration on hematite as "truth stone" can be found in Horowitz and Hurowitz, "Urim and Thummim in Light of a Psephomancy Ritual from Assur."

5. "Tasty red soil in Iran's Hormoz Island," CGTN, September 17, 2019, accessed July 27, 2022, news. cgtn.com/news/2019-09-17/ Edible-Hormoz-Island-in-Iran-K3VUuLhCLu/index.html.

6. From "Wars, Arms, Rams, Mars," in Hillman, *Mythic Figures*.

7. In Māori creation stories the first human being is "born of the red earth," and is "the origin of our menstrual cycles," as described in Montgomery-Neutze, "Ki Te Ao Mārama: Introduction to Colour Theory, a Tirohanga Māori."

8. Trungpa and Lief, *True Perception: The Path of Dharma Art*, 181.

9. From one of my favorite poems on red, the stunning "Rothko Chapel Poem," in Taggart, *On Music*, 132.

10. Rimell, *Liminal Contact*.

11. Farrelly, et al., "Competitors Who Choose to Be Red Have Higher Testosterone Levels."

12. Shinhmar, et al., "Weeklong Improved Colour Contrasts Sensitivity After Single 670 nm Exposures Associated with Enhanced Mitochondrial Function."

13. Ershoff, B.H. and Bajwa G.S., "Physiological Effects of Dietary Clay Supplements." Publication available from NASA online: https://ntrs.nasa.gov/ citations/19660023330.

14. Rogers, *Full Spectrum*, 20.

15. Franks, *Stones as Medicine*, 314.

16. Mai et al., "Characterization of cosmetic sticks at Xiaohe Cemetery in early Bronze Age Xinjiang, China."

17. Campbell, *Earth Pigments and Paint of the California Indians*, and his 2009 KPBS interview with Maureen Cavanaugh, available online: kpbs.org/news/2009/ apr/08/discovering-the-history-of-california-indian-art.

18. Excerpt from Pavini Moray and bronte valez interview, *Bespoken Bones* podcast #63, https:// bespokenbones.com/episode-63-what-happens-after-you-are-buried-alive-art-ancestors-and-black-erotic-power/.

19. In the neighboring language of goddess Anat (the vulture queen I mentioned previously, written about in Ugaritic myths), there was a cuneiform term akin to "inside-house," *kbd*, which had many meanings: entrails, organs, liver, bosom, heart, womb, mind, splendor, honor, inside, inner, innards, in the midst, within, immanent glory, heaviness, and

a welcoming. This term was associated with places in the earth, like fallow fields, where red ochre was offered down as a soothing balm. Painting red ochre down on the fields was a healing act, an attempt to nourish and pacify weather gods.

20. Thanks to Devon Deimler for insight on this point. Articulated more thoroughly in her thesis on aesthetics, "Ultraviolet Concrete: Dionysos and the Ecstatic Play of Aesthetic Experience." Here she states, "Etymologically and conceptually, 'aesthetic' harks back to the Greek *aisthesis* (to sense and perceive) and its deeper roots in the word *aisthomai* (to breathe in). This is also the meaning of the word inspire, or to be 'full of the god' ".

21. LaPointe, 2022, 209.

22. Goll, *Dreamweed*.

YELLOW OCHRE

1. Cornell, Rochelle M., and Udo Schwertmann, *The Iron Oxides: Structure, Properties, Reactions, Occurrences and Uses* (Germany: WILEY-VCH Verlag GmbH & Co., 2003), 3.

2. Hall, *Irons in the Fire*, 39.

3. Bachelard discusses iron's metaphoric, material, and imagination powers in several inspirational writings, including "The Cosmos of Iron" and *Earth and Reveries of Will*.

4. Rosen, "The Science of Climate Change Explained: Facts, Evidence and Proof."

5. Willis, "The Valor of Iron," in *The Emerald City*, 35.

6. As described by the powerful geographer Kathryn Yusoff. For more, see her talk, "Geo-Logics: Natural Resources as Necropolitics," youtube.com/ watch?v=QM8B-XZG8OQ.

7. Geist, Christopher, "The Works at Falling Creek, *Colonial Williamsburg Journal* (Autumn 2007), accessed July 27, 2022, https://research. colonialwilliamsburg.org/ Foundation/journal/Autumn07/ iron.cfm.

8. Keeler, *Edge of Morning*, 115.

9. Ochres could be associated with *ogres*, as mythologist Devon Deimler reflected to me. And there's curiously "a whole mythology [that] classes iron workers among the various categories of giants and demons," as Mircae Eliade points out in *The Forge and the Crucible*, 67.

10. Hauschka, *The Nature of Substance: Spirit and Matter*, 180. Thanks to Thomas Little for offering this esoteric nugget of insight.

11. Bachelard, *Earth and Reveries of Will*, p. 188.

12. Bucklow, *Red*, 175.

13. As noted in Heimann, R. B., 2002.

GREEN EARTH

1. For a very deep dive into place and existence, see Weiss, "*Buddhist Topology and the Practice of Lama Orgyan Khandro Norlha*."

2. Narodovskiy, "Boris and Gleb," as quoted in O'Hanlon, "Sankir—Underpainting of Flesh—Part 1."

3. An important historical note: verdaccio, like sankir, may technically also mean a formula of three non-green colors blended to become a green earth (usually indigo, yellow ochre, and lead white), or a blended green earth of various places, or a single green earth, depending on the workshop, recipe, locale, or spiritual tradition.

4. A term found in the poem "The Ruin." Translated by Patton, *Curious Masonry*, and further commented upon in Patton, *Dumuzi*.

5. "Painting the Verdaccio," Damian Osborne, accessed July 27, 2022, https://www.damianosborne.com/ painting-the-verdaccio.

6. Pastoureau and Lloyd, *The Colour of Our Memories*, 198.

7. From Vladislav Andrejev as translated to me by Nick Maione, devoted student of Andrejev's iconographic school.

8. A notable exception being Robert Fuller's dedicated chapter on green earth in: *Artists' Pigments*.

9. There's a fantastic exploration of Scheele's Green and psychology via Carl Jung in Mellick, *The Red Book Hours*.

10. Read exchanges via open-source compilation online: www. oregongeology.org/milo/archive/ Commodities/Pigments/Pigments. pdf.

11. As illustrated in Baty, *The Anatomy of Color*.

12. "The Bomb," *Time*.

13. In conversation with Dr. Morgan Williams.

14. Dário et al., "Evaluation of the Healing Activity of Therapeutic Clay in Rat Skin Wounds."

15. Alavi et al., "The Effect of a New Impregnated Gauze Containing Bentonite and Halloysite Minerals on Blood Coagulation and Wound Healing."

16. Further deep dive into uranium and local Indigenous land history can be found in Puglionesi, *In Whose Ruins*.

17. From "Wars, Arms, Rams, Mars," in Hillman, *Mythic Figures*, 135.

18. Found online at earthcitizens.net/ nuclear-guardianship/r33resp.htm.

19. As described in Hansell, 2007.

BLUE OCHRE

1. Thanks to Jason Logan for this realization, in a personal note he wrote me responding to a piece of vivianite and an early version of the "Blue Clay" essay: "Carbuncular (Blue): It's like a skin rash in various places but is communicating / connected below the surface. Almost like the way that mushrooms come up. Also a carbuncle stone is a red or blue gem that has an inner glow that you can only see from the right distance away. Also it's used as a metaphor in the novel Mrs. Bridge at the very end to describe a kind of unplaceable magical discomfort."

2. An enigmatic term used in painting workshops at the time of Rembrandt for vivianite from his regional/local Dutch bogs.

3. Harjo, *Conflict Resolution for Holy Beings*.

4. See several anecdotal accounts in Bulkeley and Bulkley, *Dreaming Beyond Death*.

5. Hillman, *Mythic Figures*.

6. Nelson, *Bluets*, 52.

7. From the documentation in Swanton, *Tlingit Myths and Texts*.

8. Thanks to Nahaan, Tlingit language educator and incredible artist, for confirming the word.

9. Fienup-Riordan, *Yuungnaqpiallerput/The Way We Genuinely Live*, 27.

10. Yup'ik artist wished to remain anonymous and asked not to have the Yup'ik spoken word written and shared, but the English translation was given permission to share. Thanks to Tilke Elkins for the connection.

11. Ancheta, "Revealing Blue on the Northern Northwest Coast."

12. Today, however, most blue pigments used on Northwest Coast objects are post-contact commercial pigments (laundry bluing, synthetic ultramarine, Prussian blue) that do not resemble vivianite materially.

13. Excerpt from "Debris of Life and Mind" in Stevens, *The Collected Poems of Wallace Stevens*, 356: "She will think about them not quite able to sing. / Besides, when the sky is so blue, things sing themselves, / Even for her, already for her. She will listen / And feel that her color is a meditation."

14. Jensen, *The Myth of Human Supremacy*, 40.

15. From a private conversation with an operator (name kept anonymous).

16. As Biskakone Greg Johnson told me in private conversation. The animals eat seasonal plants that have medicine in them, thus the fat of the animal contains medicine from the plants.

17. An accessible popular article on the grotesque side of vivianite can be found in "This Strange Mineral Grows on Dead Bodies and Turns Them Blue," sciencealert.com/ vivianite-the-blue-mineral-that-eerily-turns-buried-bodies-blue.

18. Nickles, "On the Presence of Vivianite in Human Bones."

19. For a brilliant long read on similar topics involving vivianite's cousin mineral, struvite, see Rosen, "Humanity is Flushing Away One of Life's Essential Elements," where she points out, "the writer and chemist Isaac Asimov, in a 1959 essay, dubbed phosphorus 'life's bottleneck.'"

20. Egger et al., "Vivianite is a Major Sink for Phosphorus in Methanogenic Coastal Surface Sediments," and Kubeneck et al., "Phosphorus Burial in Vivianite-type Minerals in Methane-rich Coastal Sediment."

21. Mark van Loosdrecht oversees these projects, and more can be found at wetsus.nl/european-projects/vivimag and in Prot et al., "Efficient Formation of Vivianite Without Anaerobic Digester: Study in Excess Activated Sludge."

22. To learn more about this process, see Wilfert, "Phosphate Recovery from Sewage Sludge Containing Iron Phosphate," and if you can find a physical copy of the dissertation, you'll see a painting I made for the cover, made with wastewater vivianite.

BLACK OCHRE

1. Dolgopyat, *The Quality of Time*.

2. Taylor et al., "Satellite-Altitude Horizontal Magnetic Gradient Anomalies Used to Define the Kursk Magnetic Anomaly."

3. Duncombe, "The Herky-Jerky Weirdness of Earth's Magnetic Field."

4. Chu, "Origins of Earth's Magnetic Field Remain a Mystery."

5. For an amazing account of geologic time, that covers far more than just magnetite's influence, see Bjornerud, *Timefulness*.

6. A quote from Philostratus, in Bucklow, *The Alchemy of Paint*.

7. Béguer-Pon, Mélanie et al., "Direct observations of American eels migrating across the continental shelf to the Sargasso Sea," *Nature Communications*, October 27, 2015, accessed July 22, 2022, doi: 10.1038/ncomms9705. See also Wikipedia, s.v. "American eel," last edited July 15, 2022, accessed July 27, 2022, https://en.wikipedia.org/ wiki/American_eel.

8. As outlined in Rifkin, "The Symbolic and Functional Exploitation of Ochre During the South African Middle Stone Age," 159–161.

9. Ibid.

10. For more on iron accumulation in the brain, see the in-depth overview, "Iron Oxides in the Human Brain," Faivre, *Iron Oxides*, 143–166.

11. I'm not really surprised that large-scale chemistry relies on dark metal arts. Chemistry originates from the Arabic al-kimiya, later chēmia in Greek, meaning "the blackest earth," inspired by the local Nile river, rich in magnetic mud and, in alchemy is perhaps associated with a fertile, flowing portal in the eye's pupil.

12. Bjornerud, *Timefulness*.

WHITE EARTHS

1. My use of "a dark peace" refers to a line in the poem "Rock and Hawk" by Robinson Jeffers.

2. MacDonald, "Histories of Selected Artists' Pigments."

3. Wagner and Aspenberg, "Where Did Bone Come From?"

4. Hirasawa and Kuratani, "Evolution of the Vertebrate Skeleton: Morphology, Embryology, and Development."

5. Bjornerud, *Timefulness*, 51.

6. Weinberger, *Angels & Saints*, 279–280.

7. DeLanda, *A Thousand Years of Nonlinear History*, 28. The entire first chapter of *A Thousand Years of Nonlinear History* further outlines the topic of skeletal architecture.

8. See McLeod, "Shellmounds of the Bay Area," or visit Save the Shellmounds (shellmound. org) and Sacred Land Film Project (https://sacredland. org/shellmounds-of-the-bay-area-united-states/) which both host several articles on Ohlone shellmound activism.

9. Earling, *The Lost Journals of Sacajewea*.

10. Megan Cope made a series of shellmounds for different regions; see her artist website for more: megancope.com.au.

11. See megancope.com.au/works/re-formation-part-iii-dubbagullee for more.

12. As published in Hudson, *Mana Whenua*.

13. Williams, *Erosion*.

14. Thank you to soil scientist Jessica Chiartas for helping me better understand this connection.

15. Morton, *All Art is Ecological*.

FROM ROCK TO PAINT

1. Thanks to David Cranswick for this insight. See his website for workshops: davidcranswick.com/the-alchemy-of-colour.

ACKNOWLEDGMENTS

1. Hall, *Irons in the Fire*, 41.

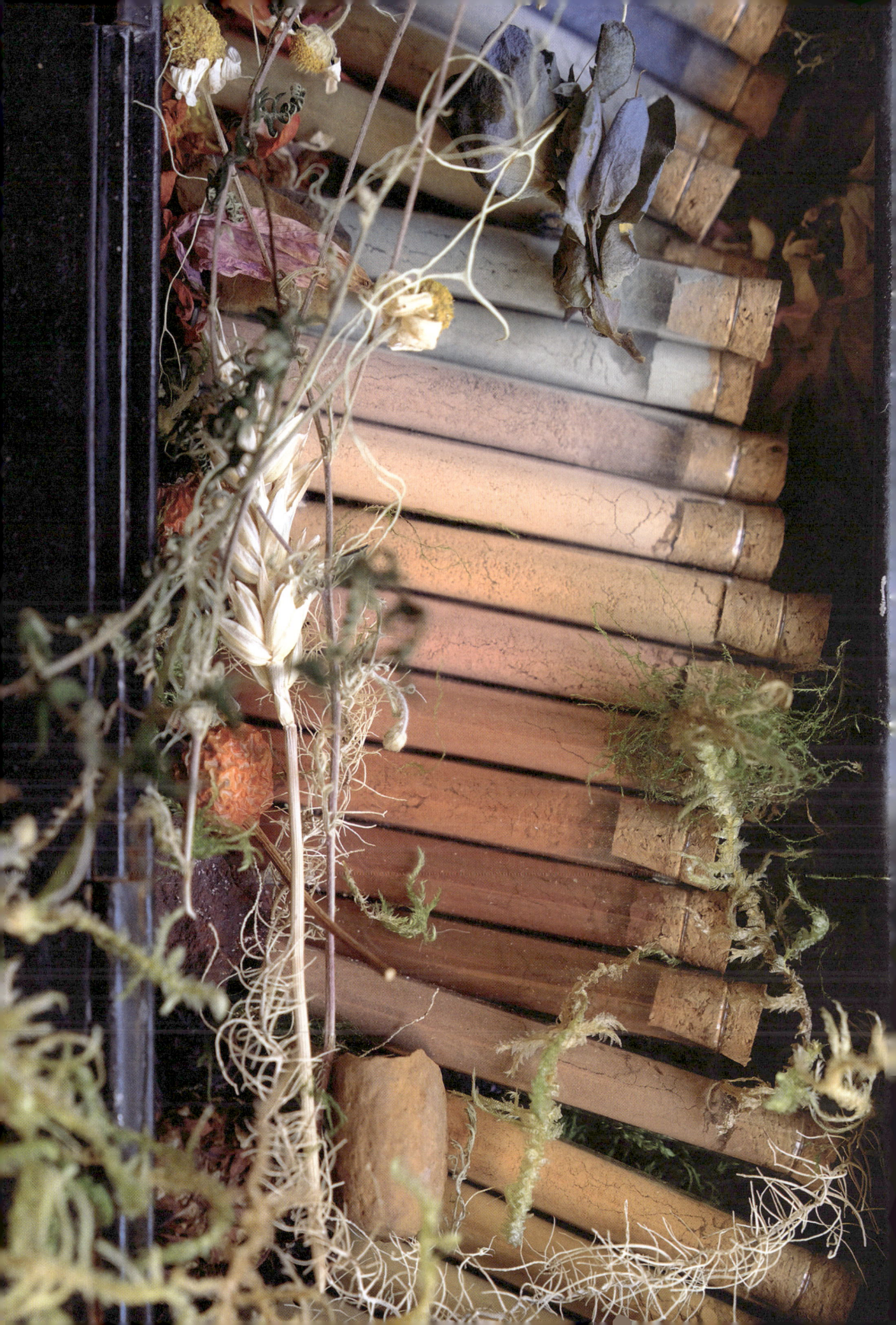

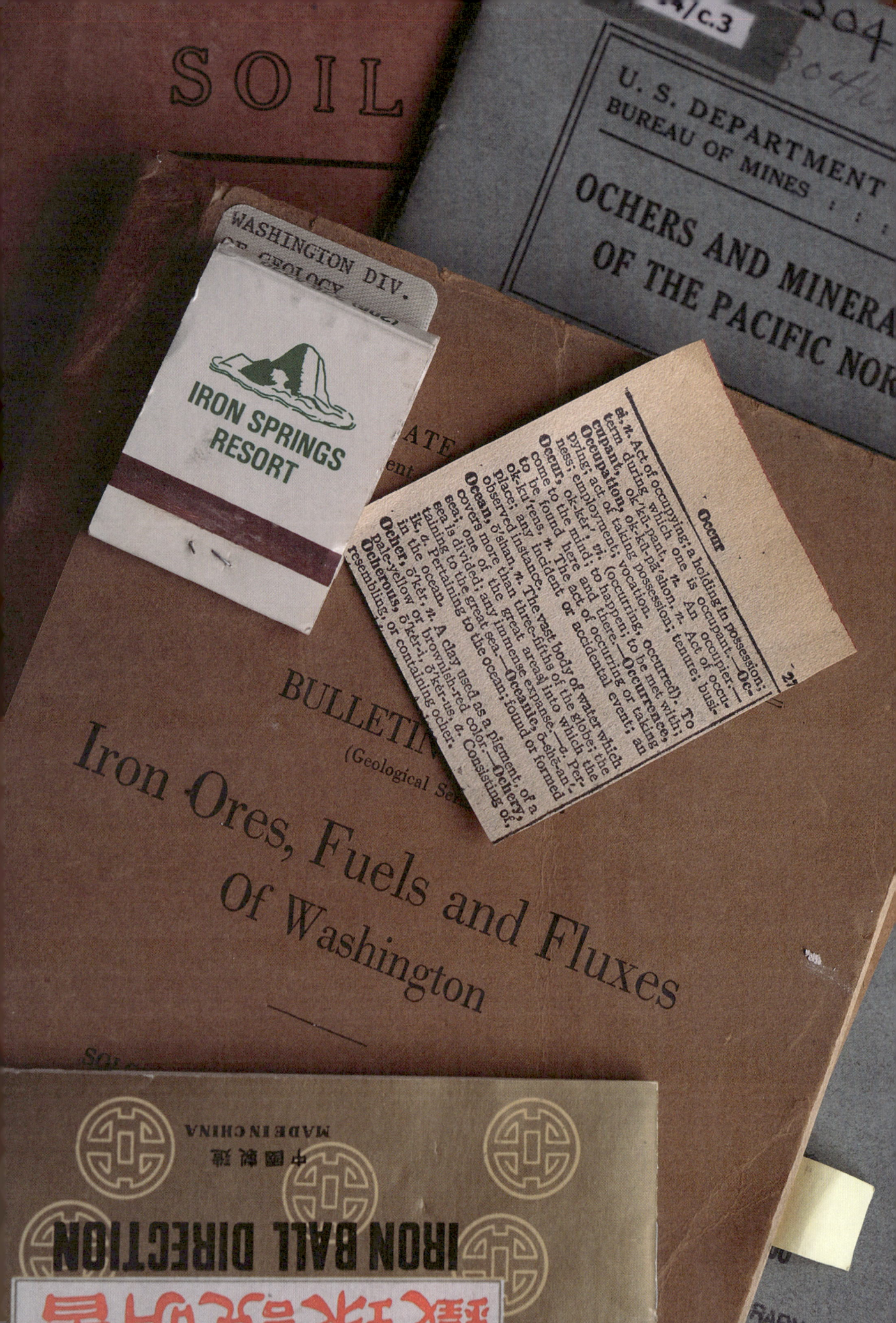

RESOURCES

Ball, Philip. *Bright Earth: The Invention of Color*. New York: Farrar, Straus and Giroux, 2002.

Campbell, Paul Douglas. *Earth Pigments and Paint of the California Indians: Meaning and Technology*. Los Angeles: Paul Douglas Campbell, 2007.

Finlay, Victoria. *Color: A Natural History of the Palette*. New York: Random House, 2004.

Kremer Pigmente Recipe Book, Germany, 2018. Select recipes available online: https://shop.kremerpigments.com/us/information/recipes/.

Logan, Jason. *Make Ink: A Forager's Guide to Natural Inkmaking*. New York: Abrams, 2018.

Massey, Robert. *Formulas for Painters*. New York: Watson-Guptill Publications, 1999.

Medlej, Joumana. *Inks and Paints of the Middle East: A Handbook of Abbasid Art Technology*. London: Joumana Medlej, 2020.

Neddo, Nick. *The Organic Artist: Make Your Own Paint, Paper, Pens, Pigments, Prints, and More from Nature*. Beverly: Quarry Books, 2015.

Rogers, Adams. *Full Spectrum: How the Science of Color Made Us Modern*. Boston: Houghton Mifflin Harcourt, 2021.

Ross, Caroline. *Found and Ground: A Practical Guide to Making Your Own Natural Paints*. Kent, UK: Search Press, 2023 (in press).

St. Clair, Kassia. *The Secret Lives of Color*. New York: Penguin Books, 2017.

Thomas, Anne Wall. *Colors from the Earth: The Artists' Guide to Collecting, Preparing, and Using Them*. New York: Van Nostrand Reinhold Company, 1980.

Webster, Sandy. *Earthen Pigments: Hand-Gathering & Using Natural Colors in Art*. Atglen, PA: Schiffer Publishing, 2013.

OCHRE & IRON OXIDES

Artists' Pigments vols. 1–4. London, 2012–2022. Vols. 1–3 available online: www.nga.gov/research/publications/pdf-library/artists-pigments-vol-1.html.

Cornell, Rochelle M., and Udo Schwertmann. *The Iron Oxides: Structure, Properties, Reactions, Occurrences and Uses*. Weinheim: Wiley-VCH, 2003.

Eastaugh, Nicholas, Valentine Walsh, Tracey Chaplin, and Ruth Siddall. *Pigment Compendium: A Dictionary and Optical Microscopy of Historical Pigments*. London: Routledge, 2013.

Faivre, Damien (ed). *Iron Oxides From Nature to Applications*. Weinheim: Wiley-VCH, 2016.

Siddall, Ruth. *"Mineral Pigments in Archaeology: Their Analysis and the Range of Available Materials."* Minerals (2018): 201. https://doi.org/10.3390/min8050201

INTERNET GEMS

ART OF SOIL
theartofsoil.com

EARLY FUTURES & OCHRE SANCTUARY
earlyfutures.com

KAUAE RARO COLLECTIVE
kauaeraro.com

NATIVE LAND DIGITAL
native-land.ca

PIGMENTS REVEALED INTERNATIONAL
pigmentsrevealed.com

WILD PIGMENT PROJECT
wildpigmentproject.org

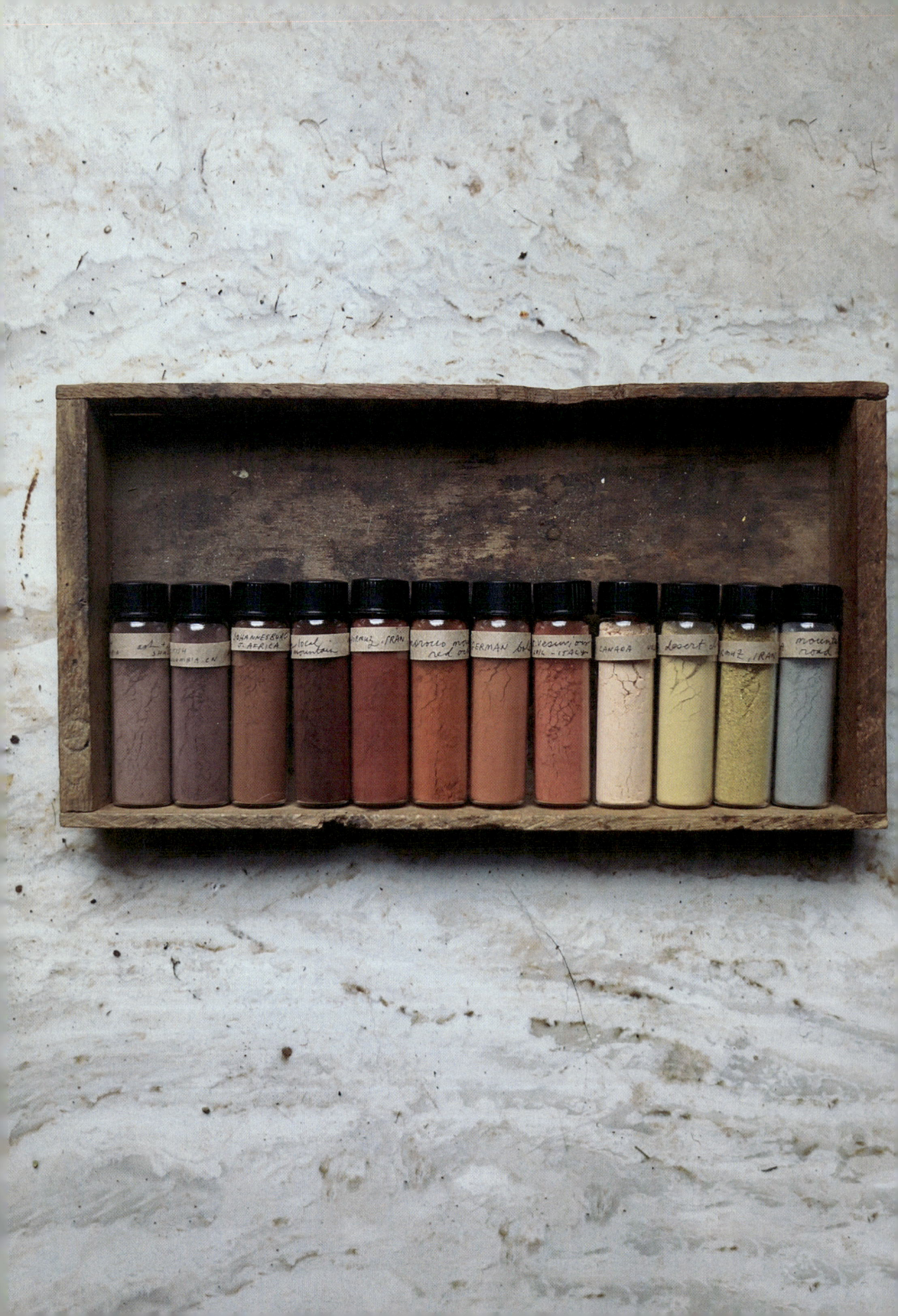

BIBLIOGRAPHY

Aeschylus Choēphoroi. *The Choephori* [The Libation Bearers]. United Kingdom: University Press, 1901.

Alavi, M., A. Totonchi, M. A. Okhovat, M. Motazedian, P. Rezaei, M. Atefi. "The Effect of a New Impregnated Gauze Containing Bentonite and Halloysite Minerals on Blood Coagulation and Wound Healing." *Blood Coagul Fibrinolysis*. 2014 Dec; 25(8):856–59. doi.org/10.1097/MBC.0000000000000172. PMID: 25004023.

Ancheta, Melonie. "Revealing Blue on the Northern Northwest Coast." *American Indian Culture and Research Journal* 43:1 (January 2019): 1–30. doi.org/10.17953/aicrj.43.1.ancheta.

Arlettaz, Raphaël, Philippe Christe, Peter F. Surai, and Anders Pape Møller. "Deliberate Rusty Staining of Plumage in the Bearded Vulture: Does Function Precede Art?" *Animal Behaviour* 64:3 (2002): F1–F3. doi.org/10.1006/anbe.2002.3097.

Athanassakis, Apostolos N., and Benjamin M. Wolkow. *The Orphic Hymns*. Baltimore: Johns Hopkins University Press, 2013.

Bachelard, Gaston. *Earth and Reveries of Will: An Essay on the Imagination of Matter*. Dallas, TX: Dallas Institute, 2002.

———. *Earth and Reveries of Repose: An Essay on Images of Interiority*. Dallas, TX: Dallas Institute, 2011.

———. "The Cosmos of Iron," *The Right to Dream*. Dallas, TX: Dallas Institute, 1988.

Baty, William. *The Anatomy of Colour: The Story of Heritage Paints and Pigments*. London: Thames and Hudson, 2017.

Bjornerud, Marcia. *Timefulness: How Thinking Like a Geologist Can Help Save the World*. Princeton, NJ: Princeton University Press, 2018.

Bordreuil, Pierre, and Dennis Pardee. *A Manual of Ugaritic*. Winona Lake, IN: Eisenbrauns, 2009.

Borg, Gregor, and Margaret Jacobsohn. "Ladies in Red—Mining and Use of Red Pigment by Himba Women in Northwestern Namibia." *Tagungen des Landesmuseums für Vorgeschichte Halle* vol. 10 (2013): 43–51.

Bucklow, Spike. *Red: The Art and Science of a Colour*. London: Reaktion Books, 2016.

———*The Alchemy of Paint: Art, Science and Secrets from the Middle Ages*. London: Marion Boyars Publishers Ltd, 2012.

Bulkeley, Kelly, and Patricia Bulkley. *Dreaming Beyond Death: A Guide to Pre-Death Dreams and Visions*. Boston: Beacon Press, 2005.

Cennini, Cennino d'Andrea. *The Craftsman's Handbook (Il Libro dell'Arte)*. Translated by Daniel V. Thompson Jr. New York: Dover, 1960.

Chan C. S., S. M. McAllister, A. H. Leavitt, B. T. Glazer, S. T. Krepski, and D. Emerson. "The Architecture of Iron Microbial Mats Reflects the Adaptation of Chemolithotrophic Iron Oxidation in Freshwater and Marine Environments." *Front. Microbiol*. 7:796 (2016). doi.org/10.3389/fmicb.2016.00796.

Chu, Jennifer. "Origins of Earth's Magnetic Field Remain a Mystery." MIT News. April 8, 2020. Retrieved February 2022. https://news.mit.edu/2020/origins-earth-magnetic-field-mystery-0408.

Collins, Billie Jean. *A History of the Animal World in the Ancient Near East*. Netherlands: Brill, 2001.

Curtis, Gregory. *The Cave Painters: Probing the Mysteries of the World's First Artists*. New York: Anchor Books, 2006.

Dário, Giordana Maciel, Geovana Gomes da Silva, Davi Ludvig Gonçalves, Paulo Silveira, Adilson Teixeira Junior, Elidio Angioletto, and Adriano Michael Bernardin. "Evaluation of the Healing Activity of Therapeutic Clay in Rat Skin Wounds." *Materials Science and Engineering: C* vol. 43 (2014): 109–116. doi.org/10.1016/j.msec.2014.06.024.

Dasen, Veronique. "Magic and Medicine: The Power of Seals." In *Gems of Heaven: Recent Research on Engraved Gemstones in Late Antiquity c. AD 200–600*. Edited by Chr. Entwistle and N. Adams. London, 2011: 69–74.

De Landa, Manuel. *A Thousand Years of Nonlinear History*. New York: Zone Books, 2019.

Deimler, Devon Erin. "Ultraviolet Concrete: Dionysos and the Ecstatic Play of Aesthetic Experience." Dissertation. Santa Barbara, CA: Pacifica Graduate Institute, 2018.

Dethier, Jean. *The Art of Earth Architecture: Past, Present, Future*. New York: Princeton Architectural Press, 2020.

Detienne, Marcel, and Jean-Pierre Vernant. *Cunning Intelligence in Greek Culture and Society*. Hassocks: Harvester Press, 1978.

Dexter Dyer, Betsey. *A Field Guide to Bacteria*. Ithaca: Cornell University Press, 2003.

Dolgopyat, Elena, trans. Richard Coombes. *The Quality of Time*, 2020. https://intranslation.brooklynrail.org/russian/the-quality-of-time/

Duncombe, J.. "The Herky-Jerky Weirdness of Earth's Magnetic Field." *Eos* 101 (2020). doi.org/10.1029/2020EO152466.

Earling, Debra Magpie. *The Lost Journals of Sacajewea*. Berkeley, CA: Peter Koch Press, 2010.

Egger, Matthias, Tom Jilbert, Thilo Behrends, Camille Rivard, and Caroline P. Slomp. "Vivianite is a Major Sink for Phosphorus in Methanogenic Coastal Surface Sediments." *Geochimica et Cosmochimica Acta* volume 169 (2015): 217–235. https://doi.org/10.1016/j.gca.2015.09.012.

Eliade, Mircae. *The Forge and the Crucible: The Origins and Structure of Alchemy*. New York: Harper & Row, 1962.

Ershoff, B.H. and Bajwa, GS. "Physiologic effects of dietary clay supplements." NASA Contractor Report, NASA-CR-65427. July 1965. Retrieved April 2022. https://ntrs.nasa.gov/citations/19660023330.

Eshleman, Clayton. *Juniper Fuse: Upper Paleolithic Imagination & the Construction of the Underworld*. Middletown, CT: Wesleyan University Press, 2003.

Farrelly, Daniel, Rebecca Slater, Hannah R. Elliott, Hannah R. Walden, Mark A. Wetherell. "Competitors Who Choose to Be Red Have Higher Testosterone Levels." *Psychological Science* 24:10 (2013): 2122–2124. doi:10.1177/0956797613482945.

Fienup-Riordan, Anne. *Yuungnaqpiallerput/The Way We Genuinely Live: Masterworks of Yup'ik Science and Survival*. Seattle: University of Washington Press, 2007.

Fox, James. *The World According to Color: A Cultural History*. New York: St. Martin's Press, 2022.

Franks, Leslie J.. *Stone Medicine: A Chinese Medical Guide to Healing with Gems and Minerals*. United States: Inner Traditions/Bear, 2016.

Goll, Yvan and Nan Watkins. *Dreamweed*. Pittsburgh, PA: Black Lawrence Press, 2012.

Goodman, Steven D. *The Buddhist Psychology of Awakening: An In-Depth Guide to Abhidharma*. Boulder, CO: Shambhala, 2020.

Hall, A. J., and E. Photos-Jones. "Accessing Past Beliefs and Practices: The Case of Lemnian Earth." *Archaeometry* 50 (2008): 1034–49. doi.org/10.1111/j.1475-4754.2007.00377.x.

Hall, Edith. "Hephaestus the Hobbling Humorist: The Club-Footed God in the History of Early Greek Comedy." *Illinois Classical Studies* 43:2 (2018): 366–87.

Hall, Nor. *Irons in the Fire*, New York: Station Hill, 2002.

Hansell, Mike. *Built by Animals: The Natural History of Animal Architecture*. New York: Oxford University Press, 2007.

Harjo, Joy. *Conflict Resolution for Holy Beings*. New York: W.W. Norton & Co, 2015.

Hashimoto, Hideki, et al. "Preparation, Microstructure, and Color Tone of Microtubule Material Composed of Hematite/Amorphous-Silicate Nanocomposite from Iron Oxide of Bacterial Origin." *Dyes and Pigments* vol. 95, issue 3 (2012): 639–43. ISSN 0143-7208, doi.org/10.1016/j.dyepig.2012.06.024.

Hauschka, Rudolf. *The Nature of Substance: Spirit and Matter*, London: Rudolf Steiner Press, 2002.

Heimann, R. B., U. Kreher, I. Spazier, and G. Wetzel. "Mineralogical and Chemical Investigations of Bloomery Slags from Prehistoric (8th Century B.C. to 4th Century A.D.) Iron Production Sites in Upper and Lower Lusatia, Germany." *Archaeometry* 43:2 (2002): 227–252. doi.org/10.1111/1475-4754.00016.

Hillman, James. *Mythic Figures*. Dallas, TX: Spring Publications, 2007.

———. "Alchemical Blue and the 'Unio Mentalis'." *Alchemical Psychology: Uniform Edition, Volume 5*. Putnam, CT: Spring Publications, 2009.

Hirasawa, Tatsuya, and Shigeru Kuratani. "Evolution of the Vertebrate Skeleton: Morphology, Embryology, and Development." *Zoological Letters* (January 2015). doi.org/10.1186/s40851-014-0007-7.

Hodgskiss, Tammy. "Ochre Use in the Middle Stone Age." *Oxford Research Encyclopedia of Anthropology*. April 30, 2020. Accessed July 27, 2022. doi.org/10.1093/acrefore/9780190854584.013.51

Holland, Heinrich D., and Karl K. Turekian, eds. *Treatise on Geochemistry*. Netherlands: Elsevier Science, 2013.

Horowitz, W. and Hurowitz, V. "Urim and Thummim in Light of a Psephomancy Ritual from Assur (LKA 137)." *Journal of the Ancient Near Eastern Society* 21 (1992): 95–115.

Houston, David, and Jamieson Copsey. "Bone Digestion and Intestinal Morphology of the Bearded Vulture," *The Journal of Raptor Research*, 28:2 (1994): 73–8.

Hudson, Sarah. *Mana Whenua*. Whakatāne: Sarah Hudson, 2020.

Jensen, Derrick. *The Myth of Human Supremacy*. New York: Seven Stories Press, 2016.

Keeler, Jacqueline. *Edge of Morning: Native Voices Speak for the Bears Ears*. Salt Lake City, UT: Torrey House Press, 2017.

Kerényi, Karl. *The Gods of the Greeks*. New York: Thames and Hudson, 1992.

Khandekar, Narayan, Georgina Rayner, and Daniel P. Kirby. "Pigments and Binders in Traditional Aboriginal Bark Paintings," in *Everywhen: The Eternal Present in Indigenous Art from Australia*. Cambridge: Harvard Art Museums, 2016.

Kimmerer, Robin Wall, *Braiding Sweetgrass*. Minneapolis, MN: Milkweed Editions, 2013.

Kolbert, Elizabeth. *The Sixth Extinction: An Unnatural History*. New York: Henry Holt and Co., 2016.

Kubeneck, Joëlle L., Wytze K. Lenstra, Sairah Y. Malkin, Daniel J. Conley, and Caroline P. Slomp. "Phosphorus Burial in Vivianite-type Minerals in Methane-rich Coastal Sediments." *Marine Chemistry* volume 231 (2021). doi.org/10.1016/j.marchem.2021.103948.

LaPointe, Sasha taqwšablu. *Red Paint: An Ancestral Autobiography of a Coast Salish Punk*. Berkeley, CA: Counterpoint, 2022.

MacDonald, B. L., D. Stalla, X. He, et al. "Hunter-Gatherers Harvested and Heated Microbial Biogenic Iron Oxides to Produce Rock Art Pigment." *Sci Rep* 9 (2019): 17070. doi.org/10.1038/s41598-019-53564-w.

MacDonald, Brandi Lee, and Nenagh Hathaway. "Histories of Selected Artists' Pigments." In Carol Podedworny et al., *The Unvarnished Truth: Exploring the Material History of Paintings (Mise à nu: une exploration de l'histoire matérielle de la peinture)*. Hamilton, ON: McMaster Museum of Art, 2015.

Macgregor, Arthur. "Medicinal *Terra Sigillata*: A Historical, Geographical and Typological Review." *Geological Society, London, Special Publications* 375: 1 (2013): 113–136.

Mai, Huijuan, Yimin Yang, Idelisi Abuduresule, et al., "Characterization of cosmetic sticks at Xiaohe Cemetery in early Bronze Age Xinjiang, China," *Sci Rep* 6 (2016): 18939. https://doi.org/10.1038/srep18939.

Marchant, Jo. *The Human Cosmos: Civilization and the Stars*. New York: Dutton, 2020.

Margalida, A., and D. Villalba. "The Importance of the Nutritive Value of Old Bones in the Diet of Bearded Vultures (*Gypaetus barbatus*)," *Sci Rep* 7 (2017): 8061. doi.org/10.1038/s41598-017-08812-2.

McClain, J. B., and K. Cooney. "The Daily Offering Meal in the Ritual of Amenhotep I: An Instance of the Local Adaptation of Cult Liturgy." *Journal of Ancient Near Eastern Religions* 5:1 (2005): 41–78. doi: https://doi.org/10.1163/156921205776137963.

McLeod, Toby. "Shellmounds of the Bay Area," Sacred Land Film Project. September 2020. Accessed January 2022. https://sacredland.org/shellmounds-of-the-bay-area-united-states/.

Mellick, Jill. *The Red Book Hours: Discovering C. G. Jung's Art Mediums and Creative Process*. Zürich: Scheidegger & Spiess, 2018.

Montgomery-Neutze, Nā Sian. "Ki Te Ao Mārama: Introduction to Colour Theory, a Tirohanga Māori." Accessed March 2022. https://www.kauaeraro.com/matauranga/ki-te-ao-marama.

Morton, Timothy. *All Art is Ecological*. London: Penguin, 2020.

Nelson, Maggie. *Bluets*. Seattle: Wave Books, 2009.

Nickles, J. "On the Presence of Vivianite in Human Bones." *American Journal of Science and Arts* (May 1856): 21, 63; *American Periodicals*, 402.

Oelkers, E. H., E. Valsami-Jones, T. Roncal-Herrero. "Phosphate Mineral Reactivity: From Global Cycles to Sustainable Development." *Mineralogical Magazine* 72:1 (2008): 337–40. Bibcode: 2008MinM...72..337O. doi.org/10.1180/minmag.2008.072.1.337. S2CID 97795738.

O'Hanlon, George. "Sankir—Underpainting of Flesh—Part 1." *Natural Pigments* (blog). 2013. Accessed May 2022. https://www.naturalpigments.eu/artist-materials/sankir-underpainting-flesh/.

Palmer, Laurie A. *In the Aura of a Hole: Exploring Sites of Material Extraction*. London: Black Dog Publishing, 2014.

Pastoureau, Michel, and Janet Lloyd. *The Colours of Our Memories*, Cambridge: Polity Press, 2012.

Patra, Jayanta Kumar, and Kwang-Hyun Baek. "Green Biosynthesis of Magnetic Iron Oxide (Fe_3O_4) Nanoparticles Using the Aqueous Extracts of Food Processing Wastes Under Photo-Catalyzed Condition and Investigation of Their Antimicrobial and Antioxidant Activity." *B: Biology*. Journal of Photochemistry and Photobiology 173 (August): 291–300. doi.org/10.1016/j.jphotobiol.2017.05.045.

Patton, Christopher. *Curious Masonry: Three Translations from the Anglo-Saxon*. Nova Scotia: Gaspereau Press, 2011.

——. *Dumuzi*. Nova Scotia: Gaspereau Press, 2020.

Photos-Jones E., C. Edwards, F. Häner, L. Lawton, C. Keane, A. Leanord, and V. Perdikatsis. "Archaeological Medicinal Earths as Antibacterial Agents: The Case of the Basel Lemnian Sphragides." *Geology and Medicine: Historical Connections.* Geological Society, London, Special Publications 452 (February 2, 2017): 141–53. doi.org/10.1144/SP452.6.

Piaget, Jean. *La Representation du Monde Chez l'enfant.* Paris: F. Alcan, 1926.

Popelka-Filcoff, Rachel S., and Andrew M. Zipkin. "The Archaeometry of Ochre "Sensu Lato": A Review," *Journal of Archaeological Science* vol. 137 (2022). doi.org/10.1016/j.jas.2021.105530.

Povinelli, Elizabeth A. *Geontologies: A Requiem to Late Liberalism.* Durham: Duke University Press, 2016.

Prot, T., W. Pannekoek, C. Belloni, A. I. Dugulan, R. Hendrikx, L. Korving, M. C. M. van Loosdrecht. "Efficient Formation of Vivianite Without Anaerobic Digester: Study in Excess Activated Sludge." *Journal of Environmental Chemical Engineering* volume 10, issue 3 (2022). doi.org/10.1016/j.jece.2022.107473.

Puglionesi, Alicia. *In Whose Ruins: Power, Possession, and the Landscapes of American Empire.* New York: Scribner, 2022.

Retsas, Spyros. "Geotherapeutics: The Medicinal Use of Earths, Minerals and Metals from Antiquity to the Twenty-First Century." *Geology and Medicine: Historical Connections.* Geological Society, London, Special Publications 452:1(2017): 133. dx.doi.org/10.1144/SP452.5.

Rifkin, Riaan F. "Ethnographic Insight into the Prehistoric Significance of Red Ochre." *South African Archaeological Society Digging Stick* 32, no. 2 (2015): 7–10.

———. "The Symbolic and Functional Exploitation of Ochre During the South African Middle Stone Age." PhD thesis, University of the Witwatersrand, Faculty of Science, 2013.

Rifkin R. F., L. Dayet, A. Queffelec, B. Summers, M. Lategan, and F. d'Errico. "Evaluating the Photoprotective Effects of Ochre on Human Skin by *In Vivo* SPF Assessment: Implications for Human Evolution, Adaptation and Dispersal." PLoS ONE 10:9 (2015): e0136090. doi.org/10.1371/journal.pone.0136090

Rimell, Bruce. *Liminal Contact: A Cognitive and Anthropological Response to the "Death" of Painting.* Lulu.com, 2016.

Roebroeks, Wil, Mark J. Sier, Trine Kellberg Nielsen, Dimitri De Loecker, Josep Maria Parés, Charles E.S. Arps, and Herman J. Mücher. "Use of red ochre by early Neandertals." *Proceedings of the National Academy of Sciences* 109:6 (2012): 1889–1894.

Rosen, Julia. "The Science of Climate Change Explained: Facts, Evidence and Proof." *New York Times.* November 2021. Accessed March 2022. nytimes.com/article/climate-change-global-warming-faq.html%20/Julia%20Rosen.

———. "Humanity Is Flushing Away One of Life's Essential Elements." *The Atlantic.* February 8, 2021. Accessed February 2022. theatlantic.com/science/archive/2021/02/phosphorus-pollution-fertilizer/617937/.

Rulandus, Martin. *A Lexicon of Alchemy.* Translated by Arthur Edward Waite. Germany: Jazzybee Verlag, 2014. E-book.

Schweitzer M. H., W. Zheng, and T. P. Cleland, et al. "A Role for Iron and Oxygen Chemistry in Preserving Soft Tissues, Cells and Molecules from Deep Time." *Proc Biol Sci.* 2013; 281 (1775): 20132741. doi.org/10.1098/rspb.2013.2741.

Shellmound.org

Shinhmar, H., Hogg, C., Neveu, M. et al., "Weeklong Improved Colour Contrasts Sensitivity after Single 670 nm Exposures Associated with Enhanced Mitochondrial Function." Science Reports 11, 22872 (2021). https://doi.org/10.1038/s41598-021-02311-1.

Siddall, R., "The Origin of Ochres: Interbasaltic Beds." The Pigment Timeline Project, UCL Blogs., 2020. Accessed June 2021. https://blogs.ucl.ac.uk/pigment-timeline/2020/04/01/the-origin-of-ochres-interbasaltic-beds/.

Sigo, Cedar. *Guard the Mysteries.* Seattle: Wave Books, 2021.

Skinner, H. C. W., and H. Ehrlich. "Biomineralization." In *Treatise on Geochemistry,* 2nd edition. Edited by Karl K. Turekian and Heinrich D. Holland. San Diego: Elsevier Science, 2014.

Smil, Vaclav. *Still the Iron Age: Iron and Steel in the Modern World.* San Diego: Elsevier. 2016.

Smith, Mark S., and Wayne T. Pitard. *The Ugaritic Baal Cycle. Volume II.* Leiden: Brill, 2009.

Turner, E. A. L. "Anvil Age Economy: A Map of the Spread of Iron Metallurgy across Afro-Eurasia." *Cliodynamics: The Journal of Quantitative History and Cultural Evolution* 11:1 (2020). doi.org/10.21237/c7clio1114.

Stein, Murray. "Hephaistos: A Pattern of Introversion" in *Soul: Treatment and Recovery.* New York: Routledge, 2016.

Stevens, Wallace. *The Collected Poems of Wallace Stevens.* New York: Vintage Books, 2015.

Swanton, John Reed. *Tlingit Myths and Texts.* United States: Scholarly Press, Incorporated, 1909.

Sykes, Rebecca Wragg. *Kindred: Neanderthal Life, Love, Death and Art.* London: Bloomsbury Sigma. 2020.

Taylor, P. T., K. I. Kis, and G. Wittmann. "Satellite-Altitude Horizontal Magnetic Gradient Anomalies Used to Define the Kursk Magnetic Anomaly." *Journal of Applied Geophysics* 109 (C). Elsevier B.V.: 133–39. doi.org/10.1016/j.jappgeo.2014.07.018.

"The Bomb," *Time*. August 20, 1945. Accessed April 2022, http://content.time.com/time/magazine/0,9263,7601450820,00.html.

Tributsch, Helmut. "Ochre Bathing of the Bearded Vulture: A Bio-Mimetic Model for Early Humans Towards Smell Prevention and Health." *Animals* 6:1 (2016): 7–17. doi.org/10.3390/ani6010007.

Trungpa, Chögyam, and Judith L. Lief. *True Perception: The Path of Dharma Art.* Boston: Shambhala, 2008.

Velliky, E.C., Hodgskiss, T., Straffon, L.M., Gustafson, H., Gollifer, A., Haaland, M.M. "The Ochre Experience Model (OEM): Towards a deep-time perspective on the earth material heritage of ochre." In *Deep-Time Art in the Age of Globalization: Understanding Rock Art in the 21st Century.* Edited by Moro-Abadía, O., M. W. Conkey, J. McDonald. Springer Nature: Cham, Switzerland, 2022 (in press).

Wagner D. O., and P. Aspenberg. "Where Did Bone Come From?" *Acta Orthopaedica* (2011): 393–98. doi.org/10.3109/17453674.2011.588861.

Watts, Ian. "Red ochre, body painting, and language: interpreting the Blombos ochre." In *The Cradle of Language*. Edited by Rudolf Botha and Chris Knight. Oxford: Oxford University Press, 2009: 93–129.

Watts et. al. "Early Evidence for Brilliant Ritualized Display: Specularite Use in the Northern Cape (South Africa) between ~500 and ~300 Ka," *Current Anthropology* (June 2016): 287.

Weinberger, Eliot. *Angels & Saints*. New York: New Directions, 2020.

Weiss, Aaron Matthew, *Buddhist Topology and the Practice of Lama Orgyan Khandro Norlha*. PhD diss. California Institute of Integral Studies, ProQuest Dis. Pub, 2021.

Wilfert, Philip. "Phosphate Recovery from Sewage Sludge Containing Iron Phosphate," Dissertation, Delft University of Technology, Netherlands, 2018. doi.org/10.4233/uuid:f3729790-0cfe-4f92-866b-eca3f2f2df24.

Williams, Morgan M., "Pedogenic Process in Engineered Soils for Radioactive Waste Containment." UC Berkeley, 2017, ProQuest ID: Williams_berkeley_0028E_19342. escholarship.org/uc/item/4zq8p8p5.

Williams, Terry Tempest. *Erosion: Essays of Undoing*. New York: Picador/Farrar, Straus and Giroux, 2020.

Willis, Daniel. *The Emerald City and Other Essays on the Architectural Imagination*. New York: Princeton Architectural Press, 1999.

Zirkle, Conway. "Animals Impregnated by the Wind." *Isis* vol. 25, no. 1 (1936): 95–130. JSTOR, jstor.org/stable/224987.

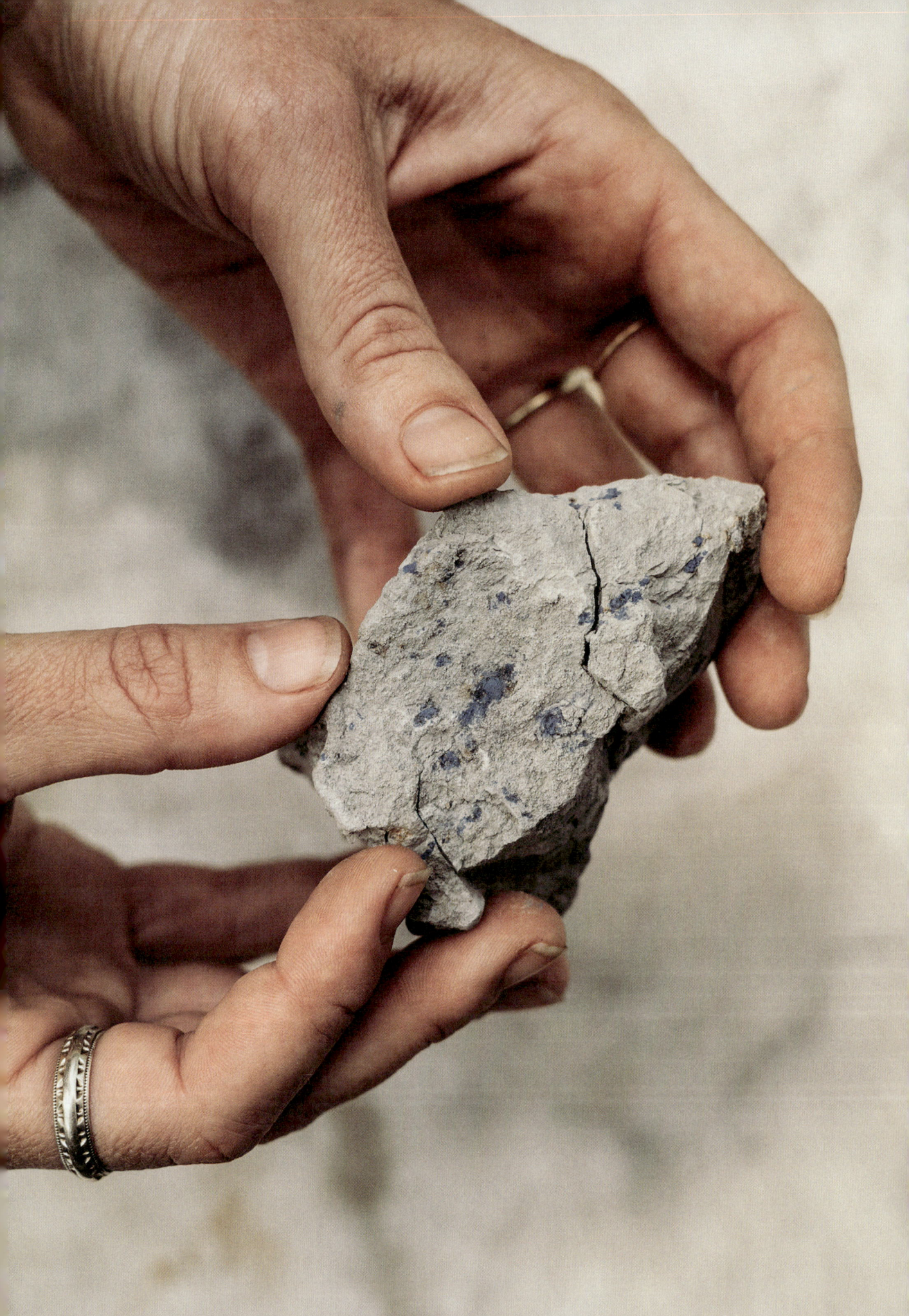

IMAGE CREDITS

ALL IMAGES
© Heidi Gustafson, unless noted below.

PAGES 6—7, 16:
© Brendan Pattengale

PAGES 10, 22, 24—25, 46, 58, 70—71, 72 (TOP), 80, 88, 98, 108 (BOTTOM), 114, 118—119, 138, 146 (TOP), 152, 154—155, 220—221:
© Brian Merriam

PAGES 12, 21, 78, 174, 218:
© Kyle Johnson

PAGE 38:
© Ben Wirtz Siegel/Courtesy of Ohio University

PAGE 43:
© Klaus Robin

PAGE 48 (BOTTOM):
© Scott Sutton

PAGE 54:
© John Sabraw

PAGE 55 (TOP):
© Matthew Hughes

PAGE 68 (TOP):
© Sheena Callage

PAGES 77, 90 (TOP), 146 (BOTTOM), 150—151, 164 (TOP), 176, 186, 190, 210, 224:
© Corwin Fergus

PAGE 96:
© Joumana Medlej

PAGE 97:
© Julia Norton

PAGE 90 (BOTTOM):
© Crtomir 'Harald' Lorencic

PAGE 108 (TOP):
© Elpitha Tsoutsounakis

PAGE 112:
© Morgan Williams

PAGE 134 (BOTTOM):
© Tim Runde

PAGES 150—151:
© Getty Images/Bloomberg

PAGE 162:
© Marta Abbott

PAGE 164 (BOTTOM):
© Karen Vaughan

PAGES 163 (BOTTOM), 167:
© Natascha Libbert

PAGES 172—173:
© Ann Gollifer

PAGE 170:
© Chiara Zonca

PAGES 180—181:
© Meghan McMackin

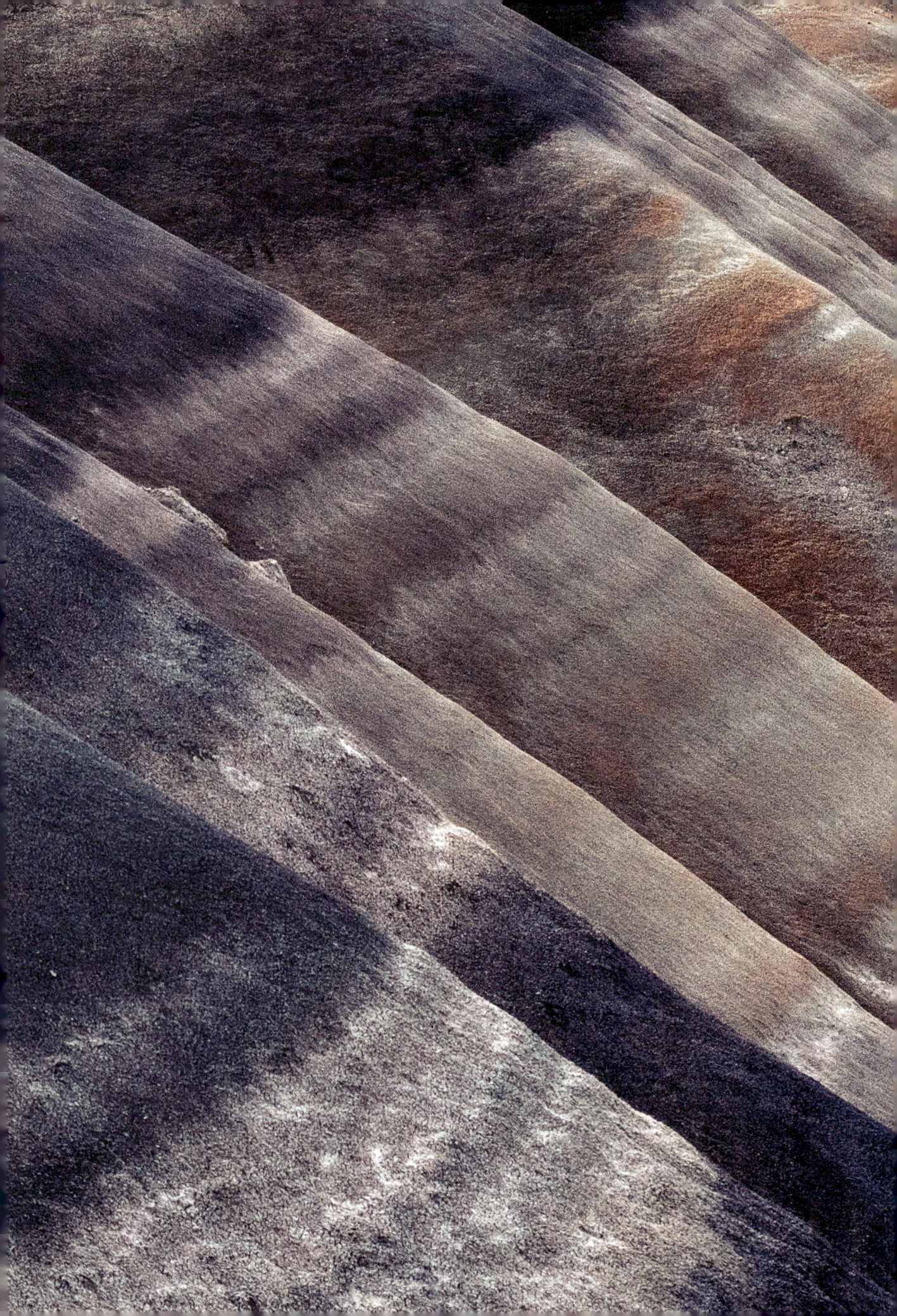

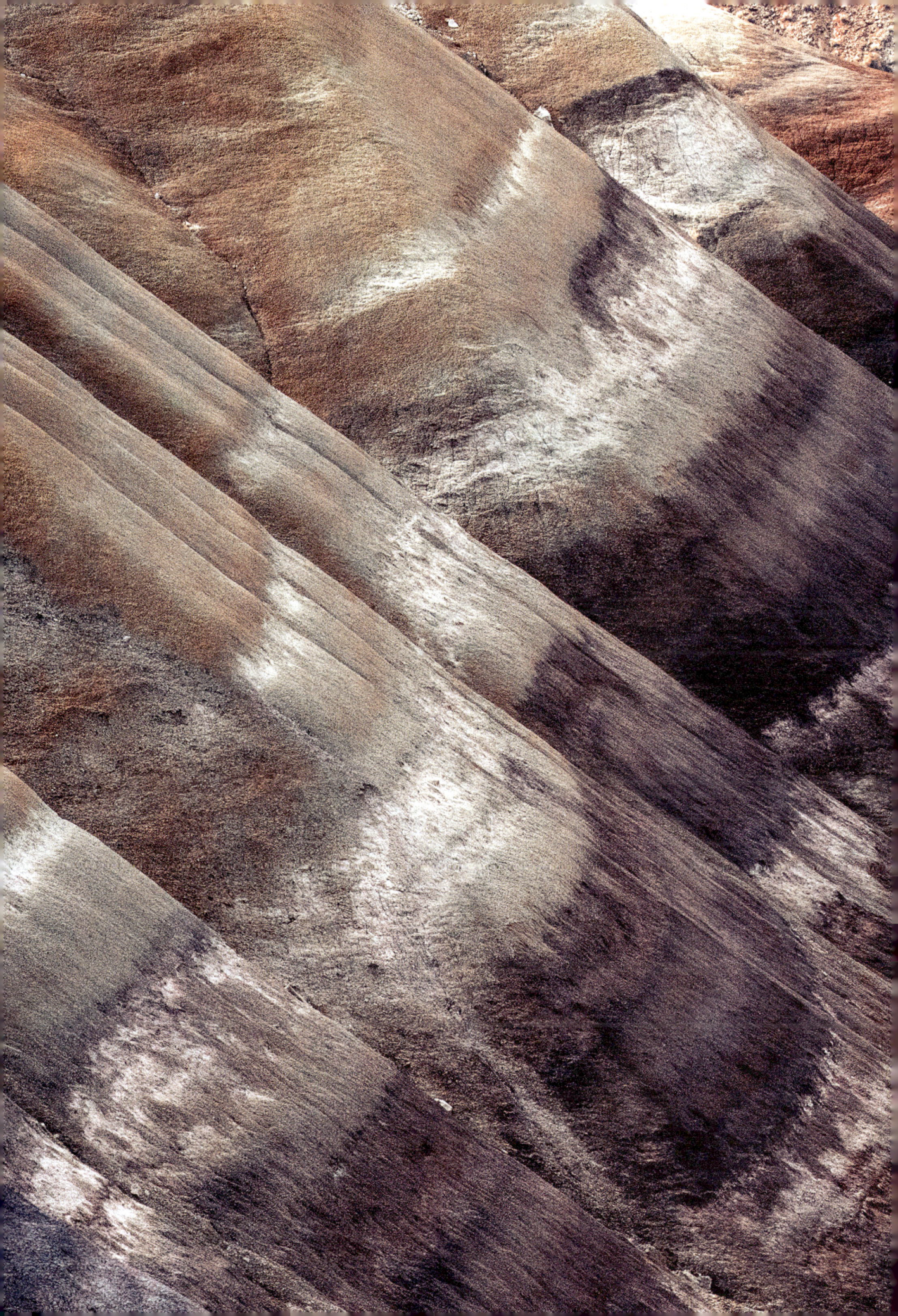

ACKNOWLEDGMENTS

As dear Nor Hall observes of women initiated by iron, "A young woman talks about the paradox of iron's command: You are forced into extreme, isolating focus by virtue of the sheer difficulty of the work and yet there is an unspoken, complete reliance on your colleagues who provide collective support. They appear to be a ragtag group."[1] Evidence of voices and insight from my ragtag connections, my incredible friends, mentors and collaborators, contributors, students and mentees, and readers are threaded throughout this book. I thank each of you.

First and foremost, this book would not be here without guidance from Ochres to whom I belong, their manifold wisdom, places, ancestors, and spirit protectors who are willing to summon, work with, and guide as they may.

Let me repeat here what I say elsewhere: I am deeply indebted to and acknowledge that Indigenous peoples, specifically Coast Salish, Duwamish, Nooksack, Methow, Lummi, Lisjan Ohlone peoples are the traditional caretakers and storytellers of the lands on which I was born, live, and work. I honor Indigenous and Native stories, traditions, technologies, knowledge, and cultures, and their continuous connection to land, water, sky, and creatures; and I celebrate and support their stewardship of land and ochre places across time and space. I offer deepest respect and well-being to the ancestors and earth protectors worldwide.

I am beyond grateful to learn from my loving, supportive, joyful, and curious inherited family—my parents, Jody and Gregg Gustafson; my brilliant sister and best friend, Claire; my nieces, Sierra and Sequoia; and my brother-in-law, Kyle: I am so lucky to learn, play, and grow with all of you. And special thanks to my extended family, close relatives, ancestors, and, in particular, my dearest cousin Karen Eisenstadt, whose generosity and soulful eye are a gift in countless ways.

This work would not exist publicly without the immense and humbling blessing of loving companionship with elders, mentors, and teachers:

Meeting, and being met by, the unwavering compassion of my (very private) spiritual companion, Steven Goodman, changed me deeply and was an incredible blessing. Your fierce heart awareness, dharma precision, laughter, and associative genius (and devotion to your own lineage teachers) remains a beacon, from whatever trickster realm you are in.

To my local mentor and ochre pigment soulmate, Melonie Ancheta, no one compares. Your friendship, bravery, brilliant observations, and straight talk are ceaseless teachings, and I am forever grateful for every generous gift you offer. And so many thanks to dear Henry Ancheta, whose supreme kindness and humor are a wellspring of inspiration. I'm beyond grateful and very honored to get to include your skillful woodcraft collaboration in the trays that carefully hold ochres throughout this book.

Cellular love to Buck "Bucasso" McAdoo: mushroom-foraging genius, luminous artist, joy of a human. You secretly lurk on hands and kneepads, below and behind, many of my rock encounters here. Your generosity, entertaining stories, and howling wolf music are an eternal lifeboat.

I also owe a huge gratitude to early mentors in Baltimore: world's greatest teacher Ms. Winnie Faye Thomas, and all the very young children we taught and learned from side by side for so many years; and an extra beloved gratitude to artist and early advisor Hugh Pocock (and sons).

My loves and innermost circle of chosen family, who are always hot springs of inspiration, and who gave keen and critical feedback, orientation, refuge, and axe-sharpening at crucial phases of my work and soulwork: I xlovex you and thank you all ways, always, Devon Deimler—my truest, oldest north song and keynote. Deep and long gratitude ever to

Elizabeth Snowden of the ever-flowing palatial poetry. Dylan Young, who somehow remains buoyant even in quicksand or whirlpool. Deborah Goldman, a lighthouse of artistic endurance and humanity.

I don't know where I'd be without the profound revolutionary oracles who I lean on and trust professionally: bewitcher and co-conspirator Jason Logan, land energy channel Morgan Williams, alchemist Thomas Little, tender tender Tilke Elkins, and wildheart of the vinegar fires Caro Ross. As well, very grateful to be alive at a time when there are incredible ochre/pigment/soil women colleagues kicking ass and sharing knowledge in academic education and research, especially dearest Tammy Hodgskiss Reynard, Elizabeth Velliky, Brandi McDonald, Larissa Mendoza Staffron, Kelsey Hansen, Elpitha Tsoutsounakis, Julia Norton, and Karen Vaughan.

Immense gratitude and splendor to fellow ochre and earth protectors, artists, and educators (and for those who I have yet to meet, I pray a ripe time arises!). Thank you to those who've shared important teachings, heart, or subtle calls, even if only for a fleeting moment: Janey Chang, Karla Sofía Claudio Betancourt, Lanae Cable, Catalina Christensen, Symeon van Donkelaar, Anita Ekman, Julie Gibney-Vamvakari and George, Ann Gollifer and her teacher Motsei Nkwemabala, Corrina and Deja Gould, Jamie Graham-Blair, Sarah Hudson, Hao-Lun Hung, Greg Biskakone Johnson and Alexandra Sulainis, Phillipa Jahn, Koichi Kuritawa, Melissa Ladkin, Camas Logue, Lee Wayne Lomayestewa, Joumana Medlej, Nahaan, Rosaura Rodríguez and family, Alan Salazar and Mona Lewis, Darin Siles, Hana Shahnavaz, Scott Sutton, and the youth involved in protecting líq'təd springs through the Urban Native Education Alliance, including Chayton and Cante Remle, Asia Gellien, Isaac Hochberg, and Tim Shay.

Deep appreciation and gratitude to the dreamcatchers who collaborated and offered photographic contributions. Especially bow deeply to Brain Merriam, whose brilliant sensibilities help this project travel further than I ever could alone (please read full photo credits for all contributing photographers).

Huge appreciation and respect to the tireless women who helped steer this project in the world—my local Northwest agent, Amy Levenson, and the supportive world-weavers at Abrams: steady editors Meredith A. Clark and Soyolmaa Lkhagvadorj, designer Heesang Lee, creative director Deb Wood, and their team, including keen-eyed Erin Slonaker, Margo Winton-Parodi, Annalea Manalili, and many others.

A deep appreciation to many folks whose notes, expertise, talks, and encouragement gave direction at twists and turns along this path: Marta Abbott, Stan Ambrose, Gary Astrachan, Patricia Belyea, Karima Cammell, Ashley Chambers, Jessica Chiartas, Kim Francis, Emily Freidenrich, Reed Garber, Magnus Haaland, Nor Hall, Hannah Hirshhorn, Matthew Hughes, Narayan Khandekar, George Mustoe, Nance Khlem, Mark van Loosdrecht, Megan Lucas, Nick Maione, George Mustoe, Tim Runde, Toby McLeod, Christopher Patton, Sabine Pinon, Rolf Pixley, Adam Robbert, Alana Seigel, Gabe Snyder (who foresaw that my first book might be a supernatural cookbook), Benjamin Turner, Aaron Weiss, Philip Wilfert, Nick Wyatt, Andrew Zipkin, and the entire board of Pigments Revealed International, and others who contribute to the Wild Pigment Project community.

Thank you to the hundreds of people worldwide I've talked to on the side of the road or in Instagram back alleys and those who've generously sent me their land and ochres and rocks (or photos) to hold and to cherish for my long-term Ochre Sanctuary project. Your notes and trust are an ongoing source of inspiration. Please visit earlyfutures.com for a partial list of contributors and supporters.

And gratitude to every other person or being whom I failed to name, and those who wished to remain fugitive, unknown, and unseen.

Lastly and firstly, Corwin Ho Fergus, my love and heart tender extraordinaire. You held me through it and carried this work as my intimate counsel, midwife, fire tender, reader of every single word, and generous partner. No amount of credit will do. Thank you.

Editor: Meredith A. Clark
Designer: Heesang Lee
Managing Editor: Annalea Manalili
Production Manager: Larry Pekarek

Library of Congress Control Number: 2022944489

ISBN: 978-1-4197-6465-3
eISBN: 978-1-64700-829-1

Text copyright © 2023 Heidi Gustafson
For image credits, see page 219
Crystallographic system illustrations by Heesang Lee

Cover © 2023 Abrams

Printed and bound in China
10 9 8 7 6 5

The activities and materials discussed in this book may be potentially toxic, hazardous, or dangerous. Any adult should approach these activities with caution. We also advise that children should only participate in said activities with appropriate adult supervision. The author and publisher do not accept liability for any accidents, injuries, loss, legal consequences, or incidental or consequential damage incurred by any reader in reliance on the information or advice provided in this book. Readers should seek health and safety advice from physicians and safety and medical professionals.

Abrams books are available at special discounts when purchased in quantity for premiums and promotions as well as fundraising or educational use. Special editions can also be created to specification. For details, contact specialsales@abramsbooks.com or the address below.

Abrams® is a registered trademark of Harry N. Abrams, Inc.

ABRAMS The Art of Books
195 Broadway, New York, NY 10007
abramsbooks.com